Large Scale Applications of Heat Pumps

Proceedings of the

3rd International Symposium on the

Large Scale Applications of Heat Pumps

Oxford, England : 25-27 March 1987

BHRA The Fluid Engineering Centre

SPRINGER — VERLAG

Berlin • Heidelberg • New York
London • Paris • Tokyo

Editor: Lorraine Grove

EDITORIAL NOTES

The organisers are not responsible for statements or opinions made in the papers. The papers have been reproduced by offset printing from the authors' original typescript to minimise delay.

When citing papers from the volume, the following reference should be used:- Title, Author(s), Paper No., Pages, Proceedings of the 3rd International Symposium on the Large Scale Applications of Heat Pumps, Oxford, England. Organised and Sponsored by BHRA, The Fluid Engineering Centre, Cranfield, Bedford, England. (25-27 March, 1987).

British Library Cataloguing in Publication Data

International Symposium on the Large-Scale Applications of Heat Pumps (3rd)

Proceedings of the 3rd International Symposium on the Large-Scale Applications of Heat Pumps.
1. Heat Pumps
I. Title II. Grove, Lorraine III BHRA
621.402'5 TJ62

ISBN 0-947711-24-4 BHRA, The Fluid Engineering Centre, Cranfield, Bedford, MK43 0AJ
ISBN 3-540-17750-7 Springer-Verlag Berlin Heidelberg New York
ISBN 0-387-17750-7 Springer-Verlag New York Heidelberg Berlin

ACKNOWLEDGEMENTS

The valuable assistance of the Organising Committee and Panel of Referees is gratefully acknowledged.

ORGANISING COMMITTEE

(Chairman)	
Prof. F.A. Holland	University of Salford
Dr. C.A. Bailey	University of Oxford
R. Gluckman	March Consulting Group
Dr. D. Hodgett	Electricity Council Research Centre
Dr. M.B. King	University of Birmingham
J. Masters	British Gas
Dr. R.L. Powell	ICI Mond Division
D.A. Reay	International Research & Development Co. Ltd.
Prof. I. Smith	Cranfield Institute of Technology
Symposium Organiser	
Lorraine Grove	*BHRA*

OVERSEAS CORRESPONDING MEMBERS

Prof. T. Berntsson	Chalmers University of Technology, Sweden
P.E. Frivik	University of Trondheim, Norway
J.A. Knobbout	T.N.O., Netherlands
Prof. P. Worsøe-Schmidt	Technical University of Denmark, Denmark
P. Zegers	Commission of European Communities, Belgium

3rd International Symposium on the

Large Scale Applications of Heat Pumps

Oxford, England : 25-27 March 1987

CONTENTS

ABSORPTION SYSTEMS II

POSTER DISPLAY

3rd International Symposium on the
Large Scale Applications of Heat Pumps
Oxford, England : 25-27 March 1987

PAPER A1

EXPERIENCES FROM PERFORMANCE TESTING OF LARGE HEAT PUMPS

P.O Fahlén, K.O Lagerkvist

The authors work in the Department of Energy Technology at the Swedish National Testing Institute, Box 857, S-501 15 BORÅS, Sweden.

SYNOPSIS

Experience from performance testing of 20 large heat pump plants has demonstrated the importance of such tests and the importance of clear guarantee conditions. The heat pump plants which were evaluated in this investigation consisted of 26 heat pump units, each with a thermal output in the range of 10-30 MW. All plants are connected to municipal district heating systems using treated sewage water, lake water, industrial waste water or geothermal ground water as the heat source.

The tests show that stability criteria for temperatures and flowrates, as required by the test method developed by the National Testing Institute, could be fullfilled in all cases. The typical error of measurement is in the range of 2-4 % and the average deviation of the measured COP from guaranteed values is -4 % for full load operation.

NOMENCLATURE

Roman letters

P_1 = heating power
P_2 = cooling power
P_e = total electric power consumed excluding pumps for heat source and heat sink flows
COP = coefficient of performance = P_1/P_e
COP_c = Carnot efficiency of heat pump cycle
EB = energy balance = $\frac{P_1-P_2-P_e}{P_1}$

t_1 = heating water temperature out from heat pump
t_2 = heating water temperature in to heat pump
t_3 = heat source temperature in to heat pump
t_4 = heat source temperature out from heat pump
q_1 = heating water flowrate
q_2 = heat source flowrate
ρ_1 = heating water density
c_{p1} = heating water specific heat

Additional subscripts

m = measured value
g = guaranteed value
o = operating conditions

INTRODUCTION

Good operating economy and short pay back times have resulted in a large number of heat pumps being installed for base load operation in district heating systems in Sweden. In May 1986 approximately 55 large units were installed or contracted with a total output exceeding 1100 MW. Unitary turn-key installations with outputs up to 30 MW per unit exist and the units can be combined to cover any desired load. The most common heat sources are purified sewage water and lake or sea water but industrial waste water, geothermal ground water, untreated sewage water and outdoor air are also used.

In order to demonstrate the profitability of a heat pump plant accurate calculations have to be performed when the plant is contracted. How profitable the plant will be is decided by the size of the investment, the availability of the installation, the cost of operation and maintenance and the relative energy prices.

The main part of the variable costs consists of the electricity bills for the driving power of the compressor. This is considered in contracts by stating guaranteed values for heat output and coefficient of performance for one or several sets of operating conditions. In most cases bonuses or penalties are stipulated depending on how well the supplier is able to conform to contracted conditions. Thus there are strong economic incentives for both vendor and purchaser to verify the performance of the plant in the most accurate possible way. How strong these incentives can be is illustrated by the following example.

Using the performance data given in table 1 the annual extra energy consumption to produce the yearly guaranteed amount of thermal energy can be estimated.

Annual extra power consumption
Part load: 1554 MWh
Full load: 2069 MWh

In this example it has been assumed that the heat pump has operated under part load conditions during four summer months and under full load conditions during the rest of the year.

Using current Swedish high voltage rates

Held at St. Catherine's College Oxford, England. Organised and sponsored by BHRA, The Fluid Engineering Centre, Cranfield, Bedfordshire, MK43 0AJ England.

for electricity (0,183 SEK/kWh in May-September, 0.221 SEK/kWh the rest of the year and 305 SEK/installed kW including all taxes and other charges) the following annual extra cost can be determined.

Annual extra cost
Part load: 313 kSEK
Full load: 503 kSEK
Power rates: 244 kSEK
Total: 1060 kSEK
(Approx. £ 103000)

This example is just intended to give an idea of the changes in the operating costs that can be caused by seemingly minor deviations from guaranteed values. Of course these deviations may work both ways as will be shown later in this paper.

Despite the moderate discrepancy between measured and guaranteed coefficients of performance in this example operating costs are increased by more than 1 million SEK per annum. A fairly large part of this increase is due to the power rates for the extra power consumption (305 SEK/kW). Using a real interest rate of 5 % and a pay back time of 5 years the present value of this extra cost will amount to 4.7 million SEK. Thus a performance test can carry a fairly sizeable cost.

For this type of commercially contracted heat pump installation the National Testing Institute of Sweden has designed a test method and used it on more than 20 large heat pump plants. The test method describes stability criteria and accuracy of measurement.

DESCRIPTION OF THE TESTED HEAT PUMPS

The units considered for this particular evaluation all operate on municipal district heating systems. Of the 20 plants 13 utilize purified sewage water as heat source, 3 operate on lake water, 2 use industrial waste water and 2 are supplied with geothermal ground water.

All units are furnished with turbo compressors powered by high voltage (11 kV) electric motors. Due to the high thermal outputs these plants carry fairly sizeable refrigerant charges. Using a rough estimate of 1 ton of refrigerant per MW thermal output a plant of 3 x 13 MW would have at total refrigerant charge of 39 tons.

Sewage water

The principle layout of a heat pump plant using purified sewage water as heat source is shown in figure 1. Sewage water is pumped from the treatment plant to a well and then pumped through the evaporator (2), normally a shell and tube design.

The typical design uses a two stage turbo compressor (3) and a two step expansion process by means of a high pressure (11) and a low pressure (13) control valve. Flash gas is separated in an economizer vessel (12). To achieve a maximum amount of sub cooling the district heating water is separated into two flows through the condensor (6) and sub cooler (7) respectively. These flows are joined again before returning to the district heating system through the delivery flow meter (q1).

In table 2 a list is given of the various sewage water heat pumps giving their noninal output, type of refrigerant and the year they were taken into operation.

Lake water

In figure 2 the layout of a lake water heat pump is illustrated. The principle is very similar to the sewage water design apart from the evaporator. Due to the very low water temperatures during winter when lakes are frozen the evaporators are of a special falling film design. Water is sprayed on the outside of the evaporator making it easy to clean the heat transfer surfaces from algae and ice.

During winter water is taken at a substantial depth through the winter inlet (2). In summertime the warm surface water is utilized by means of another shallow summer inlet (1).

In table 3 a list of lake water heat pumps is given.

Waste water

Two installations using industrial waste water have been tested (see table 4). The principle design is the same as for sewage water, as described in figure 1.

Ground water

Geothermal ground water is special by being highly aggressive, therefore requiring the use of titanium tubes in the evaporator. Otherwise the plant configuration is similar to the sewage water design. Two installations are given in table 5.

TESTED PERFORMANCE DATA

In most cases contracts prescribe guaranteed performance data for several sets of operating conditions. This is to enable tests to be carried out with varying temperatures and flowrates for the heat source and heat sink since these parameters normally can not be controlled. Furthermore many installations are designed to operate with reduced output and therefore part load performance is often guaranteed.

In order to give an idea of the performance to be expected from these types of large heat pump installations, measured values for heat output and coefficient of performance are given in tables 6-8. Since guaranteed operating conditions are different for every individual contract the figures given in the tables can not be compared with each other. The incoming heat source temperature (t_3) and out going heating water temperature (t_1) are therefore given to designate the operating conditions as (t_3/t_1). Only one performance point is given for each heat pump unit. Note that COP values are fairly high in spite of very high forward feed temperatures. Typical COP values range between 60-65 % of the Carnot efficiency.

COMPARISON WITH GUARANTEED PERFORMANCE DATA

Guaranteed performance data can be given in the form of diagrams but is generally stated for one or several sets of fixed operating conditions. This requires a correction to be made either to the test results or to the guaranteed data since tests can not generally be carried out at the exact operating conditions of the guarantee. This will be discussed in the section on guarantee problems.

In the present investigation the measured thermal output (P_1), electric power con- sumption (P_e) and coefficient of perfor- mance (COP) have been compared with cor- rected guaranteed values. On the average measured values of P_1 for full load operation exceeded guaranteed values by 4.0 % with a standard deviation of 6,3 %. The maximum differences from guaranteed values of P_1 were +12.9 % and -24.0 % respectively. For part load operation the average difference is +3,8 % with a maximum deviation of +6.4 %. If P_1 on the average has exceeded expectations so has P_e also. For full load operation the average difference is +6.4 % with maximum differences of +16.8 % and -17.6 % respectively. Corresponding figures for part load operation are +13.3 % mean difference and a highest value of +25.0 %.

These discrepancies for P_1 and P_e result in COP values being 4 % lower than expected on the average for full load operation. The standard deviation is 5.5 % with high and low values of +26.7 % and -10.5 % respectively. The distribution of deviations for COP values is illustrated in figure 3 for full load operation. It can be seen in the diagram that the bulk of the tests lie between +2 % and -6 % deviation between actually measured and guaranteed COP calues.

Sometimes only the cooling power (the "free" energy) is guaranteed and then it might be argued that COP values are unimportant. This however
is only true when the heat pump operates as base load with electric heating to cover the peak. Otherwise the extra electric energy consumption and particularly the extra power tariff will prove costly.

In figure 4 the deviations from guaranteed values of P_1 and COP are compared with calculated values of the error of measurement. This diagram will show the occurence of penalties or bonuses being paid. Deviations smaller than the error of measurement are normally not considered which further emphasizes the importance of accuracy of measurement.

A common figure for the size of a bonus or penalty is 0.25 % of the total contract sum for each % deviation of P_1 and P_e. For a typical installation this would amount to something like 40000 SEK (approximately £ 4000) for each % deviation resulting in 160000 SEK (£16000) for a typical deviation of 4 %. This is excluding the error of measurement, making each % of this error equally costly. Therefore an improvement of the error of measurement by 1 % is worth 40000 SEK. This also makes the estimation of measuring errors extremely important.

DISCUSSION OF GUARANTEES

In many cases contracts guarantee that the heat output (P_1) should not be lower than a certain value and that the electric power consumption (P_e) should not exceed a certain value. Sometimes this will prove disadvantageous for the heat pump manufacturer. If the heat pump has a higher output than guaranteed the power consumption may exceed guaranteed levels and even though COP values may be better than promised a fine will have to be paid. This example shows the importance of how a contract is laid out.

The recommended practice would be to guarantee heat output and COP. However particular problems arise in the case of part load operation. Two separate situations occur.

In the first case part load operation is induced by a reduction of the heat demand. This being the case part load performance should be defined in terms of achieving a guaranteed COP value for a defined percentage of the full load output for the same set of operating conditions.

The second case however is more complicated. This occurs when the heat pump output has to be reduced due to design limitations in terms of excessive motor currents, extreme heat source or heat sink temperatures etc. Then the guaranteed operating range of the heat pump in terms of design limitations may be far more important than any possible deviations in efficiency. For instance in the case of a lake water heat pump the economic losses from not being able to extract the full amount of energy from the water due to a low lake water temperature is of much greater importance than COP values. In this case premature operation of protection devices should be checked carefully. Even fractions of degrees are important. Selection of the full capacity rating points should be chosen carefully to avoid close proximity to any of the heat pump operating limits. It is generally very difficult to get meaningful results when different types of protection devices start to affect heat pump performance.

Another important aspect concerns the recalculation of measured data to make them comparable to guaranteed values in spite of deviations in the operating conditions. Normally the manufacturer uses computer programs to calculate data for any set of operating conditions. It has to be stated in advance how large deviations from the design point in temperatures and flowrates that can be tolerated without invalidating the computer program. Prior to the performance test either curves showing how performance data vary with temperature and flowrate or several calculated values for different sets of operating conditions (with only one parameter changed at a time) should be supplied by the manufacturer.

In figure 5 there is a diagram showing the deviations from the design heat source temperature for the present number of tests. The diagram indicates that most of the tests have been within ± 2 K and a large portion within ± 1 K of the heat source temperature of the guarantee conditions. Figure 6 demonstrates the same thing for the heat sink temperature. It is evident that in this case the scatter is much larger. However most temperatures lie within ± 3 K of guaranteed conditions. The mean deviation from guarantee conditions of the heat source temperature is -0.4 K and for the heat sink temperature -1.5 K. Extreme deviations are +3.2 and -3.8 K and +3,4 K and -9.8 K respectively.

Flowrates have differed considerably both for the heat source and the heat sink. These discrepancies normally stem from too little knowledge of the heat source and the district heating network by the purchaser. Heat source flowrates have varied between a maximum of +98 % and a minimum of -18 % from design values with a mean value of +12.5 %. Similarly heat sink flowrates have varied between +108 % and -19 % with a mean deviation of +9.4 % from design conditions.

As discussed in the previous section accuracy of measurement is of great importance. The principle for calculating errors and the errors ascribed to installation deficiencies and calibration should preferably be decided and agreed upon in advance.

MEASURING PROBLEMS

Stability criteria

Heat pump performance is greatly affected by temperature and flowrate. This can be designated by

$$P_1 = P_1\ (t_1,\ t_3,\ q_1,\ q_2)$$
$$P_e = P_e\ (t_1,\ t_3,\ q_1,\ q_2)$$

Uncertainty in determining the exact operating conditions during a test will result in an additional error, ΔP_{1o}.

$$\Delta P_{1,o}\quad \frac{\partial P_1}{\partial t_1} \bullet \Delta t_1 + \frac{\partial P_1}{\partial t_3} \bullet \Delta t_3 + \frac{\partial P_1}{\partial q_1}\ \Delta q_1 + \frac{\partial P_1}{\partial q_2} \bullet \Delta q_2$$

It is therefore important not only to have operating conditions close to design values but also to keep them constant. Variations will unavoidably increase Δt and Δq in the expressions above thereby increasing $\Delta P_{1,o}$.

The test method used in these tests require heat source temperatures to be stable within ± 0.5 K, heat sink temperatures within ± 1 K and flowrates to be stable within ± 5 % for a stabilizing period of 1 hour and a measuring period of 0.5 hours. It has been possible to fullfill these requirements on all occasions. The mean variation for t_1 is 0.4 K with a maximum of 1 K. For t_3 the mean variation is 0.1 K with a maximum of 0.5 K. Flowrates are normally very stable with mean variations of 1.2 % and 1.8 % for q_1 and q_2 respectively.

The error due to instability of the operating conditions is typically 0.5 %. In some installations it has reached a high value of 1.5 % although variations have still been within the limits prescribed by the test method.

Accuracy of measurement

A prerequisite for obtaining accurate results is the use of high quality equipment with up to date calibration status. It is important that calibration has been performed for the actual operating conditions of the application. Flowmeters for instance are mostly calibrated using cold water where as the heat output flowmeter operates in the range 50-100 °C. For instruments with several output signals (e.g. frequency, current etc) only the output used in the calibration process should be used.

The above conditions are necessary but unfortunately not sufficient. In most of the tests in this report estimated errors because of installation of measuring equipment deviating from ideal conditions make a major contribution to the total error. This applies to flowmeter installations in particular.

The error due to measuring uncertainties is calculated from

$$P_1 = q_1 \bullet \rho_1 \bullet c_{p1}(t_1 - t_2)$$

Logarithmic differentiation yields

$$\frac{\Delta P_{1,m}}{P_1} = \frac{\Delta q_1}{q_1} + \frac{\Delta \rho_1}{\rho} + \frac{\Delta c_{p1}}{c_{p1}} + \frac{\Delta (t_1 - t_2)}{(t_1 - t_2)}$$

Accuracy of measurement of the individual parameters will determine the total error of measurement.

Some particular problems are listed below.

- Temperature.
  To stand the high temperature and pressure of the district heating system, thermometer wells have to be used giving fairly slow response. Normally 2-4 PT100 transducers are used in a measuring plane. Thermal stratification has not been experienced as a problem in pipe measurements although this has been difficult to check in practice.

  In some installations flowrates have been very high compared to contracted values giving low temperture differences, thus increasing the relative error. For lake water it may sometimes be difficult to get a representative value due to large gradients around the inlet. The total uncertainty for the temperature difference is normally less than 0.2 K (calibration of instrument better than 0.005 K).

- Flowrate.
  In most installations electromagnetic flowmeters have been used with good results. Unfortunately the heat delivery meter has often been installed with insufficient straight lengths of pipe before and after the meter. A typical installation is shown in figure 7. For this type of installation the uncertainty of the meter performance is the single most important contribution to the total error.

  In one installation electrode leakage in the meter housing led to readings being consistently 10 % low. This was detected through the energy balance check.

  In a number of installations ultrasonic flowmeters have been used for the heat source flowrate with poor results. Deviations have ranged from +7 % to -17 %.

  In one installation problems with interference were experienced when connecting the frequency output of the meter to the test instrument. This caused the heat pump to shut down due to faulty flow wignal.

- Electric power.
  For measurement of electric power 2 or 3-phase measuring instruments are employed. The most difficult problem is to get the power meter connected to the high voltage measuring transformers with short notice. On one occasion the power consumption had to be measured in a substation 1,5 km away from the heat pump plant and therefore transmission losses had to be considered.

  Sometimes commercial kWh-meters have been connected to the same set of measuring transformers loading these enough to derate the transformer accuracy class one step.

Total accuracy

Total accuracy is normally derived by adding the errors quadratically and stated at a confidence level of 95 %

$$\Delta P_1 = \sqrt{\Delta P_{1,m}^2 + \Delta P_{1,o}^2}$$

In figure 8 results for the total error of measurement are shown. The average value is ± 2.8 % with a maximum of ± 4.7 % and a minimum of ± 1.4 %. This is the range of accuracy values to be expected in this type of testing using first class instrumentation.

Whenever possible an energy balance calculation should be performed. The heat output (P_1) should equal the heat input (P_2) plus the electric power input (P_e) minus any losses to the surroundings.

$$EB = \frac{(P_1 - P_2 - P_e)}{P_1} \, 100 \, \%$$

Results from these tests are illustrated in figure 9. It can be seen that in most cases the balance agrees within ± 2 %. All the extreme values are due to inferior quality flowmeters being used on the heat source flow. For tests using the same quality of equipment for both the heat source and heat sink the average balance is -0.3 % with the largest deviation being 3.2 %. Thus if a heat balance shall be of use high quality instrumentation must be used all around.

OPERATING PROBLEMS

Since performance testing for the purpose of guarantee evaluation is carried out during the course of 1 or 2 days there are normally few problems with the operation of the heat pump. Long term experience has also returned annual availability figures in excess of 90 %, sometimes approaching 100 % (VAST, 1986). Some examples of problems which have affected performance testing are given below.

On one lake water installation there was a problem of too low water temperatures due to an unsually mild autumn. This delayed ice coverage causing larger than usual heat losses from the lake. Due to this climatic problem it was impossible to operate the heat pump on full load.

On some sewage water installations there have been problems with leaves blocking screens and with unstable temperatures and flowrates due to flushing of filters. Furthermore tube vibration in the evaporator caused a few tubes to crack requiring these tubes to be plugged in one installation.

Insufficient or unappropriate instrumentation has delayed testing in a few installations. There have also been problems in using existing flowmeters directly on a few occasions due to the control system of the heat pump. If the flowsignal is affected the flow control unit may stop the compressor.

Factors directly affecting performance have been for instance too little refrigerant charge, refrigerant level control in condensors and evaporators, hot gas distribution in condensors and control of hot gas by pass in connection with potential pumping of the compressor. Estimation and control of the actual refrigerant level in the system seems to be a generic problem. Level indicators have degraded due to heat and vibration and since there is no accumulator capacity for refrigerant in the system the charge has to be changed and optimized according to the operating conditions of each individual installation. This makes the function of the monitoring system all the more important.

Factors affecting heat pump performance on a long term basis have been tube vibration in heat exchangers, refrigerant leakage and poor control of cooling of oil and electric motors. In particular problems with refrigerant leakage through seals constitutes a general problem. Also the oil has sometimes been degraded through pollution by refrigerant which has decayed due to high temperatures when it has leakaed through bearings. The degradation of the oil has caused one or two bearing failures.

Finally one compressor failed on starting up due to too tight clearances between impeller and housing and one system lost a sizeable amount of refrigerant when a safety valve opened. This was due to accidentally passing too hot water through the condensor of a unit not in operation while trying to control the condensor temperature of another unit for testing purposes.

It must be emphasized that most problems are very specific and have happened once or twice. In general these types of heat pump plants have proved very reliable. The only generic problem of importance concerns refrigerant leakage.

CONCLUSIONS

This survey of a number of large heat pump installations has demonstrated that on average performance compares well with guaranteed values. It has also been indicated that due to the economic consequences accuracy of measurement is vital in this type of performance testing. In some cases money has been saved in the first cost of measuring equipment and its installation leading to unnecessarily high inaccuracies in the evaluation of the heat pump.

In order to provide unambiguous conditions for evaluation of guaranteed performance data the method of test, the method of error calculation, the data to be supplied prior to testing, choice and calibration status of equipment and last but not least clear definitions of the guaranteed performance should all be stated or referenced in the contract. The relevant guarantees should normally be comprised of heat output and COP for full load operation, COP at a defined percentage output for normal partload operation and the operating range for part load operation due to design limitations.

REFERENCES

The Swedish National Testing Institute, 1985. SP METOD ET 1985:1 "Performance testing of large scale heat pump installations". Borås. (in Swedish).

VAST Information, 1986. No 86:52, 1986-05-25. Stockholm. (in Swedish).

Elteknik, 1984. "Testing heat pump performance". 1984:2. Stockholm. (in Swedish).

TABLE 1. Example of performance test data

	Guaranteed	Measured
Full load		
- P_1, MW	16.6	18.3
P_e, MW	5.0	5.8
COP	3.3	3.1
Part load		
- P_1, MW	8.4	9.4
P_e, MW	2.7	3.6
COP	3.1	2.6

TABLE 2. List of sewage water heat pumps

Nominal output (MW)	Year taken into operation	Type of refrigerant
13.5	1982	R12
3x13	1982	R12
13	1982	R12
13	1983	R12
27	1983	R12
12	1983	R12
14	1984	R12
12	1984	R12
15	1984	R12
7	1984	R12
29	1984	R12
9	1986	R500
2x15	1986	R12

TABLE 3. List of lake water heat pumps

Nominal output (MW)	Year taken into operation	Type of refrigerant
10	1982	R12
3x26	1986	R22
2x20	1986	R500/R22

TABLE 4. List of waste water heat pumps

Nominal output (MW)	Year taken into operation	Type of refrigerant
12	1984	R12
2x14	1984	R22

TABLE 5. List of ground water heat pumps

Nominal output (MW)	Year taken into operation	Type of refrigerant
19	1985	R500
19	1986	R500

TABLE 6. Full load performance of sewage water heat pumps

P_1 (MW)	COP	(t_3°C/t_1°C)
12.9	3.7	(14.7/61.2)
13.3	3.5	(17.8/68.5)
13.5	3.4	(12.9/60.4)
13.5	3.4	(14.9/67.0)
13.3	3.3	(12.1/67.3)
12.0	3.1	(4.1/68.4)
29.5	3.1	(13.6/77.2)
11.6	3.3	(11.1/67.1)
14.0	3.4	(11.2/58.8)
11.3	2.8	(3.6/73.1)
12.4	3.4	(9.2/65.1)
4.3	3.0	(10.7/79.1)
28.2	2.7	(12.1/87.8)
9.3	2.8	(7.5/74.7)
31.4	3.4	(8.3/61.2)

TABLE 7. Full load performance of lake water heat pumps

P_1 (MW)	COP	(t_3°C/t_1°C)
11.0	2.8	(2.4/61.6)
78.6	3.4	(4.1/57.0)

TABLE 8. Full load performance of waste water and ground water heat pumps

P_1 (MW)	COP	(t_3°C/t_1°C)
11.0	3.5	(20.4/64.7)
14.3	3.4	(11.2/62.4)
45.1	3.2	(20.8/70.1)

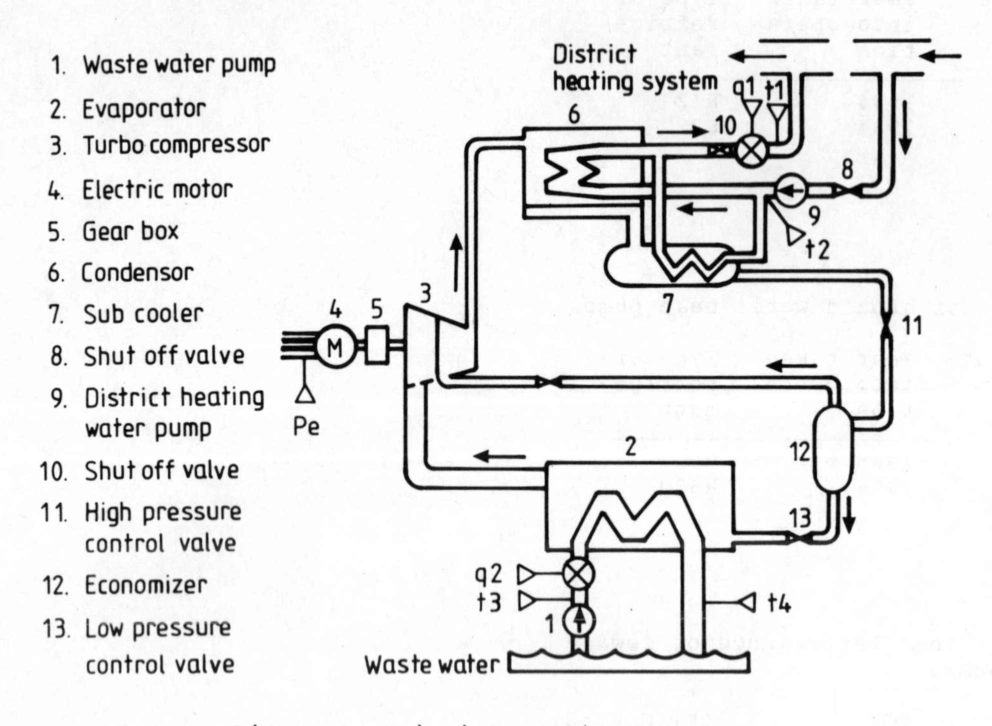

Figure 1. Principle design of a waste water heat pump (ASEA STAL) with the necessary measuring points included.

1. Summer inlet
2. Winter inlet
3. Lake water pump
4. Panel evaporator
5. Turbo compressor
6. Electric motor
7. Gear box
8. Condensor
9. Sub cooler
10. Shut off valve
11. District heating water pump
12. Shut off valve
13. High pressure control valve
14. Economizer
15. Low pressure control valve

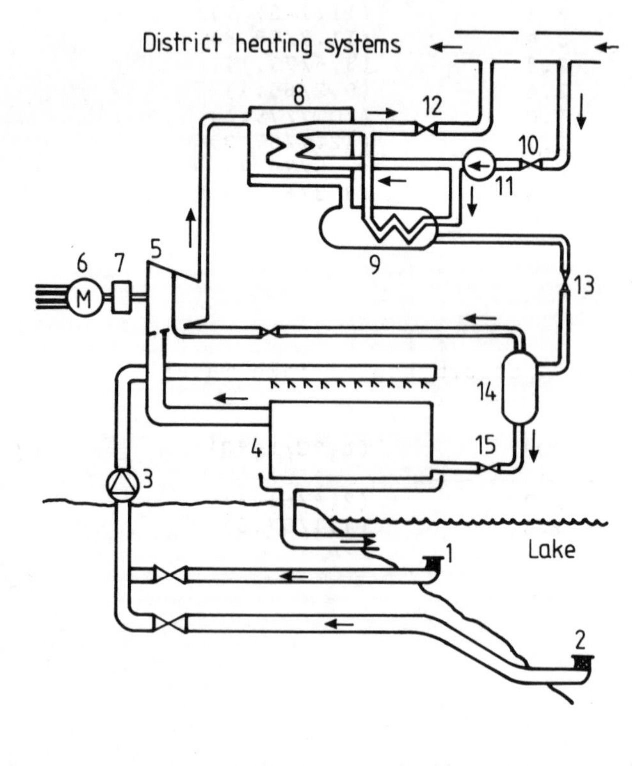

Figure 2. Principle design of a lake water heat pump (ASEA STAL).

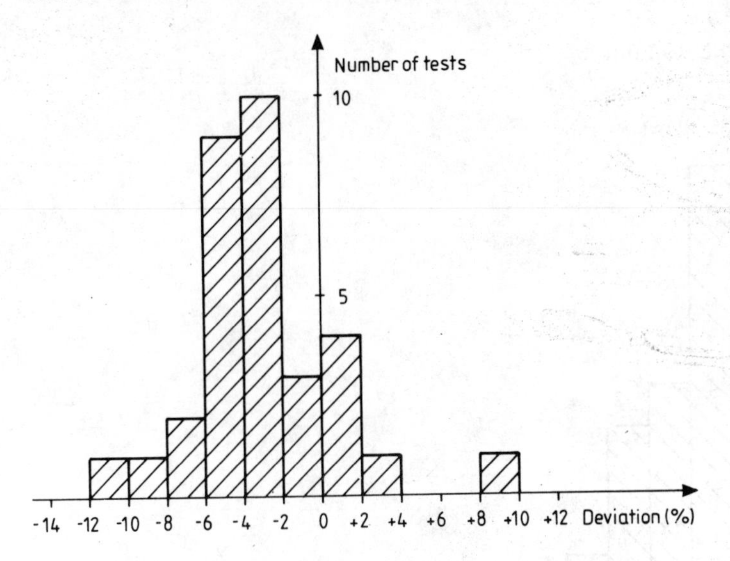

Figure 3. Number of tests with a certain deviation between measured and guaranteed COP in 2 % intervals.

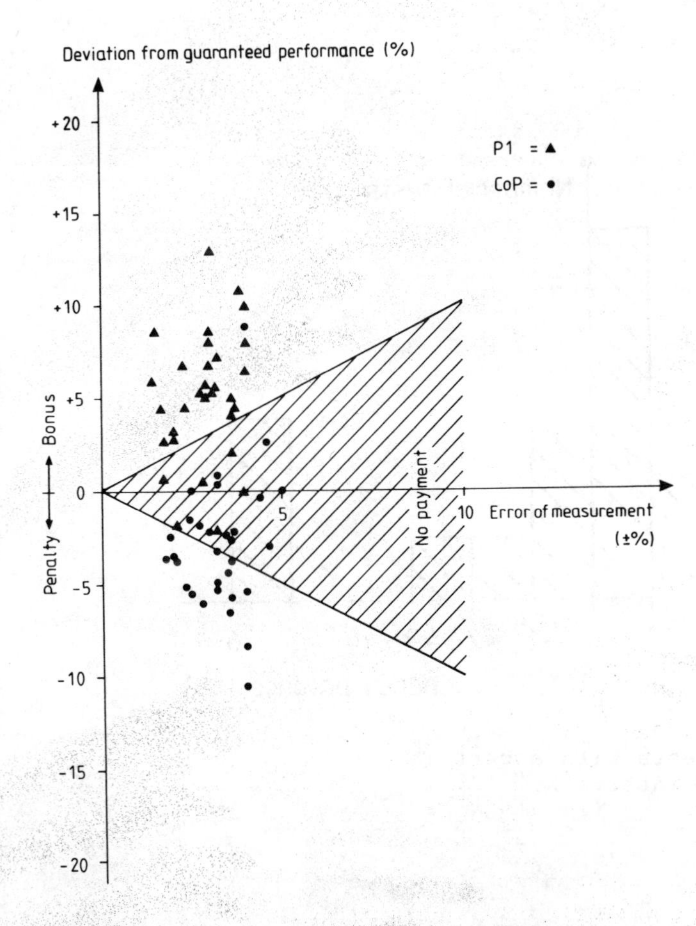

Figure 4. Diagram indicating the influence of error of measurement on payment of penalty or bonus.

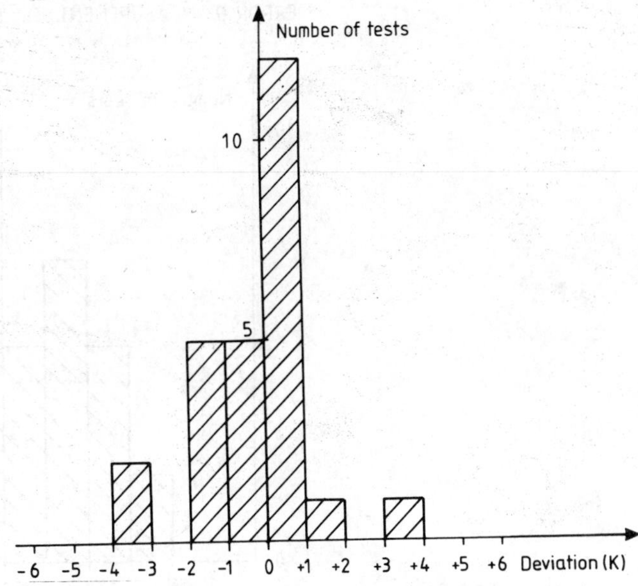

Figure 5. Number of tests with a certain deviation between the measured heat source temperature and the temperature for guaranteed conditions in 1 K intervals.

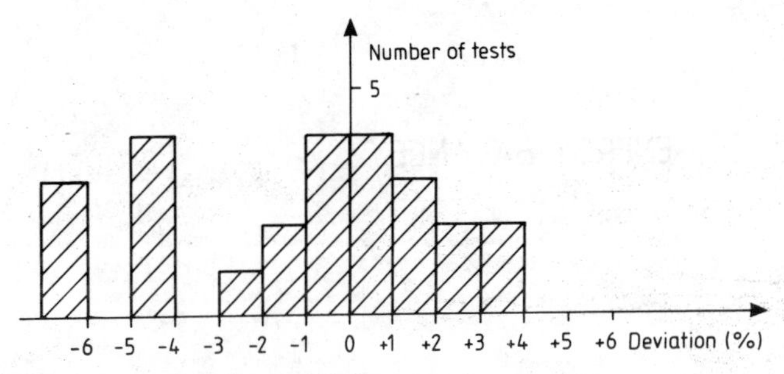

Figure 6. Number of tests with a certain deviation between the measured heat sink temperature and the temperature for guaranteed conditions in 1 K intervals.

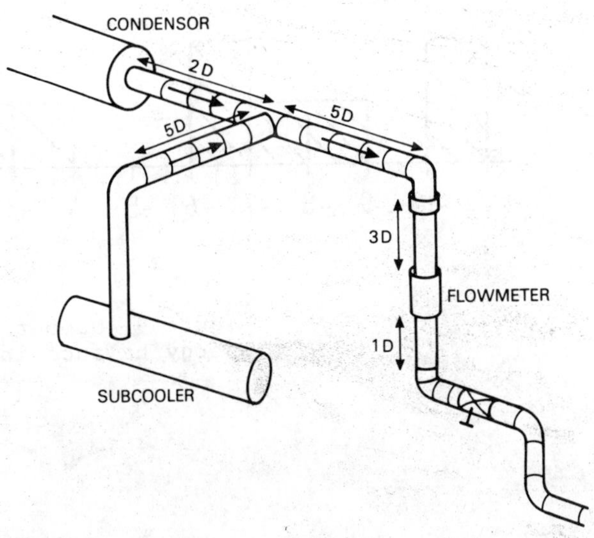

Figure 7. Typical flowmeter installation in a large heat pump unit.

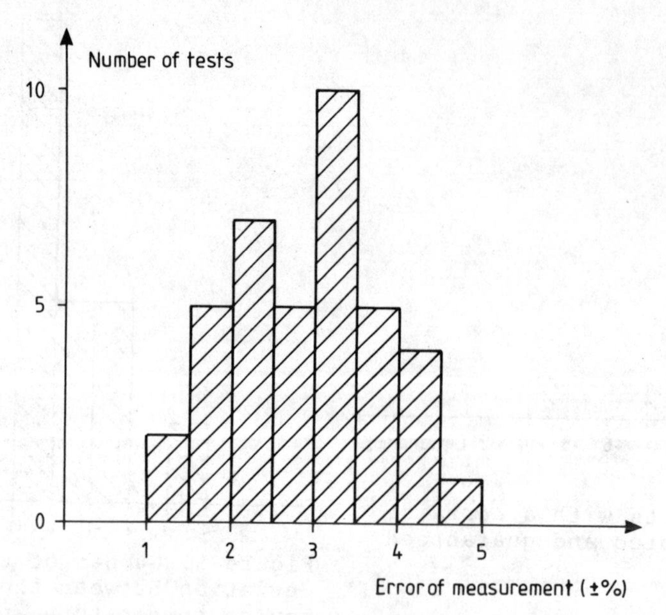

Figure 8. Number of tests with a certain calculated error of measurement for the heat output in 0,5 % intervals.

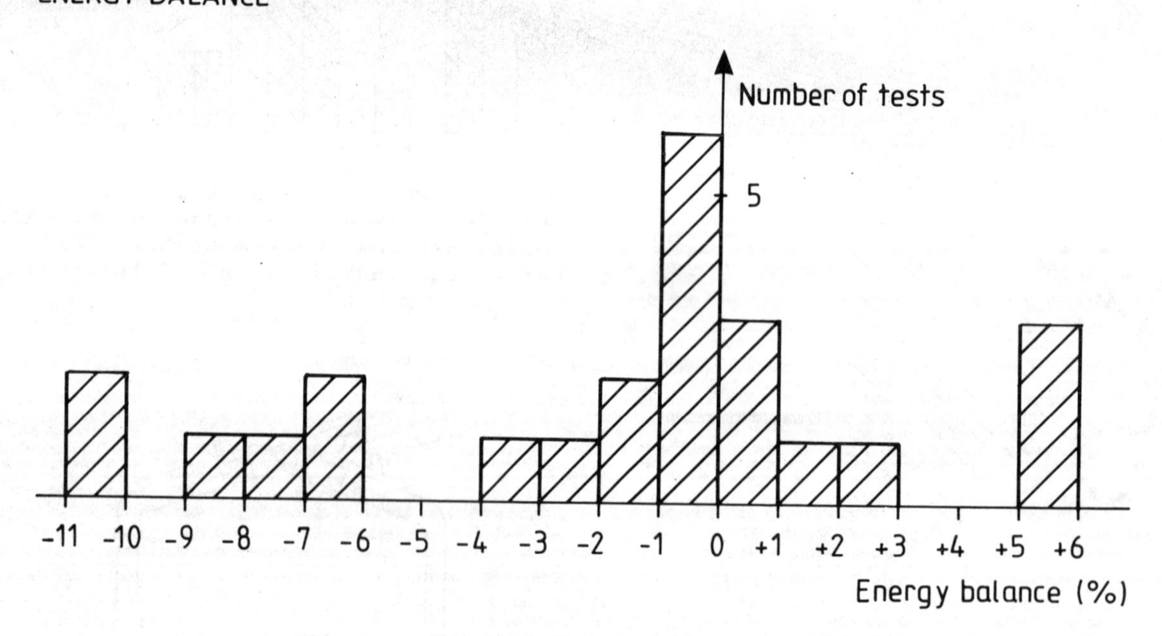

Figure 9. Number of tests with a certain energy balance in 1 % intervals.

3rd International Symposium on the

Large Scale Applications of Heat Pumps

Oxford, England : 25-27 March 1987

PAPER A2

TITLE : APPLICATION OF HEAT PUMPS BASED ON WASTE HEAT IN A CORRUGATED CARDBOARD FACTORY.

AUTHOR : A.C. GILLET
The author is an energy consultant in an engineering and architecture company
ATELIER "D" S.C.
Bellevoie 23 - B. 1300 Wavre - Belgium.

SYNOPSIS

A demonstration project has been planned and put in action with the financial support of the E.E.C. in a corrugated cardboard factory : Les Papeteries CATALA S.A. in Drogenbos, Belgium.

Previously vented outside, radiant heat from boilers as well as convection heat emitted and humidity vaporized during manufacture of cardboard are absorbed and/or condensed on evaporator batteries hanging just above the sources. Energy so retrieved without any forced ventilation is transfered by refrigerant gas to nine compressor-condenser groups, each of 300 KW heating capacity, and distributed by luke warm water through destratifying aerotherms hanging from the roof in the various parts of the factory.

On the basis of a general energy audit, drastic reduction of the heating load has been done before to match the deperdition and the recuperation possibilities.

Difficulties encountered as well as technical and economical results are described.

I. INTRODUCTION

In 1982, the old oil based heating system of the CATALA factory needed refurbishment.

Our company made an energy audit. It was obvious that besides the renewal of the heating system, the factory needed thermal insulation and controled ventilation. Two actions were summarized in fig. 1 :

1. Drastic reduction of the heating load through thermal insulation of the roofs and reduced ventilation.
2. Substitution of light fuel oil burnt in decentralized aerotherms scattered in the factory, by a centralized system based on retrieval of the excess heat from the process. At that time, this was lost through ventilation to the atmosphere of radiant heat from three steam producing boilers and by convection of humid air emitted by three corrugators.

Recovery and transfer of the excess heat was most economicaly feasible by a thermodynamic system the originality of which was in the use of static evaporators without any forced ventilation system hanging under the roof just above the heat/humidity producing machinery.

The newness of the proposal and the results predicted were quite attractive but the magnitude of the investment was such that the manager hesitated to make a decision. So the project was presented to the E.E.C. ad hoc Committee and finaly accepted and funded in December 1983.

II. GENERAL AIMS OF THE PROJECT

Classically, the recovery of waste heat from the exhauste air of industrial processes makes use of active air handling. Air is forced through relatively compact exchangers where it is chilled and/or dried and the recovered energy is transfered to secondary air flows which transport it to where it is needed.

About four cubic meters of air are needed to transfer, as sensible heat, the same quantity of heat as with only one liter of water or some cubic centimeters of refrigerant, as latent heat.

Drawbacks of such air handling systems are then :

- large section of air ducts,
- large volume of building for to same,
- high auxiliary energy for air moving.
  Fans need four to five times more energy than pumps for the same result. Extra energy needed by compressor is compara-tively negligible.

So, one of the aims of the project was to demonstrate the feasibility of static evaporators and their performance when used in natural convection air flows above excess heat producing industrial machines from an energy efficient point of view.

The audit made the point that a monovalent thermodynamic system was conceivable if the heating load of the building to be heated was reduced so as to equilibrate the existing recovery possibilities.

Actions were then taken as to :

- recover 38 % of the thermal transmission through the roofs;
- control natural as well as forced ventilation losses.

Held at St. Catherine's College Oxford, England. Organised and sponsored by BHRA, The Fluid Engineering Centre, Cranfield, Bedfordshire, MK43 0AJ England.

This was done by November 1983.

The thermodynamic system covering about 20 % of the residual thermal load was ready by November 1984. Data acquisition and a scientific follow-up were in action by that time and have been sustained until now.

III. DESCRIPTION OF THE PROJECT

1. Thermal insulation of the roofs

After detailed study, the following actions were taken :

1.1. Installation of an underroof mat of 5 cm thick bakelized glass wool with an aluminium sheeting underneath as radiation barrier;

1.2. Doubling of all the glass sections with double polycarbonate extruded sheeting (thickness : 1 cm).

2. Control of the natural and forced ventilation losses

Scrap and cuttings of cardboard are removed and transported to the cyclones by an pneumatic system. This resulted in a depression of air pressure inside the building and an aspiration of cold air from outside through every opening in the envelope.

Beside the purchase of automatic quick opening and closing doors and a carefull inspection of all the envelope for air entry, air ducts opening outside on the roof were mounted near each aspiration mouth. Thanks to these ducts outside air is substituted to inside air for the transportation of scrap and cuttings. A tremendous saving of heat was experienced.

3. Thermodynamic retrieval of lost industrial heat and environmental heat back-up

(See general plan n° 248/82/1 (fig. 2) and working diagram 2.3.a (fig. 3)).

3.1. Lost industrial heat

374 static evaporators (sketch 4.3.3.) (fig. 4) are suspended under the roof in the heat producing rooms. They have a total of 9,350 m2 of heat exchange area and absorb about 1,800 kW computed on the basis of a pilot test in the project.

Refrigerant vaporized in this array is compressed by three independant compressor-condenser groups with respectively 1, 4 and 4 parallel compressors. Water for the heating system is pumped through the condensers.

The whole represents a huge air to water heat pump with a overall heat output of about 2,700 kW.

The predominant part of this is obtained from the condensation of water vapor onto the evaporators cooled by vaporizing refrigerant. A gutter net drains the condensed water to the sewers.

3.2. Environmental heat back-up

When the production is stopped and no waste heat is available, e.g. during week-ends and holidays, a environmental back-up is obtained from the Senne waters flowing in sewers underneath the floor of the factory and nearby (minimal temperature recorded 5°C).

Two evaporator batteries (fig. 5) were immersed in special chambers having a gate, the closure of which overfloods the battery when needed (fig. 6).

For a 2.5 k temperature decrease and a water flow of 400 m3/h, the capacity is about 1,290 kW, sufficient to keep the premises from freezing.

Liquid refrigerant is fed to these batteries (total head exchange area : 530 m2), instead of to the static air evaporators, and compressed by the two four compressors groups which act then as water to water heat pumps.

4. Transfert and emission of heat

The first proposal forsaw a distribution of energy by mean of the high pressure refrigerant gas from the compressors directly to heat exchangers scattered in the rooms to be heated (air to air heat pump; fig. 7).

The risks of such a system with pipes extending over more than 200 m radius, have forced us to prefer a classical system of luke warm water distribution through a net of parallel piping to and from the condensers of the nine groups. Total flow of water is about 450 m3/h. Piping is made of high density polyethylene, uninsulated, hung from the steel framework of the premises. 73 heat exchangers have a total heating capacity of 2,900 kW for 15°C air and 45°C entering water temperatures. Position near the ridge of the roof ensure a good thermal destratification.

5. Regulation and data acquisition

The needs of the measurement program required by the E.E.C. support, have constrained us to use a system for both control and data acquisition, as follows.

The premises were divided on the basis of their needs into :

- 14 heated zones
- 5 heat producing zones
- 2 environmental heat producing places.

A central programmable controller regulates the whole system. It is equiped with :

- 14 analoge probes measuring ambient temperature in the heated zones
- 9 ditto and
- 6 analoge probes measuring temperature and relative humidity in the heat producing zones
- 4 analoge probes measuring water temperature in the sewer heat sources
- 6 analoge probes measuring luke warm water temperature in and out of the distribution network
- 1 controlling system which chooses the appropriate cold sources and regulates the work of compressors, pumps, fans and the like, on the basis of a program interrogating the above sensors.

The thermal balance and the heat pumps' performances are obtained by three flow measurements combined with the temperature data and four independant recording electrical energy meters.

The safe operation of the compressors is ensured by the use of six pressure transducers inserted in the various high and low pression piping, monitored by the controller which has a built-in alarm system.

Averaged data are printed on paper every hour.

IV. PROBLEMS AND SOLUTIONS

There were no problems with thermal insulation done between September and November 1983.

Mechanical ventilation was controlled without difficulties during October 1984.

A first phase of the thermodynamic system was planned for late 1983. It was postposed for lack of a refrigerant tank for the first group adjacent to boilers.

The whole system was only in action by the 3rd of December 1984, in time anyhow for the heating season. Autumn was very mild that year and combined effects of insulation, ventilation control and thermal destratification were sufficient to ensure the comfort in November.

Difficulties were quickly apparent.

1. Evaporator batteries in Senne water

Fouling of the batteries occurs rapidly. We were aware of the fact that this was a weak point for the system and railing was thus provided at the entrance of the sewers. Unfortunately an overflow, in November, carried along a heap of dead leaves which clogged the batteries.

It was then decided :

- to halve the batteries by pulling out an alternate layer of pipes;
- to control the level in the sewers by automatic gates set at the entrance by the river;
- to assemble the removed pipes into a third battery immersed directly in an old arm of the river flowing nearby.

One of the batteries was disassembled in December, the other during Spring 1985, and the whole system was ready for the 85-86 heating season.

2. Polyethylene piping net

During the first trials of the heat pumps, some defects in the jointing and in the general stability of the piping became apparent.

We had to repair these by ourselves. The contractor for this part of the work was a newcomer in polyethylene piping and he had gone bankrupt.

3. Compressor-Condensor groups

After some weeks of work electrical problems appeared : clinging of the power contactors and fire in three of the nine electrical circuits. The coil of one motor was even damaged and had to be replaced.

A bad connection of the signal wiring to the controller was responsible for these accidents resulting in a lack of a time-delay between the on and off positions of the motors.

4. Refrigerant piping network

4.1. As already stated,late delivery of liquid refrigerant tanks had caused troubles. It appeared also that they were improperly cleaned. The refrigerant piping network was heavily polluted by dust and even scraps of solder. Fortunately we had inserted cartridges filters in the outlet of these tanks. Cartridges were changed several times before remaining clean. It seems anyhow that dust not stopped by the filters may be partly responsible for a rapid wear of the mechanical parts of the compressors described below.

4.2. Rigidity of the junction between network and compressors was responsible of some breakage and refrigerant leakage. We had to replace faulty connections by flexibles.

4.3. During 85-85 winter, an unfortunate modification of the controller program caused "liquid slugs" the repetition of which damaged compressors. Two of them were even out of service for the end of the season. An abnormal wear was detected during the check-up, ascribed to lack of lubrication and "refrigerant fire".

V. RESULTS AND DATA

1. Overall results

1.1. Reduction of heat load

As stated before, thermal insulation of the roofs was ready as soon as November 1983, while control of ventilation could only influence consumption by November 1984.

Reduction in purchase of light fuel is the only index of the effect of insulation between these dates and before the beginning of thermodynamic heat emission.

Average consumption of oil previous to the audit was about 1,000 m3/year. In 1984, the factory bought 397.5 m3 thus saving more than 60 %, a result greatly in excess of the target of 40 %.

We think this can be ascribed, for the most part, to a significant modification of the staff's and workers' mind. The audit itself and the quasi permanent presence of consultant engineers on the spot have promoted a sense of duty translated in a general hunt for waste and a reduction of the comfort requirements of men themselves. Many small steps were taken which combined to give a net economical improvement.

1.2. Substitution of oil by thermodynamic heat

1.2.1. Further comparison of purchases between 1984 and 1985, which were climaticaly quite different, gives a first idea of the saving ascribed to this substitution.

We must first state that the old system had to be put in action again during January and February 1985 as a back-up, for some hours on Monday mornings, to ensure minimal comfort. These winter months were exceptionnally rigorous (20 % colder than normal).

Purchases of oil in 1985 went down to 316 m3 that is 81.5 m3 less than the preceeding year and 684 m3 less than average consumption before the audit.

The counterpart of this is logically an increase of electricity purchase. 1985 consumption run to 5,624 MWh compared to 4,700 MWh in 1984. The difference of 976 MWh cannot entirely be ascribed to the substitution as some modifications of the manufacturing machinery and programmes may have influenced the consumption.

It is interesting to point out that the systematic use of compressors during night has resulted in a decrease of the peak hours part of the bill and of the mean price of the kWh (Table 1).

TABLE 1

	Peak hours consumption	FB/kWh
1984	85.4 %	3.673
1985	82,8 %	3.557
1986	80.2 %	3.320

1.2.2. Detail analysis of the end use of oil purchased and the connection of independant electric energy meters in 1984 allow us to give a better comparison between 1985 and 1984.

Oil consumption for heating only went from 210 m3 in 1984 to 115 m3 the next year with a saving of 962 KFB.

Electricity consumption for the same purpose was 732.5 MWh of which 516.2 MWh were consumed by the compressors themselves with an extra cost of 1,740 KFB.

1.2.3. In conclusion, for 1985 only, gross result of the project, by comparison to average previous years, is a net gain of 7,776 KFB. But the substitution itself of thermodynamic heat to oil resulted in net loss of 778 KFB.

We show below that this situation is partly ascribable to an abnormaly high electric consumption of the groups.

2. Results of data acquisition

2.1. Difficulties met

The measuring campaign has started with the installation of the controller and its program in November-December 1984. The data collected cover only the 1984/85 and 1985/86 winters without any comparison with previous winters.

The installation of the apparatus and its connection with counters, probes, gauges, flowmeters and the like by inexperienced own staff have led to many difficulties (bad wiring, electric failure, long delays for repair, etc.). Consequently gaps exist in the data collected and unseasonable stops in the heating occur.

The flowmeters, on the other hand, were immediately faulty once (disagreement of voltage) and had to be replaced with very long delays. It is not until November 1985 for Group I and April 1986 for Group II that we collected reasonnable data, but not yet entirely reliable, because they were working at end of scale nearly all the time. Correction of this systematic bias is discussed later.

Flowmeter of Group II was only ready in July 1986 too late to be useful for the present report.

2.2. Thermal comfort, regulation and degree-days

In accordance with the factory council, the following temperature levels were used in each zone during winter 1984/85 :

13°C zone heat exchangers and their circulators on,
14°C stop of the heat emitting system.

This levels were raised by 1°C for winter 1985/86. It must be remembered that internal industrial heat covers more than 80 % of the heating load of the premises. Production works from 0600 to 2200 hours five days a week. As a result, relatively large variations of temperature occur normally. The two hourly diagrams appended (fig. 8) illustrate these variations before (29/11/84) and after (01/02/85) the substitution of thermodynamic heat, for relatively mild days.

A diurnal variation of 2 to 3 Kelvin is thus not uncommon for the measuring period. It would have be more significant before thermal insulation.

The call for heating in any one zone (temperature going down under fixed level) alerts the controller which puts the heat pump system in action. This is programmed for the cascade starting of compressors in the three groups and full working until water temperature out of the network reaches 45°C. Afterwards, modulation of the number of compressors on keeps this level.

Heat emission by the heat exchangers is so directly related to the thermal production by the groups.

This schedule goes on as long as waste industrial heat is available from boilers and/or corrugators.

This condition is fullfilled :

- in boiler room : as soon as temperature at a hight of 1,5 m reach 15°C, with stop under 10°C;
- over corrugators : as soon as difference of temperatures between attic and the height of 1,5 m reaches 10 K, with stop under 5 K.

From the beginning of the 1985/86 winter, if the call for heating in at least one zone goes on and no more waste industrial heat is available, Group II and III turn over to the batteries immersed in Senne water. This modification of the program has allowed an increase in the off-peak and week-end operation of the system without having yet to ensure the temperature asked for after a prolonged pause of production during a cold spell, especially on Monday morning.

A back-up heating system is planned for next winter to overcome this difficulty, heat being drawn from a neighbouring conventional central system through an exchanger.

Figure 9 gives, for a relatively long period, the variation of the mean daily temperatures, inside and outside, as registered by the controller.

One must notice first that the mean outside temperature registered by the controller differs somewhat, especially on week days, from the same registered, some kilometers away, at the Uccle Observatory. The origin of this difference is in the position of the probe on the wall of the factory, protected from insolation but not from hot air flowing out the nearby quick opening door of the corrugator room.

During production, temperatures registered are then biased upwards.

Totalization of mean daily temperature differencies between heated premises and outside gives the number of degree-days for each heating season biased to the low side. They are given in Table 2.

TABLE 2

Winter	1984/85	1985/86
November	-	189
December	148.7	125.6
January	358.6	251.1
February	262.2	280
March	242.9	212
April	-	131.6
Total	1012.4	1189.3

2.3. Electricity consumption and coefficient of performance

As can be seen, the increase of 1 K of the temperature required in the premises and the lengthtening of the heating season combined to give a increase of more than 17 % of the heating load.

The absence of backing up by oil burning during the second winter has led to a further increase in consumption of electricity by the thermodynamic system. The resulting consumption for the two seasons is given in Table 3.

The difference between the two successive winters (+ 62 %) cannot be explained anyhow by the reasons quoted before. It must be sought in an appreciable lowering of the system performance.

Unfortunately, we do not have complete and reliable data on the water flows through the condensers and hence rather no computation of the overall coefficients of performance of the groups.

During the period when the flowmeters were active, for Group I and III, we have computed the apparent average monthly coefficient based as said before, on end of scale functionning of the flowmeter (Table 4).

TABLE 3

	1984/85 kWh	%	1985/86 kWh	%
Auxiliaries (circulators and heat exchangers)	159,152	25.7	277,117	27.6
Group I (boiler room)	45,183	7.3	95,821	9.5
Group II (corrugators 18-20)	200,470	32.3	257,732	25.7
Group III (corrugator 21)	214,990	34.7	373,741	37,2
Total season	620,395	100	1,004,411	100

TABLE 4

Month	Dec.	Jan.	Feb.	Mar.	Apr.	Mean
Group I	2.92	2.61	2.90	3.00	2.39	2.80
Group III						
air to water	-	-	-	-	1.86	
water to water	-	-	-	-	2.58	2.24

Instant measurements with a differential manometer on the primary pump of Group III enable us to evaluate to about + 20 % the bias between the actual flow and the flow metered by the controller.

The averaged Coefficient of Performance, computed at the end of the season, must have been about 2.7. Knowing that the auxiliaries use 26 to 27 % of the total consumption of electricity, the seasonal Coefficient of Performance must have been in the order of 2.1, clearly under the 2.5 figure forseen at the initiation of the project.

Many hypotheses, alone or combined, may explain this state of affairs :

a. optimistic predictions;
b. early wear of the compressors;
c. lack of lubrication;
d. lack of refrigerant at the evaporators and evaporation temperatures too low.

Check-up and group repairing are being done at the time of writing this report. Hopefully they will tell us which of these hypothesis was predominant.

VI. CONCLUSIONS

Taking into account the novelty of many parts of the system and the hazard normally associated with any demonstration project, two conclusions may be drawn from results and costs so far :

a. Technically, the possibility to retrieve energy lost in the atmosphere of industrial premises by a static thermodynamic system has been demonstrated. A Coefficient of Performance of heat pump in the order of 3 may be expected for an optimised installation in good order.

b. Economically, in the actual conditions of the market (July 1986), the proposal is not paying back. Results show that the cost of the energy saved (11,846 FB/Ton Equivalent Petroleum) is superior to the cost of one ton of oil subsituted. The economical viability of the system is questionable.

At the very least, the combination of the chosen actions, reducing of the thermal load and substitution of thermodynamic heat for oil burning, gave a net profit during the two seasons reviewed.

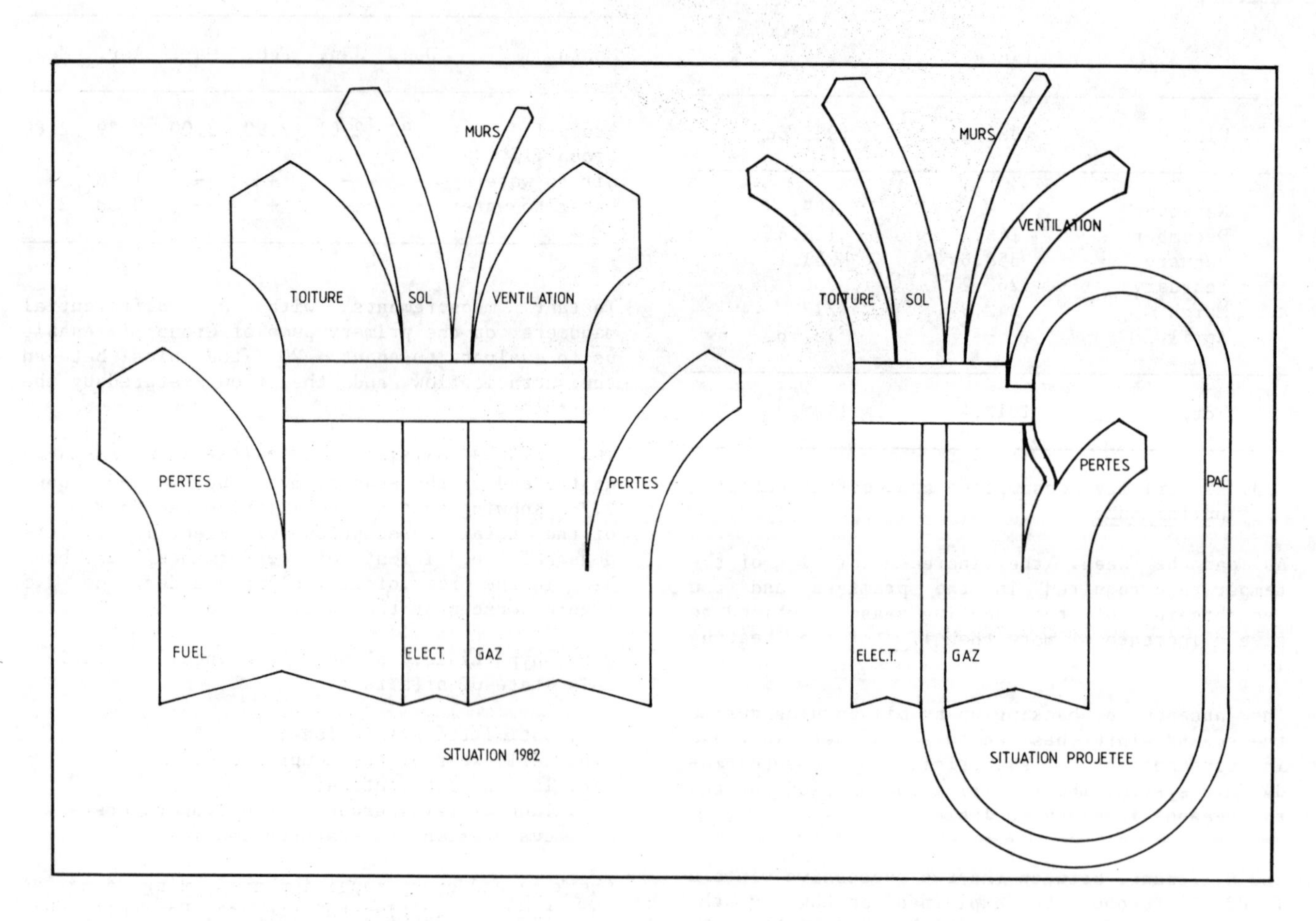

Fig. I Thermal flows demonstrated and suggested by the I982 audit

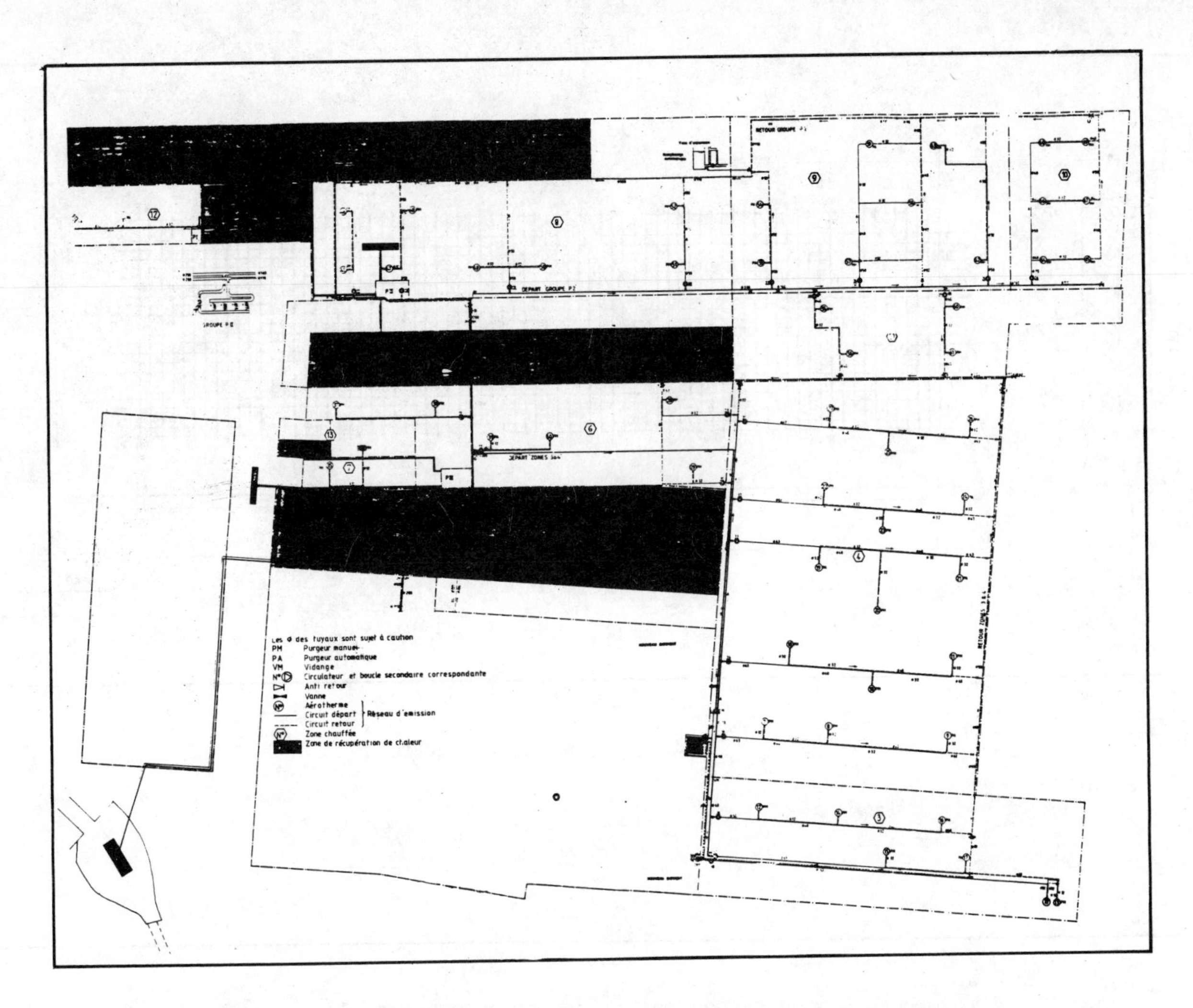

Fig. 2 Ground plan of the factory Darkened: Zones where retrieval of heat is possible

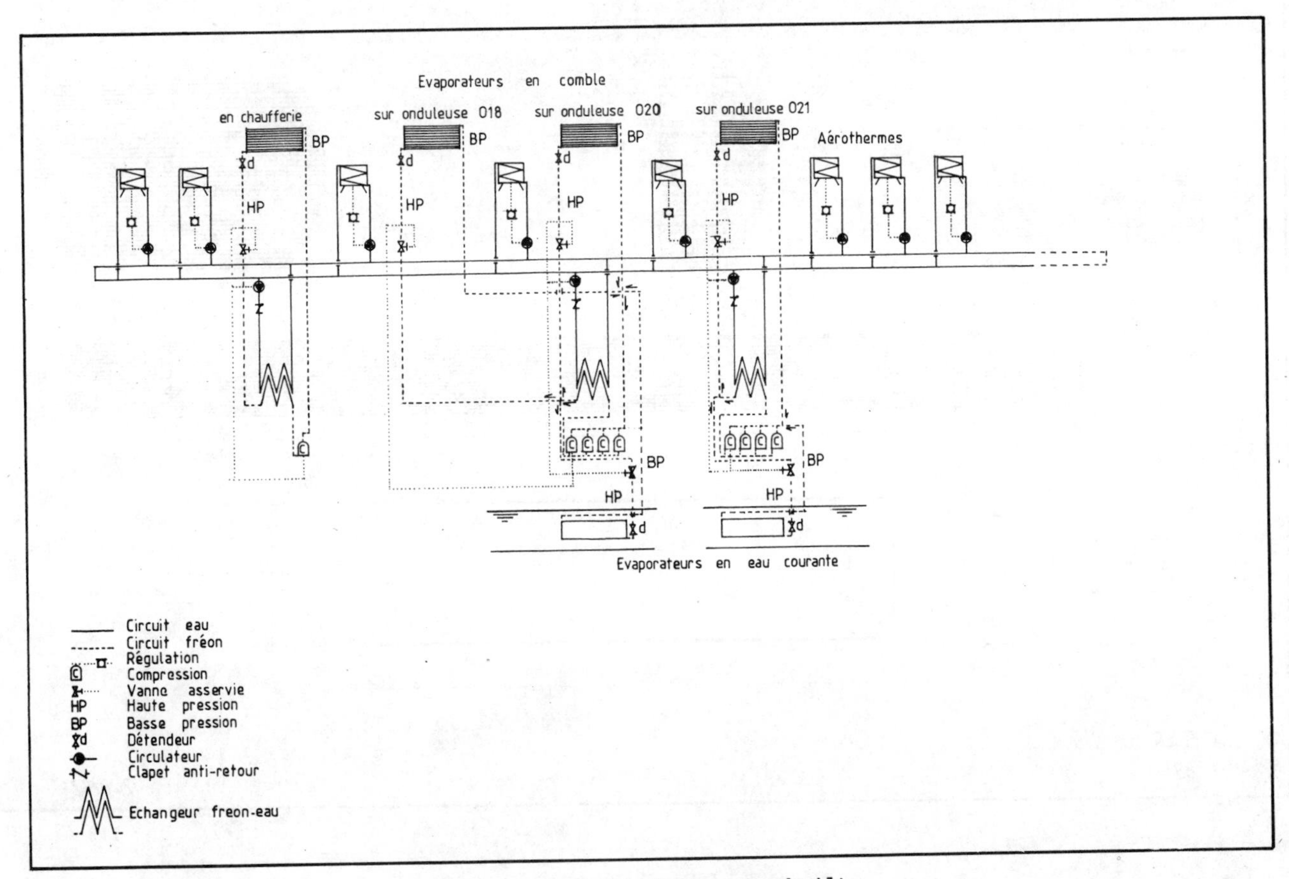

Fig. 3 Working diagram as built

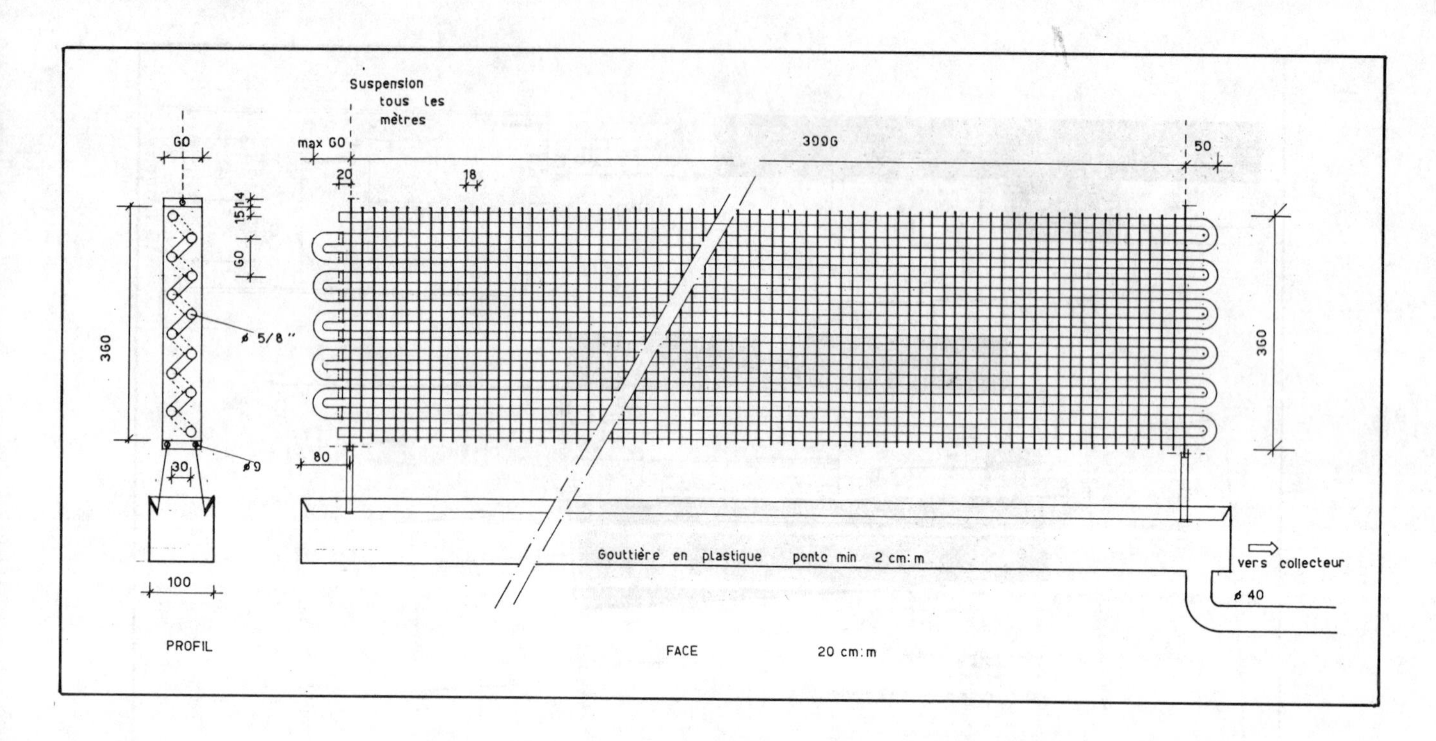

Fig. 4 Static evaporator and gutter

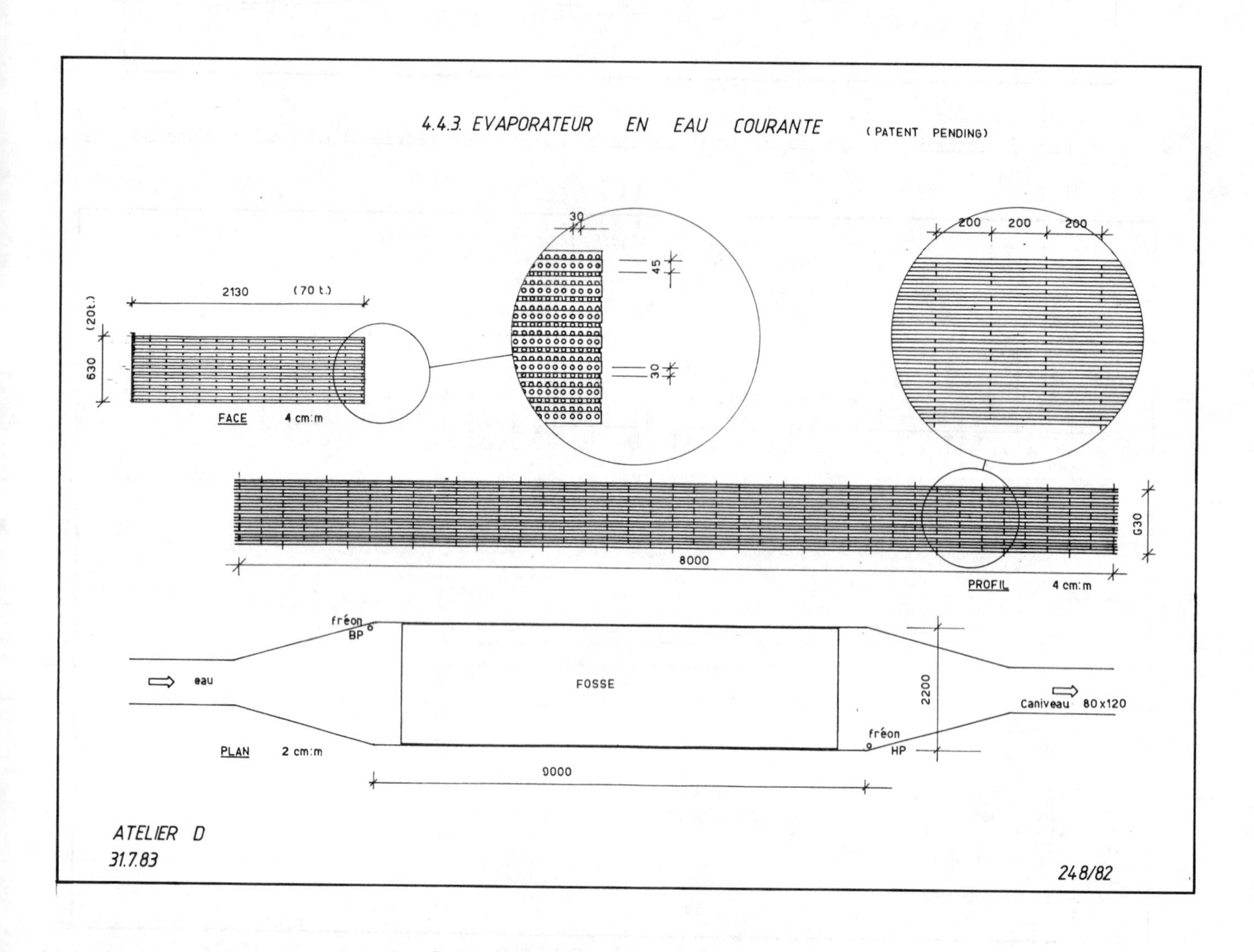

Fig. 5 Evaporator battery and trench scheme

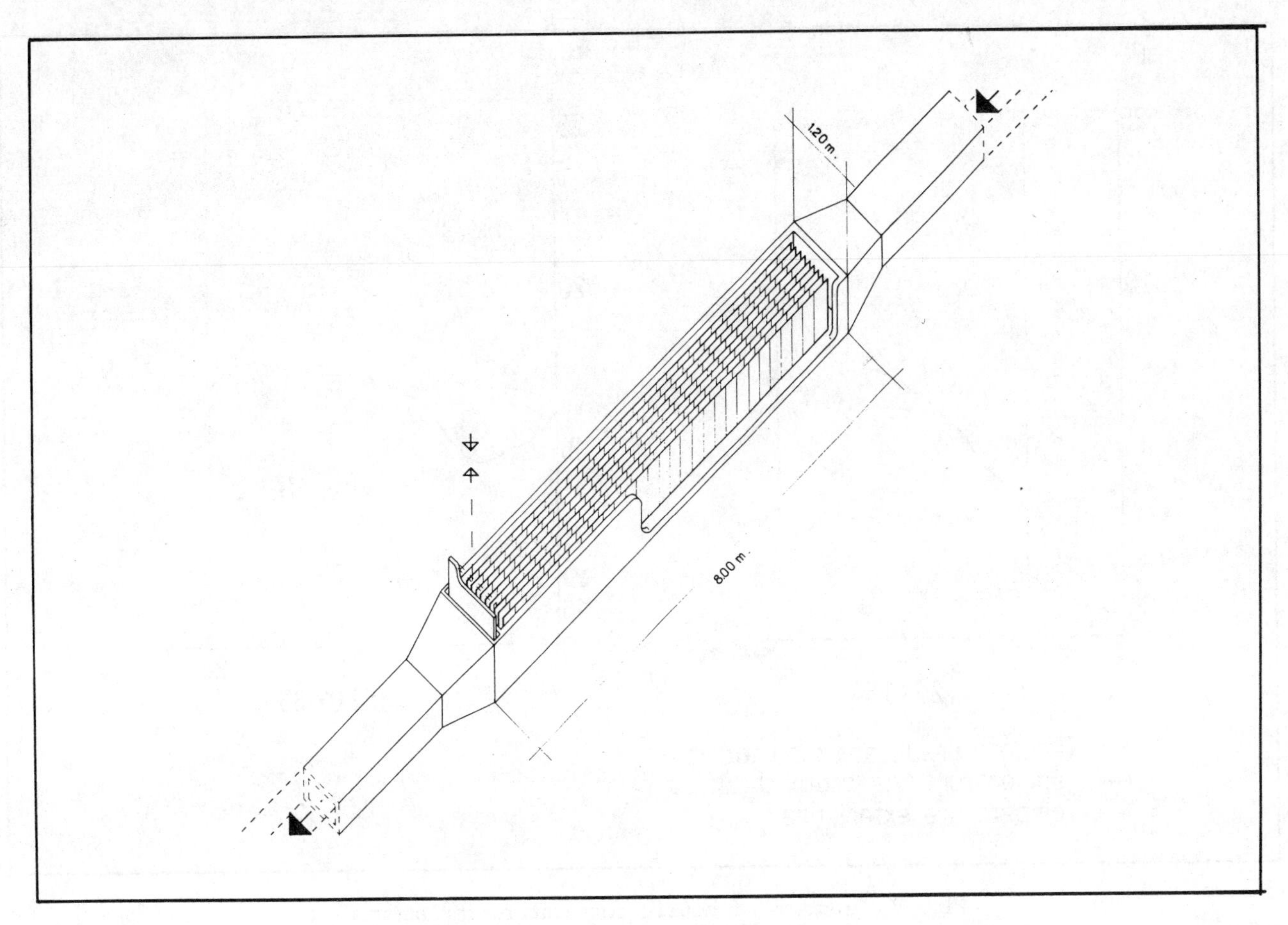

Fig. 6 Perspective view of evaporator in its trench

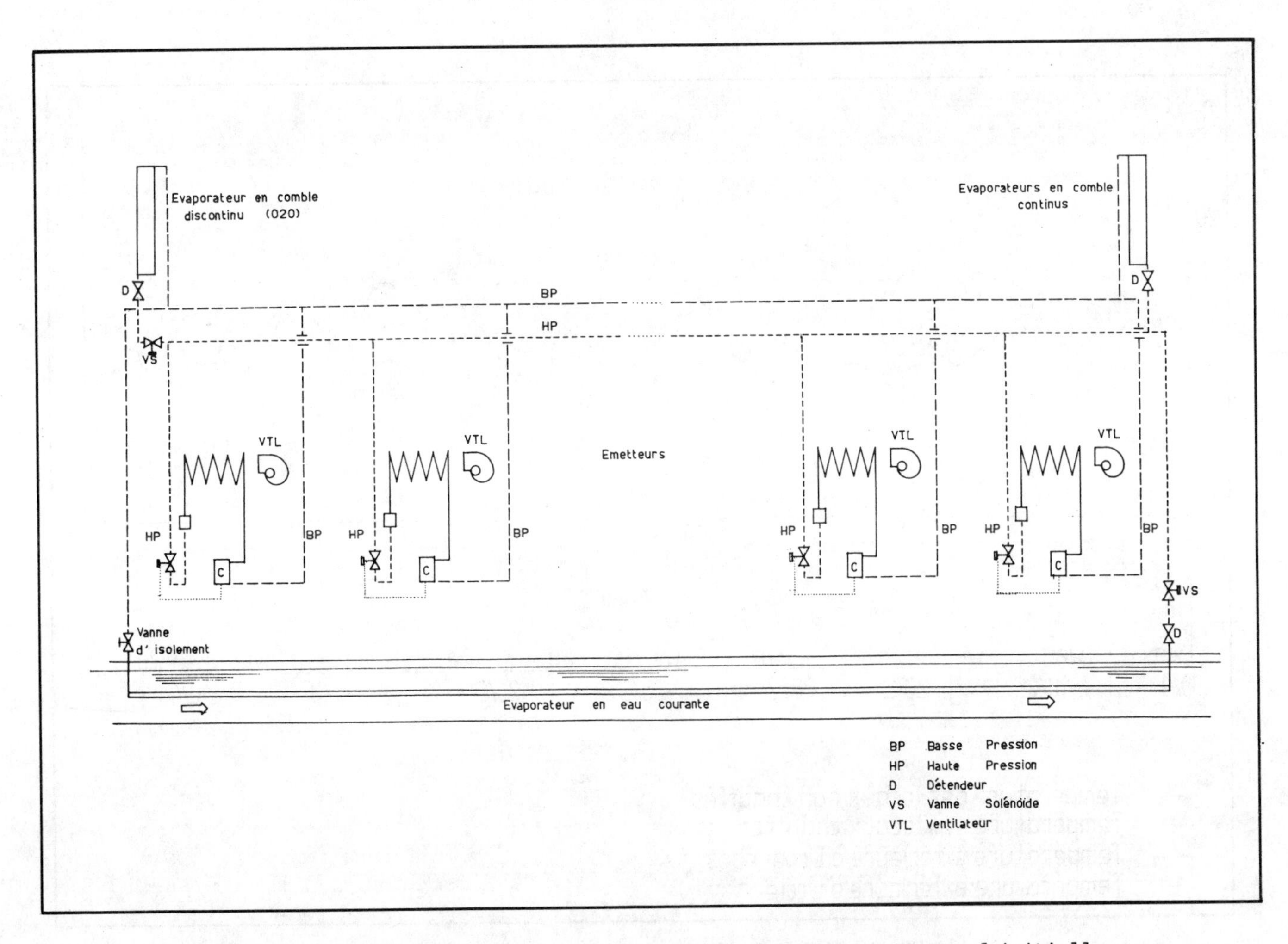

Fig. 7 General scheme of an air water to air heatpump as proposed initially

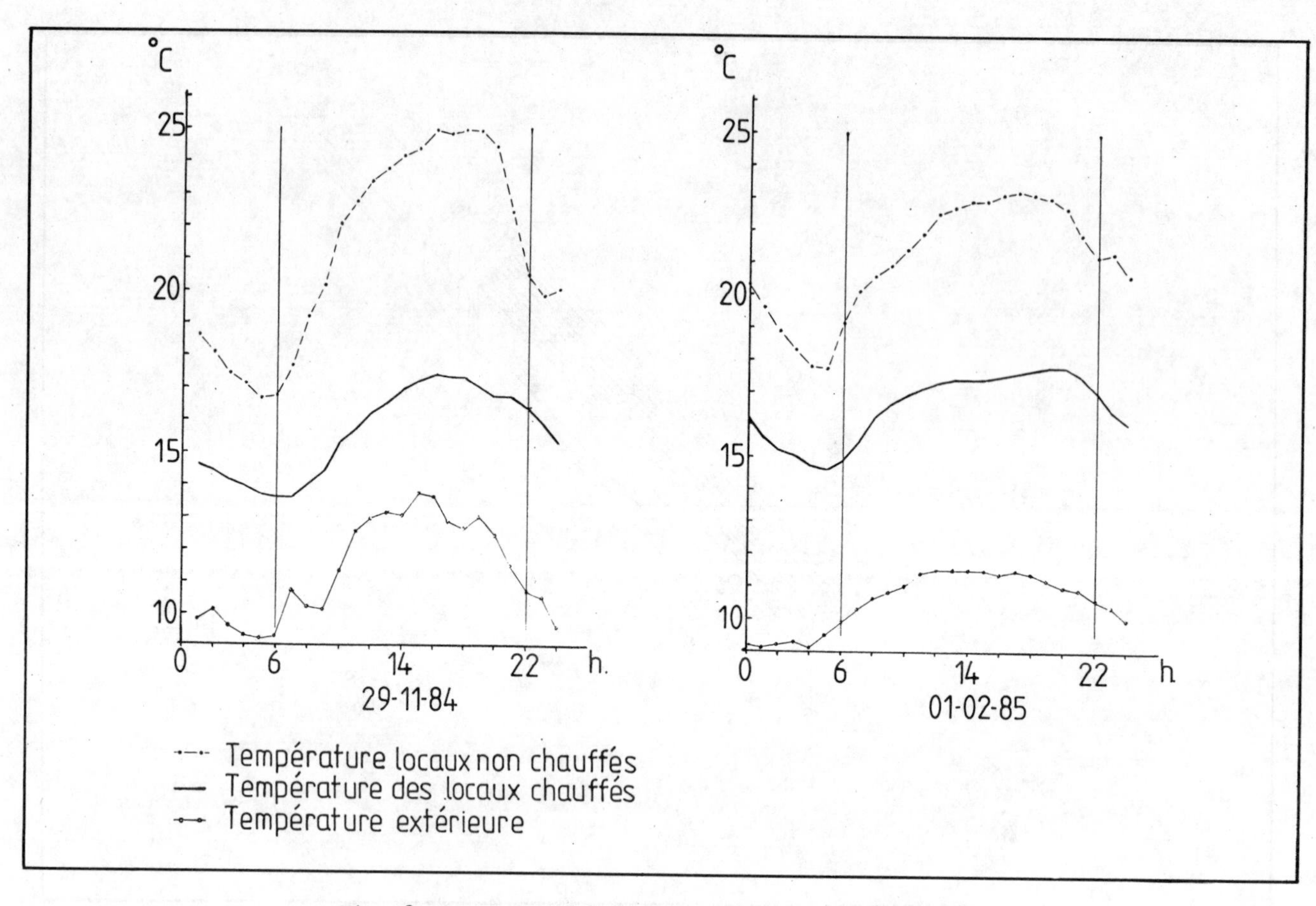

Fig. 8 Diagrams of hourly temperature variations

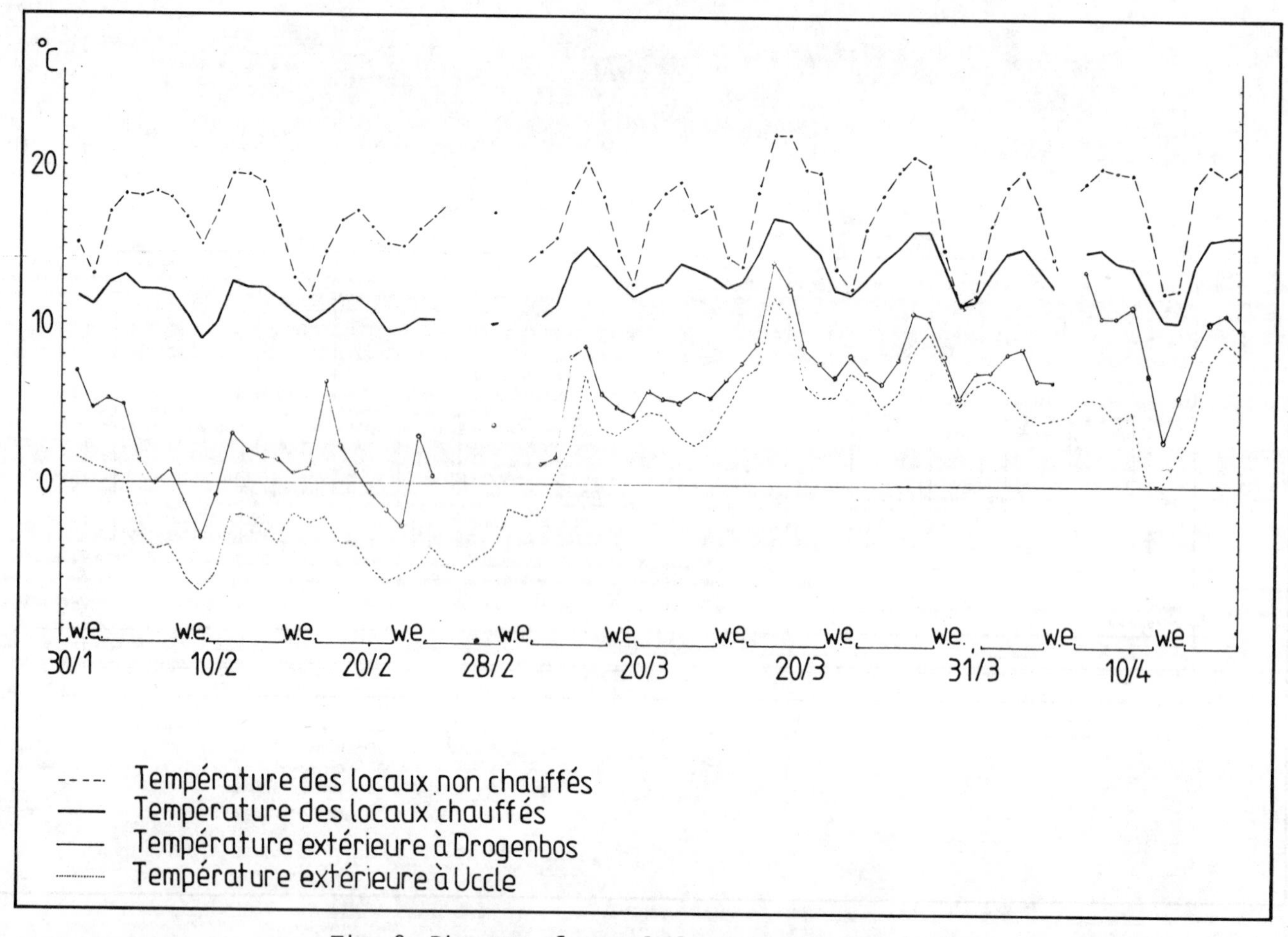

Fig. 9 Diagrams of mean daily temperature variations

3rd International Symposium on the

Large Scale Applications of Heat Pumps

Oxford, England: 25-27 March 1987

PAPER A3

A LARGE ABSORPTION HEAT PUMP
Measurements on and experiences of the first installation in Sweden

M. Wimby
P-Å. Franck
T. Berntsson

The authors are M.Eng. Martin Wimby, M.Eng. Per-Åke Franck and Prof. Thore Berntsson. M. Wimby is in the Department of Chemical Engineering Design, P-Å. Franck and T. Berntsson in the Department of Heat and Power Technology at Chalmers University of Technology, Göteborg, Sweden.

ABSTRACT

In November 1984 the first large absorption heat pump (AHP) in Sweden was installed. The AHP delivers 7 MW heat to a district heating system, using excess steam as drive energy and cooling water as heat source. The steam and the cooling water originate from a near by chemical plant as waste energy. The heat pump is manufactured by SANYO. It is one of the largest AHP (type I) in the world so far.

The generator is heated with 10 bar steam and in the evaporator, water is cooled from 60°C to 30°C. The heat pump raises the temperature of the district heating water from 60° C to 80° C. The working pair used is Lithium bromide-water. A single control loop reduces the steam flow when the temperature of the district heating water exceeds 83°C.

In this paper the energy balances and internal behavior of the heat pump are presented. The energy balances were continuously determined from temperature and flow recordings every other minute in the external streams of the AHP. The internal behavior - such as heat transfer - have been studied by means of surface temperature sensors.

The COP was found to vary between 1.3 and 1.8 with a mean value of 1.6. Since startup the availability has been 100 % . The AHP is very flexible with load variations between 0 % and 100 % without complications. Corrosion has been avoided by means of inhibitors and a purge pump system.

INTRODUCTION

In the beginning of the eighties, the district heating company of Trollhättan (TFAB) was offered to buy heat from Kema Nord, an electrochemical industry situated close to the district heating central at Trollhättan. The offer comprised heat at two temperature levels: 150°C and 30°C. This created an ideal situation for a type I absorption heat pump (AHP). TFAB accepted the offer and invested in a SANYO type I AHP. In November 1984, the 7 MW AHP was installed. At that time, the AHP was the largest one in the world.

The AHP at Trollhättan was the first large-scale unit in Sweden. In 1985 and 1986, 3 more type I AHP have been installed, one by SANYO and two by Hitachi.

TFAB has received financial support for the AHP-project from the Swedish Council for Building Research, who also sponsors the independent evaluation of the project presented in this paper.

THE HEAT SINK AND THE HEAT SOURCES

The AHP is one of several production units in the district heating network. The AHP operates in the low temperature range, typically heating the district heating water from 57°C to 75°C.

The high temperature heat source is 10-bar steam, which is produced in a hydrogen-fired boiler. The hydrogen is a by-product in the electrochemical process. Due to security reasons, the hydrogen has to be destroyed and since the need for high temperature heat in the process is very small, the steam is regarded as waste heat.

The low temperature source is cooling water from crystallizers. The temperature ranges from 60°C to 30°C.

THE CYCLE

The AHP is a single-stage one with Lithium-bromide/water ($LiBr/H_2O$) as working fluid. Figure 1 shows a process diagram over the heat pump cycle. In the evaporator, water evaporates and the cooling water from Kema Nord is cooled down. The water vapour is absorbed by the LiBr-solution in the absorber, and thereby the district heating water is heated. The solution is pumped from the absorber to the generator via the solution heat exchanger and the condensate cooler. In the generator the absorbed water is evaporated. The concentrated solution is then cooled down in the solution heat exchanger and throttled into the absorber. The vapour condenses in the condenser and is throttled back to the evaporator.

THE MEASUREMENTS

To suit the requirements of the measurement program, the AHP was equipped with a measurement system which records the temperatures in all streams and the flows in the external streams, every second minute. Hourly mean values are stored on a floppy disc. Platinum resistance thermometers (PT 100) with a maximum error of ±0.03 °C in the external streams and ±0.3 °C in the internal streams are used. The district heating water flow

Held at St. Catherine's College Oxford, England. Organised and sponsored by BHRA, The Fluid Engineering Centre, Cranfield, Bedfordshire, MK43 0AJ England.

and the cooling water flow are measured with magnetic inductive flow meters with a maximum error of ±0.25 %. The condensate flow is measured with a turbine flow meter with a maximum error of 1 %. The average heat balance error is 2 % with ±2 % hourly variations.

The measurements were started in March 1985 and they will continue until April 1987.

THE HEAT EXCHANGERS

As can be seen in Figure 2, the absorber, the condenser, the generator and the evaporator are horizontal tube heat exchangers with the external streams on the tube side. The absorber and the evaporator are placed in the same shell. They are both of the falling-film type and the evaporator part is equipped with a recirculation pump, which ensures the wetting of the heat exchanger tubes. The liquid from the condenser is sprayed on the evaporator tubes. The condensate that does not evaporate is collected in an open container and pumped back to the spray nozzles. If condensate accumulates, the open container overflows and the water flows into the absorber and dilutes the LiBr-solution without heating the district heating water. This arrangement works as a concentration controller that effectively prevents crystallization in the absorber. In order to create a more counter-current-like flow, the external streams are passed four times in the evaporator and in the absorber.

The condenser and the generator are not built in the same shell, but they are connected with six wide and short tubes in order to minimize the pressure drop. The cooling water is recirculated from the outlet to the inlet of the evaporator.

The mixing lowers the temperature, hence this arrangement causes irreversible losses, but since the evaporator only is capable of working at one temperature, the loss is inevitable. The advantage gained is a higher water velocity in the evaporator, and hence a higher heat transfer coefficient.

The condensate cooler is necessary since the condensate system at Kema Nobel is held at atmospheric pressure. Temperature measurements show that 40 % of the solution from the absorber passes through the condensate cooler. An alternative solution would be to cool down the condensate with the outlet of the district heating water.

The circulation ratio, defined as massflow of solution pumped to the generator divided by massflow of condensate from the condenser, has been estimated to lie between 10 and 20 at full load. Since the solution pump is not controlled, the circulation ratio increases as the heat load decreases. The LiBr concentration in the absorbent circuit varies widely depending on temperatures in the external streams. From temperature measurements, the concentration interval has been estimated to 40 - 62 %.

To keep the low pressure part of the AHP free from air, a purge pump is connected to the evaporator. The pump is in operation one hour once a week.

The heat pump is controlled with a throttle valve on the steam pipe. The valve closes automatically when the district heating temperature reaches 83°C. It also closes when the generator temperature exceeds 145°C.

DISTRIBUTION OF TEMPERATURE DIFFERENCES

The concentration controller described in the previous paragraph introduces an extra degree of freedom, since it allows the absorber/evaporator to work independently of the generator/condenser. This makes the heat pump robust in the sense that the low temperature energy source can be shut off without any serious implications. Without the cooling water, the heat pump works as a heat exchanger which heats the district heating water with steam.

The concentration controller also affects the distribution of the available temperature differences. As can be seen in Table 1, the logarithmic mean temperature differences (LMTD:s) are lower in the evaporator and the absorber than in the generator and the condenser. The difference between the saturation temperature of the solution in the absorber and the saturation temperature of the water in the evaporator increases as the LiBr concentration in the absorber increases. Since the concentration is limited, the temperature gap between the absorber and the evaporator is limited. This implies that the only way to increase the gap between the low temperature heat source and the heat sink is to lower the logarithmic mean temperature differences in the absorber and the evaporator. If it were possible to operate at higher LiBr concentrations, the heat pump could be optimized with respect to heat exchanger surface. Such an optimization would probably result in a more even distribution of the temperature differences.

OPERATIONAL EXPERIENCES

The extensive measurement program has enabled a careful evaluation of the heat pump. The original design values are given in Table 2. During the first winter (-84 and -85) it was found that the temperature of the inlet district heating water was too high, which decreased the heat output drastically. The average output during this period was approximately 3 MW. In June -85, the maximum concentration was adjusted by replacing some of the water with LiBr. The temperature controller on the district heating water was set on 83°C instead of the earlier 75°C. Approximate design values after this change are given in Table 2 . After the adjustment, high temperatures in the district heating network have caused no problems.

The COP, defined as heat delivered to the district heating system divided by heat from steam and condensate, shows small variations. As can be seen in the COP duration curve in Figure 3, the average COP value is 1.6. The low COP-values arise from the situation when the cooling water is shut off. COP values higher than 1.8 should be neglected since they are caused by inaccurate temperature difference measurements at part load.

During one summer month, the heat pump is shut off due to the low energy consumption. This can be seen in the duration plots in Figure 4. During approximately 1000 hours per year, the heat pump operates at part load, mainly because of shortage of steam caused by steam demand peaks in the chemical plant. Due to operation disturbances the cooling water is turned off for some hundred hours per year. In Figure 4, it can be seen that the output is higher than 6 MW for 5500 hours per year.

Generally speaking, heat pumps tend to decrease their efficiency when they are most needed . As is shown in Figure 5, this is not the case here. The figure shows monthly energy amounts from the heat sources and to the heat sink. The reason is that the required temperature lift is smaller than the design temperature lift.

Absorption heat pumps are known to be reliable. The AHP at Trollhättan is no exception, it can not be blamed for any stops. The reliability is important in complex systems like this, since disturbances propagate to both the heat sources and the heat sink.

Though evaporator temperatures lower than 25 °C have been measured, no serious air leaks have been observed.

CONCLUSIONS

The absorption heat pump is reliable. During our measurements the availability was 100 %.

The average COP was 1.6.

The absorption heat pump operated at full load for 5000 hours per year.

ACKNOWLEDGEMENTS

The authors wish to thank Dr. Leif Nilsson and Dr. Kjell Schroeder for setting up the measurement system and Mr. Gunnar Fasth of TFAB for interesting discussions and for his hospitality. They also wish to thank the Swedish Building Research Council for financial support.

	Abs.	Gen.	Evap.	Cond.
Inlet ext. sream	65.9	154	50.1	73.7
Outlet ext.	73.7	154	34.3	79.9
Inlet int.	81.1*	139.9	34.1	81.4
Outlet int.	72.6	139.9	34.1	81.4
LMTD	7.1**	14.1	3.6	3.8

* Calculated value, based on generator, condenser and evaporator temperatures.

** Counter flow is assumed.

Table 1. Temperatures and logarithmic mean temperature differences (LMTD) in the different heat exchangers at full load (86-01-01, 11 pm).

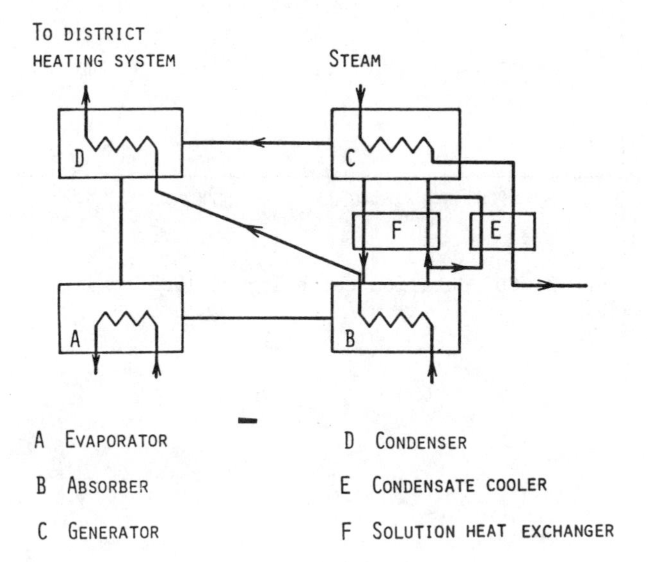

Figure 1. The absorption heat pump.

	Original	After adjustment
Temperatures (°C)		
District heating, in	57	65
District heating, out	75	83
Cooling water, in	40	50
Cooling water, out	30	35
Steam	180	180
Heat loads (MW)		
Cooling water	2.8	2.8
Steam	4.2	4.2
District heating	7.0	7.0

Table 2. Design values before and after June 1985.

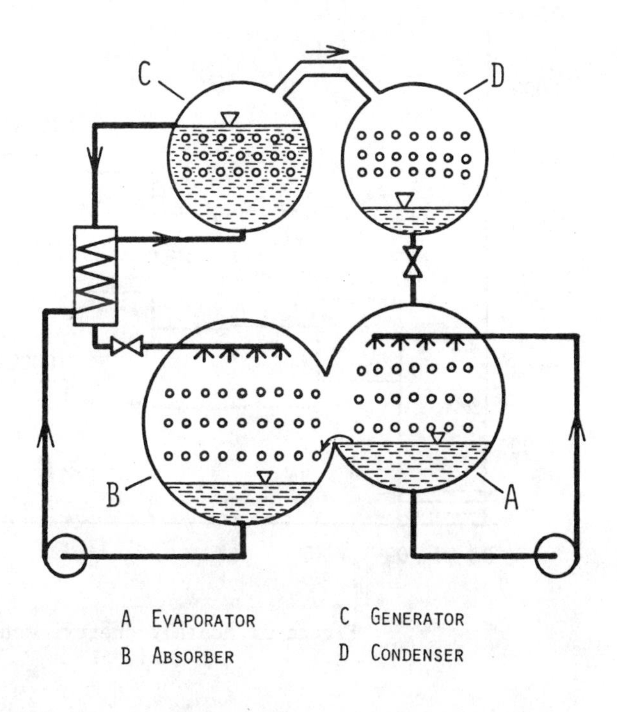

Figure 2. A cross-section of the absorption heat pump.

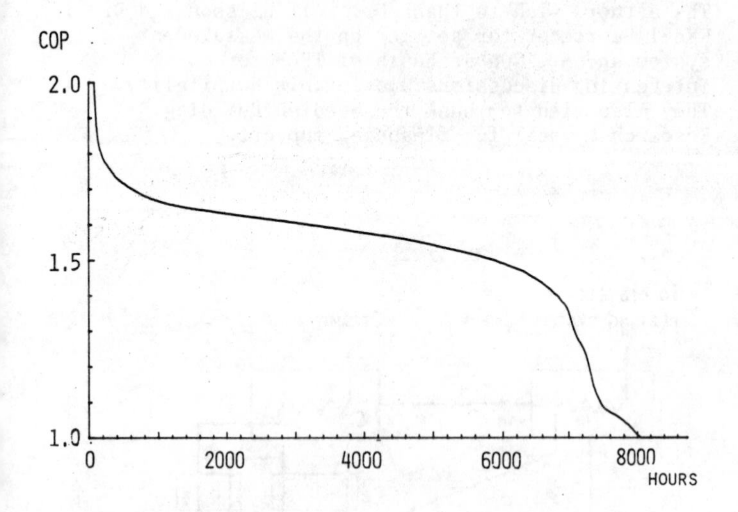

Figure 3. COP duration curve August 1985 to July 1986.

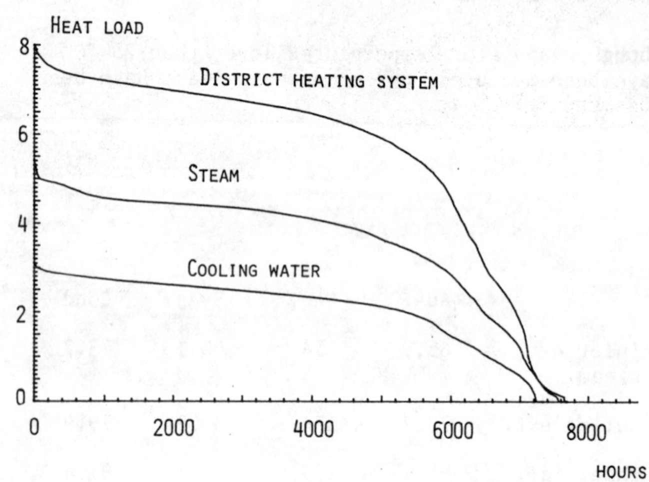

Figure 4. Heat load duration curves August 1985 to July 1986.

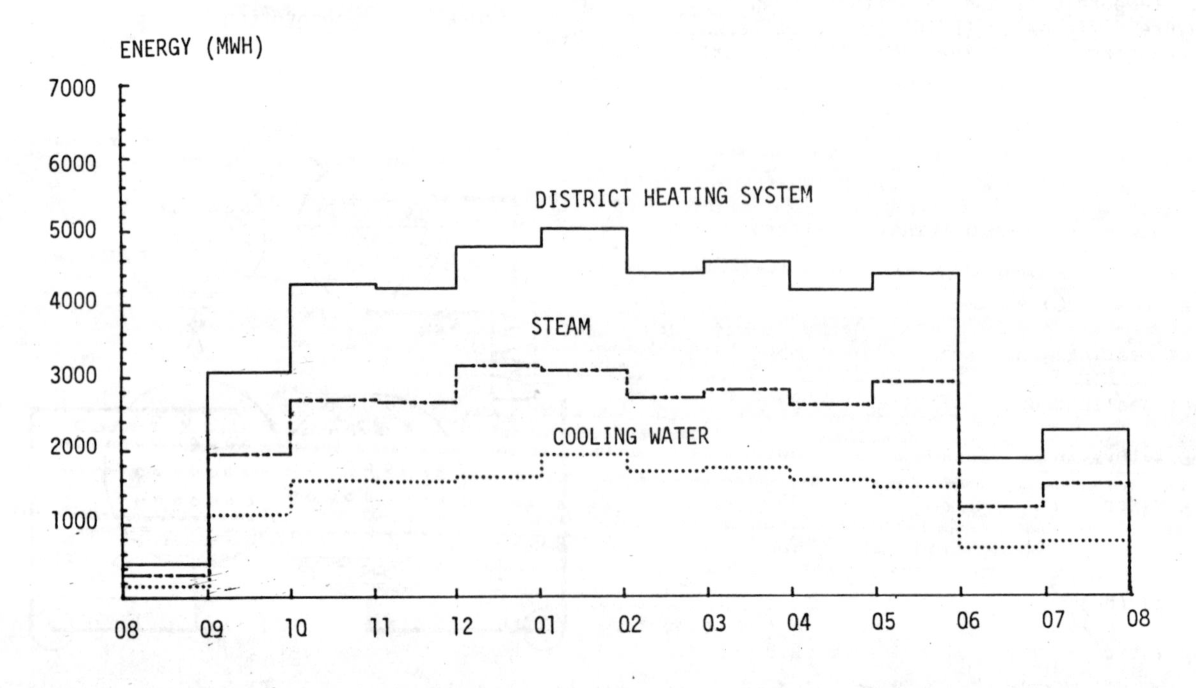

Figure 5. Monthly energy amounts August 1985 to July 1986.

3rd International Symposium on the

Large Scale Applications of Heat Pumps

Oxford, England: 25-27 March 1987

PAPER B1

REVIEW OF FIVE INDUSTRIAL HEAT PUMP DEMONSTRATION PROJECTS

By A W Deakin and R Gluckman

Dr A W Deakin is a project officer in the Energy Technology Support Unit AERE Harwell.

Mr R Gluckman is a director of the March Consulting Group.

SYNOPSIS

The performance of five industrial closed cycle heat pump projects is reviewed from both the technical and economic viewpoint. Lessons from the projects, along with recommendations for future installation are given.

Market prospects for heat pumps in the UK are reviewed in the light of the recent decreases in fossil fuel prices. To increase the competitiveness of heat pumps work is required to markedly improve performance. Particular care should be given to selecting applications to ensure firstly that the system will give net benefits to the whole process and secondly to ensure the heat pump operates at the high COP's necessary for economic viability.

1. INTRODUCTION

During the past seven years the Energy Efficiency Office of the UK governments Department of Energy have supported a number of heat pump projects under the Energy Efficiency Demonstration Scheme. This review will concentrate on the performance of five closed cycle heat pump projects. These are:

(a) Heat Pump 1 - A steam turbine driven unit in a food processing company that recovers heat from a cooling tower circuit and delivers heat to boiler feedwater.

(b) Heat Pump 2 - An electric heat pump installed in a dairy, producing UHT milk, that recovers heat from steriliser cooling water and delivers heat to various users at 70°C including boiler feedwater, process water and space heating.

(c) Heat Pump 3 - A gas engine driven heat pump at a textile finishing works used to recover heat from dyehouse effluent and supply heat to process water and boiler feed-water.

(d) Heat Pump 4 - a gas engine driven heat pump at a food processing plant used to recover heat from the exhaust of a dryer in order to preheat incoming air to the unit.

(e) Heat Pump 5 - a gas engine driven heat pump at a food processing plant operating in a similar way to Heat Pump 4.

2. PERFORMANCE

Table 1 summarises some general points about each heat pump. Further details of the five heat pumps can be obtained from ETSU, AERE Harwell, Oxen. OX11 ORA

Economic Performance

The most important criterion for measuring the success of any heat pump is the financial return. The only reason for buying an expensive and complex device like a heat pump is to reduce energy costs. There are many ways to present the economic performance of any energy saving investment. Ideally one should use discounted cash flow methods but this is difficult when comparing units in five different companies (each with its own accounting methods). For this reason we shall apply simple payback periods.

In all cases, except Heat Pump 2, the heat pumps were retrofit ie alternative heating methods were already installed. In the case of Heat Pump 2, although the heat pump was designed as part of a new plant, it was not possible to make any offset against the capital cost of the heat pump; full size boilers were required for start-up and standby. Thus in all five demonstration projects the full capital cost had to be recovered by energy cost savings.

Table 2 summarises the financial details of each heat pump. It is interesting to note the following points:-

(a) Capital costs varied between £92/kW and £463/kW. The low cost of Heat Pump 2 is because the unit is electrically driven and it is based

Held at St. Catherine's College Oxford, England. Organised and sponsored by BHRA, The Fluid Engineering Centre, Cranfield, Bedfordshire, MK43 0AJ England.

on large, packaged, cetrifugal heat pumps. These units were developed from mass produced American air conditioning chillers, hence their low cost. The price of Heat Pump 3 is artificially high at £463/kW because half the heat duty is from passive heat exchangers. However, the heat pump is still relatively expensive because it is very much smaller than the other units.

(b) The price difference between Heat Pump 4 and 5 partly reflects the effect of site specific installation cost. The heat pumps are similar in many respects and yet Heat Pump 5 is much more expensive. However, Heat Pump 4 is twice the size of Heat Pump 5 hence economy of scale contributes to this difference.

(c) None of the projects achieved the projected savings. The two dryer heat pumps (Heat Pumps 4 and 5) have performances reasonably close to design after teething troubles were sorted out. The other three units have all performed very badly with payback periods between 200% and 400% of design values.

The key issue raised by these figures is whether the actual achieved performances will be repeated in new installations. If they cannot be improved then the potential market for heat pumps is small. If lessons can be learned from these projects to enable second generation heat pumps to attain good performance, then the technology of heat pumps should be encouraged. In order to review this issue it is necessary to examine the technical performance of the demonstration heat pumps from two viewpoints:

- why did the systems fail to meet their design performance
- could the actual design have been improved.

Technical Performance

Rather than describe the performance of each unit in detail this section will describe the type of problems experienced, the probable causes and possible improvements.

One of the main reasons why three out of five of the installations did not meet their projected performance was due to oversizing. In particular heat pump 2 was almost twice as large as it needed to be. The oversizing problem was further excacerbated for heat pumps 1 and 2 because of poor performance at part load. Mismatching problems of this type could have been avoided at the design stage.

Heat pump 4, after overcoming a number of problems such as heat exchanger corrosion, failures of reciprocating compressor shaft seals and off loading gear, performed within 10% of the projected design values. In contrast heat pump 5, which in many respects is similar to heat pump 4, has suffered from a series of reliability problems and was eventually scrapped. The list of problems covers almost all parts of the system and exemplifies the difficulties that 'first generation' heat pump designers have had. In retrospect some of the problems are obvious and could have been avoided with more thought at the design stage. Many of the problems stemmed from two basic causes:

1. The gas engine air intake was located inside the plant room. Leaking R12 was drawn into the engines and caused severe corrosion in the engine and exhaust system. Air intakes must be ducted outside the engine room to a position where they cannot be affected by refrigerant leakage. Flourocarbons break down into highly corrosive products at the high temperatures generated in an engine.

2. No adequate provisions were made to reduce the vibration between the engine and compressor. Torsional vibration resulted in a number of failures of engine components (eg timing gears, magnetos). Design of engine/compressor mountings and couplings needs specialist attention on new heat pumps.

Another problem encountered on the gas engine driven machines has been the larger than expected maintenance costs. This may in part be due to plant engineers lack of understanding but is also due to the inherent complexity of gas engine driven heat pumps. Since operators are concerned to reduce the need for skilled labour on site and to reduce the maintenance time to a minimum every effort should be made to minimise heat pump maintenance costs.

3. THE MARKET POTENTIAL FOR HEAT PUMPS IN THE UK

Analysis of the market potential for heat pumps is very complex. Heat pumps include a wide range of different types and system options. Heat pump types include closed cycle compression, open cycle compression, absorption and reversed absorption machines. As well as these basic cycles there are many options with regard to choice of operating fluids, compressor type, compressor drive etc.

The type of situation where heat pumps might be applicable are very diverse, some of the more obvious are:

Water Heating

The main categories of water heating are:

- o boiler feedwater
- o process water
- o washing water
- o space heating water
- o water as an intermediate heat transfer fluid

Boiler feedwater is often required when processes use a lot of steam directly, without any condensate return. In these circumstances water must be heated from about 10°C to 90°C.

Process water covers a wide range of possibilities which are very site specific. Typical heating temperatures are similar to boiler feedwater although in some cases temperature requirements are much lower.

Washing water is often used in large quantities in many factories. Typical temperature requirements are 60°C. Modern trends are towards using cooler water together with chemicals, instead of very hot water.

Water is sometimes used as an intermediate heating fluid, for example to heat air in dryers. Often systems will be pressurised to allow water delivery temperatures of, say, 120°C. Unlike the other examples above the water is not heated from cold. The return temperature is usually only 5°C or 10°C below the delivery temperature.

Evaporation Processes

Evaporation and concentration processes are one of the best potential markets for open cycle and absorption heat pumps. Many examples of such processes are to be found in the food and chemical industries.

Distillation Processes

Another boiling process that is applicable is distillation.

Drying

In a number of convection drying processes heat pumps are applicable. In this situation the heat pump uses moist exhaust air as a heat source as in the malt dryers described earlier in section 2.

Combined Heating and Cooling

Applications with simultaneous requirements for refrigeration and heat at up to, say, 60°C will often lead to good heat pump applications.

Space Heating

Factory heating is often a heat user in heat pump systems although limited hours of operation reduces the economic potential.

The various applications described above are relevant to a wide variety of industrial processes. ETSU have reviewed the UK market potential for industrial heat pumps on an industry by industry basis taking into account the typical applications described above. A more recent look at the chemical industry by Mercer (1986) has identified a greater potential for open cycle heat pumps on evaporation and distillation processes than that of the earlier ETSU study. Table 3 shows the combined results of these studies. The overall energy saving potential for industrial heat pumps is about 24 PJ/yr.

The above estimates took no account of the recent findings of process integration on the correct placement of heat pumps. To be of benefit to the overall process heat pumps must be placed across the process pinch. Taking account of this principle will undoubtedly reduce the potential savings given above. Using evidence from fourteen completed process integration studies it seems likely that about a third of the above potential applications would be incorrectly placed. However, this still leaves a substantial potential of around 16 PJ/yr. As an example if an average industrial heat pump saved about 1.0 MW for 6,000 hours/yr the above potential implies the need for about 750 heat pumps in UK industry.

Market Penetration

Although considerations of the most appropriate placement has reduced the market for heat pumps there is still a large potential to be tapped. Whether current technology can penetrate the market is largely dependent on economics. However, another important and complementary factor is the markets perception of heat pump reliability. From the results of the projects outlined earlier it is clear that closed cycle heat pumps are high risk investments and UK industry would typically expect such projects to have a two year payback.

To try and estimate the market penetration of heat pumps requires some estimate of payback period. In simple terms payback is given by:

$$\text{Payback period} = C/(H(F/e - E/COP))$$

where C - Capital cost/kW of heat output
H - Operating hours/y
F - Price of fuel being replaced
e - Efficiency of original fuel usage
E - Energy price for heat pump drive
COP - Coefficient of performance

This equation is displayed graphically in figure 1 and can be used for any heat pump type. The chart assumes the efficiency of use of the original fuel (eg in a boiler) to be 70%. The simple payback period predicted by the above equation will be slightly optimistic since it relates only to energy cost savings and does not include maintenance cost etc.

The effects of fuel prices on payback period are dramatic. Taking typical UK energy prices prevailing in April 1985, a site burning interruptible gas (1.0p/kWh) could install a typical electric heat pump, operating 6,000 hrs/y, and achieve a payback of around 2.5 - 3 years assuming an installed cost of £100/kW. At Summer '86 prices of around 0.6p/kWh for interruptible gas, the payback period for the same installation is around 11 years.

Table 4 shows the effect on payback of various ranges in both capital cost, running hours and performance. The fuel prices assumed were those prevailing in the Spring of 1986.

Closed Cycle Systems

From the earlier arguments it seems likely that in the UK a two year payback will be required for closed cycle systems, although slightly longer paybacks might be acceptable for electric motor driven systems. It is clear from Table 4 that only those electric heat pump opportunities offering COPs greater than six and running hours of over 8,000 hours/y will be able to approach a two year payback. It should be emphasised that very few closed cycle heat pumps currently installed in the UK can achieve a COP of six, a more typical figure would be between four and five. Also in many of the potential markets such as the food industry there are relatively few processes that operate 8,000 hour/yr, 6,000 hours being more typical. The case for gas engine systems is even worse, they are uneconomic even in the most favourable circumstance at Spring '86 price levels.

For closed cycle heat pumps to penetrate the heat recovery market requires several changes, to take place, either singly or together. The most obvious and significant effect would be for industry to increase the acceptable payback period required for heat pump projects. Although even then, the payback required to achieve signficant market penetration would probably have to be at least 5 years.

Another possible option is to improve the performance of the available system at little or no extra installed cost. Table 5 shows this effect for an electric heat pump costing £150/kW and operating for 6,000 hours/y with fuel at the prices given in Table 4.

It is evident from this table that increasing the COP has marked effects on the payback period. Research and development is therefore required to bring about such increases. One way of doing this would be to ensure that heat pumps are operated across small temperature lifts, this could be achieved by developing heat exchangers to operate at much smaller temperature differences than those currently used. It is also important to ensure the heat pump is placed correctly to ensure minimum temperature lifts, this might be accomplished by considering heat pumps at the front end of process design rather than as an afterthought added to an already existing process where the likely opportunities for operating at small temperature lifts will be considerably restricted.

It seems clear that at Spring 1986 fuel prices the potential market for closed cycle heat pumps in the UK will be small and restricted to processes that can achieve high COPs and running times in excess of 8,000 hours. Unless fuel price differentials change dramatically, to say early 1985 levels, or significant improvements in performance are obtained the situation will not change in the short to medium term.

Open Cycle Systems

The case for installing open cycle systems looks considerably better particularly in those situations where only a compressor needs to be installed. In practice the number of such installations is likely to be small and a more typical example would include the installation of extra heat transfer surface as well as the compressor. The returns in many instances will still be attractive.

In some cases the open cycle heat pump option is only considered when current equipment, such as an evaporator, is being replaced. In this instance the payback is calculated on the difference in capital costs of a system including an open cycle heat pump and a conventional system, therefore making the economic justification much easier.

At the fuel prices assumed in Table 4 the prospects for open cycle systems look reasonably good. However, if fossil fuel prices fall much below 0.8p/kWh they become increasingly more marginal and at around 0.6p/kWh only those systems running 8,000 hours/y are likely to be economic.

4. THE WAY FORWARD

In terms of the number of sites at which heat pumps can provide useful energy savings, the potential market for heat pumps is large. However, two major factors are at present stopping companies investing:

(a) the reputation of heat pump technology is poor in terms of mechanical reliability and achievement of design performance.

(b) payback periods are too long.

Reliability

The programme of first generation heat pumps has been disappointing, but if we examine the reasons for failure they fall into two main areas.

Firstly, and most importantly, the installed systems have not properly matched the host process. In particular, gross oversizing occurred. These matching problems could have been avoided if more care had been taken to assess the process needs at the design stage.

The second type of problem is more typical of any new technology and could not have been predicted in advance; this is the area of detailed design. Many detailed engineering design lessons have now been learned and there is no reason that we need to learn such lessons again.

Bearing these two categories of problem in mind it is possible, now in 1986, to design and build heat pumps that will not suffer the problems encountered in the past.

Payback Period

The question of economic viability is more difficult, at Spring 86 fuel prices closed cycle heat pumps are uneconomic and open cycle systems are marginal. However, if it is assumed that fossil fuel prices will, at some time, return to mid 1985 levels then heat pumps will again become attractive. In order to bring forward this date research and development is required.

This effort should not be to investigate esoteric new designs, it should be carried out to improve the performance and reduce the capital cost of existing designs. In particular work on improved compressor design and improved part load performance is required.

The Implementation of Good Heat Pumps

All the problems encountered in the projects described in section 2 could have been avoided at the design stage. Not all could have been avoided in 1980 but with hindsight this statement is correct. To ensure second generation heat pumps do not suffer these faults it is necessary to do just two things:

(a) allocate adequate money for proper design

(b) as part of the design exercise, use the experience of the last five years.

The design exercise should be split into two parts. Firstly, the overall system should be considered and then detailed design of components should take place.

System Design

The system design involves matching the heat pump to the host process and the selection of the best cycle configuration. This is an engineering appraisal and must be done by someone who will not favour a particular type of system for commercial reasons. Most importantly, the system design must involve an in depth understanding of the host process. The elements of the system design exercise include the following:

(a) Measurements of performance of host process. For both the heat source and heat user it is necessary to build up a complete thermal and physical picture. This includes temperature, heat flows, velocities, flow constituents (eg corrosive products). The assessment of the host process must include time related variations that occur. Such variations may be throughout the operating period or may be intermittent due to start-up/shutdown, seasonal variations or unusual process requirements. Many of the practical problems that have occurred in the past are due to the inability of the heat pump to cope with conditions different from average design figures. Variations must include possible changes in factory output.

(b) Choice of heat pump size. The most important lesson learned from the Demonstration heat pumps was not to oversize. A heat pump only gives good economics if it operates at 100% load for all available hours. Hence, the unit should not be sized to supply peaks and transients. The heat pump should be designed as a base load plant. Signficant loss of COP can occur on part load.

(c) Choice of cycle configuration. Probably the most difficult part of the system design is the choice of cycle configuration. This includes decisions such as:-

- what basic heat pump cycle should be used (eg open cycle, closed cycle, absorption).

- choice of operating temperatures and heat exchanger temperature differences.

- what type of compressor drive (electric, gas, engine, etc).
- what type of compressor.
- how many compressors, and in what configuration.

An even more basic decision should be included:-

- is a heat pump truly applicable?

These design decisions are not easy and can only be done well through experience and effort. Some general rules that emerge from the demonstration heat pumps are:

- use passive heat recovery where possible.
- use electric drive unless COP's are low.
- subcoolers can be very beneficial.
- installation costs can be very high.
- the simplest cycle will not necessarily lead to lowest capital cost.

The Role of Process Integration

The technique of Process Integration (PI) is an important tool to use in a heat pump system design. PI shows quite clearly the circumstances in which a heat pump is worthwhile.

When a PI study is carried out for a process, a significant temperature level, the "pinch temperature", is identified. The rules of PI show that a heat pump is only of benefit if it either:

(a) transfers heat across the pinch.

(b) modifies the pinch temperature in such a way that the process target energy requirement is reduced.

In general, a heat pump which operates completely above the pinch (ie heat source and heat user both hotter than the pinch temperature), does not conform to the above rules, so no benefit is obtained. The same applies for heat pumps below the pinch.

The value of Process Integration in heat pump system design emphasises that the heat pump should not be considered in isolation. It is part of a complete industrial process and must be assessed in relation to the whole process.

Detailed Design

The second part of the design exercise is detailed design. It is at this point that all the practical lessons from first generation heat pumps can be applied. The detailed design should include:

(a) Compressor design - valves, bearings, shaft seals, mounting, unloading mechanism.

(b) Prime mover design - air intake, exhaust system materials, lubricating oil, cooling water, mounting, ignition systems, exhaust pressure drop.

(c) Refrigeration system components - seals, oil separators, flexible pipes, expansion valves, suction line pressure drop.

(d) Heat exchangers - material of construction, pressure drop, oil separation, refrigerant distribution, provision for cleaning.

(e) Controls - off design operation, system protection.

5. CONCLUSIONS

The performance of five industrial scale demonstration heat pumps has been disappointing. Only one out of the five systems achieved its projected economic performance.

Oversizing and underutilisation where the main reasons for the poor performance. Although several of the systems also had mechanical problems. With hindsight most of these problems could have been avoided.

Given Summer 1986 fuel prices current heat pump technology, both open and closed cycle, is marginal in economic terms. Open cycle systems will become attractive with only a relatively small rise in fossil fuel prices whereas closed cycle systems will require a return to something like mid 1985 fuel prices to become economic.

When the economics of heat pumps improves it should be possible to design, build and operate reliable systems that perform as specified. To do this it is essential that great care is taken in both the matching of the heat pump to the process and in the detailed engineering design of the chosen system.

Research and development could bring forward the date when heat pumps become economically viable. Effort in this area should not be aimed at esoteric new designs but concentrated on improving the performance and cost effectiveness of existing designs.

REFERENCES

Mercer, A C (1986), 'A Review of the energy efficiency demonstration scheme in the chemicals industry', To be published.

	HEAT PUMP 1	HEAT PUMP 2	HEAT PUMP 3	HEAT PUMP 4	HEAT PUMP 5
HEAT PUMP OUTPUT kW	1 000	870	110 (217)*	2300 (2750)*	1 200
HEAT SOURCE	Cooling water 30°-25°C	Cooling water 53°-26°C	Cooling water 33°-26°C	Moist air 28°-20°C	Moist air 27°-21°C
HEAT USER	Boiler Feed 10°-80°C	Process water 50°-70°C	Process water 30°-45°C Boiler feed 45°-75°C	Dry air 10°-70°C	Dry Air 26°-70°C
COMPRESSOR TYPE	Single screw	Centrifugal	Reciprocating	Reciprocating	Reciprocating
COMPRESSOR DRIVE	Steam turbine	Electric motor	Gas Engine	Gas Engine	Gas Engine
DESIGN COP_H	4.1	4.5 and 6.7	4.3	4.0	4.1
DATE INSTALLED	Dec. 1982	Aug. 1979	Sep. 1982	Mar. 1981	Mar. 1982
OPERATING HOURS PER YEAR	6 000	4 000	6 500	8 000	8 000

* These systems include passive heat recovery. The bracketed figures are total system heat output

TABLE I SUMMARY OF HEAT PUMP DATA

	HEAT PUMP 1	HEAT PUMP 2	HEAT PUMP 3	HEAT PUMP 4	HEAT PUMP 5
CAPITAL COST £	200 000	80 000	51 000	426 000	324 000
SPECIFIC CAPITAL COST £/kW	200	92	463 (235)*	185 (155)*	270
PROJECTED ANNUAL SAVINGS £	80 000	40 000	18 000	120 000	46 000
PROJECTED PAYBACK PERIOD - Years	2.5	2	2.8	3.5	7
ACTUAL ANNUAL SAVINGS £	20 000	15 000	9 000	108 000	38 600
ACTUAL PAYBACK PERIOD - Years	10	5.3	5.6	3.9	8.4

* These systems include passive heat recovery. Bracketed figures = capital cost/total system heat output.

TABLE 2 FINANCIAL DATA

Sector	Potential Saving PJ/yr	No of Sites
Engineering	1.9	1,000+
Food	10.9	450
Chemicals	7.1	200
Textiles & cloth care	1.3	750
Paper and board	1.3	100
Building products	0.9	175
Other industries	1.0	500
TOTAL: (rounded)	24.4	3,200

TABLE 3 HEAT PUMP MARKET POTENTIAL

Heat Pump Type	Drive	Typical Cost £/kW	Typical PER (COP)	PAYBACK (years) (MIDRANGE)(2)	(WORST)(3)	(BEST)(4)
Closed Cycle	Electric	100 - 200	1.2 - 1.8 (4 - 6)	6.2	10	2.4
Closed Cycle	Gas engine (firm gas)	200 - 300	1.0 - 2.0	10	10	4.5
Open Cycle retrofit MVR	Electric	40 - 60(5)	1.8 - 6 (6 - 20)	1.0	2.9	0.6
Open cycle MVR	Electric	150 - 250(6)	1.8 - 6 (6 - 20)	4.0	10	2.0

(1) Assumes fuel prices (p/kWh) as follows: 0.8p, I.Gas 0.8p, F.Gas 1.1p, electricity 3.6p
(2) Average capital cost and PER, running 6,000 hrs/y.
(3) Most expensive capital cost smallest PER, running 4,000 hrs/y.
(4) Cheapest capital cost best PER, running 8,000 hrs/y.
(5) Assumes only compressor and pipe work required.
(6) Assumes complete installation, that is (5) plus evaporator.

TABLE 4 TYPICAL PAYBACK OF INDUSTRIAL HEAT PUMPS (I)

COP	Payback Years
5	6.2
6	4.8
7	4.2
8	3.7
10	3.3

TABLE 5 EFFECT OF COP ON PAYBACK FOR AN ELECTRIC HEAT PUMP

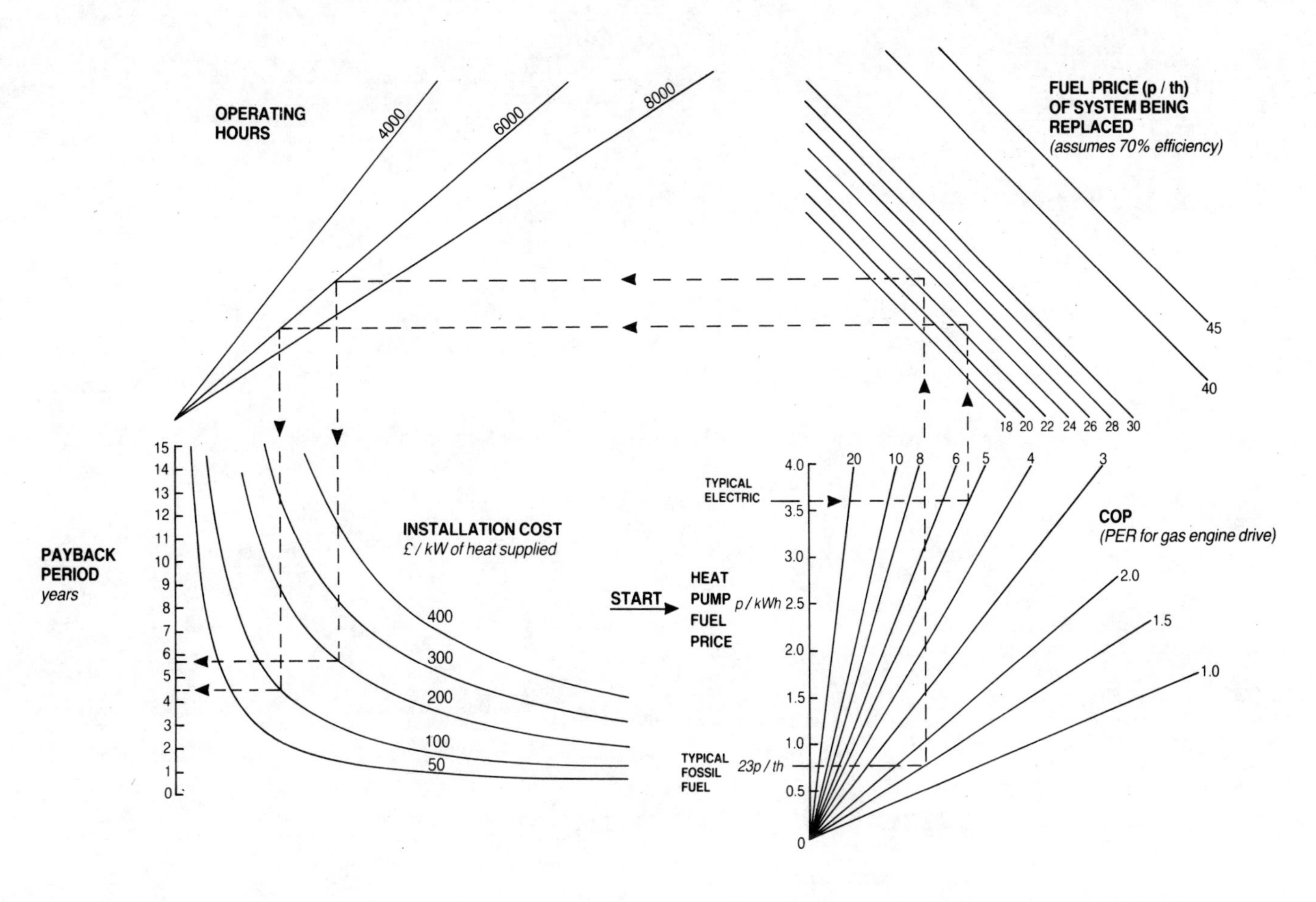

FIG. I ESTIMATE OF PAYBACK PERIOD

3rd International Symposium on the

Large Scale Applications of Heat Pumps

Oxford, England : 25-27 March 1987

PAPER B2

HEAT PUMP ASSISTED WATER PURIFICATION SYSTEMS

M.A.R. Eisa*, R. Best and F.A. Holland
Department of Chemical and Gas Engineering
University of Salford, M5 4WT, U.K.

SYNOPSIS

An evaluation of possible heat pump systems to provide a continuous flow of pure distilled water from salt or involatile contaminants has been made. A heat pump can be used to maintain a temperature difference between two vessels containing pure and salt water respectively. If the temperature of the salt water vessel is higher than the temperature of the pure water vessel by an amount sufficient to create a higher vapour pressure in the salt water vessel than in the pure water vessel, pure water vapour will flow continuously from the salt water vessel to the pure water vessel.

Heat pump assisted water purification systems have a wide range of potential applications ranging from the production of drinking water to the elimination of solids in geothermal brine to avoid the scaling of heat transfer surfaces. Scaling is a serious problem in power and other plants operating on geothermal brine.

* Mahmoud Abdel Rahman Eisa, (Ph.D), Cairo, Egypt.

NOMENCLATURE

(COP) = coefficient of performance, dimensionless
(CR) = compression ratio P_{CO}/P_{EV}, dimensionless
(FR) = flow ratio, dimensionless
H = enthalpy per unit mass of vapour, kJ kg^{-1}
P = vapour pressure, mm Hg or bar
ΔP = pressure difference, mm Hg
Q = heat load, kW
T = temperature, ^oC or K
ΔT = temperature difference, K
(VPH) = volume of water vapour per unit latent heat, $m^3 kJ^{-1}$
W = work rate delivered to the shaft of the compressor, kW
X = salt concentration, weight per cent
ϕ = entropy, kJ kg^{-1} K^{-1}

Subscripts

AB absorber
BS solution from absorber
C condensed liquid
CMVC closed cycle mechanical vapour compression
CO condenser
CW water from condenser
D evaporation
DF difference
EHW enthalpy based for heat pump assisted water purification
ETW enthalpy based for heat transformer assisted water purification
EV evaporator
EW working fluid from evaporator
F feed
GB geothermal brine
GE generator
GS solution from generator
GW working fluid from generator
HT heat transfer
LF lift
OMVC open cycle mechanical vapour compression
RCO saturated at condenser temperature
S condensation
SEV saturated at evaporator temperature
SN solution
SS superheated
SW sea water
V vapour
W working fluid

INTRODUCTION

An economic water purification system, which can operate on a minimum energy input could have a wide range of applications. The availability of fresh water throughout the world is far from uniform and the problem is aggravated by the rapid increase in population. Many of the areas suffering from a lack of fresh water are in the vicinity of sea water. In industry, vast quantities of water are used which contain ingredients that cause scaling on heat transfer surfaces. This enormously reduces the efficiency of industrial processes. Distillation is the most widely used process for desalting water. It consists of successive evaporations and condensations and is relatively energy intensive. The energy requirements can be minimized by recycling the heat produced by condensation using a heat pump cycle.

THERMODYNAMIC CONSIDERATIONS

The vapour pressure of a water/salt solution is less than for pure water by an amount that depends on the temperature and the salt concentration. Natural solutions like sea water or geothermal brines differ in composition and salt concentration from place to place. A typical sea water has a salt concentration of 3.4% (Ref.1). A typical geothermal brine has been taken as having about two thirds the salt concentration of a typical sea water (Ref.2).

Figure 1 is a plot of the vapour pressure of pure water and the vapour pressure lowering for a salt solution against temperature for a typical sea

Held at St. Catherine's College Oxford, England. Organised and sponsored by BHRA, The Fluid Engineering Centre, Cranfield, Bedfordshire, MK43 0AJ England.

water and a typical geothermal brine. The curves for a typical sea water and a typical geothermal brine are similar. The vapour pressure lowering increases with temperature and the rate of increase increases with temperature.

If a container of pure water is connected to a container of water/salt solution at the same temperature, the lower vapour pressure above the water/salt solution will cause water vapour to flow from the vessel containing pure water to the vessel containing water/salt solution. If the temperature of the water/salt solution is raised to the point where the vapour pressure of the water/salt solution is higher than the vapour pressure in the vessel of pure water, then water vapour will flow from the vessel containing water/salt solution to the vessel containing pure water. The distillation rate, heat loads and sizes of equipment in the distillation process are functions of the thermophysical and thermodynamic properties of the solution.

If T_D is the evaporation temperature in the vessel containing water/salt solution and T_S is the condensation temperature in the vessel containing pure water, then $(T_D - T_S)$ is the temperature difference between the two containers. $(T_D - T_S)$ determines the vapour pressure difference ΔP_{DF} between the two vessels and the rate of the distillation process.

Figure 2 is a plot ΔP_{DF} against salt concentration X of a typical sea water for different values of $(T_D - T_S)$ for evaporation temperatures $T_D = 60^oC$ and $T_D = 100^oC$. The vapour pressure difference ΔP_{DF} increases with the evaporation temperature T_D and decreases with the salt concentration X. The rate of decrease of ΔP_{DF} with X increases with an increase in evaporation T_D. A heat pump can be used to recycle the heat of condensation produced from the condensed vapour.

HEAT PUMP ASSISTED WATER PURIFICATION

1. Mechanical vapour compression systems

An open or a closed cycle mechanical vapour compression heat pump can be used to recycle heat in a water purification system. Open cycle mechanical vapour compression systems have been used for many years in the purification of water or for concentrating solutions. This is based on the principle that if the vapour produced in an evaporator is compressed, its saturation temperature is raised so that it can be used as the heating medium in the same evaporator.

In both the open and the closed cycle systems a mechanical compressor is used. However, in a closed cycle system the working fluid is normally a fluorocarbon.

Figure 3 is a schematic diagram for a closed cycle mechanical vapour compression system for water purification using an external working fluid. In this system the condenser of the heat pump supplies the heat required to raise the temperature and vapour pressure of the impure water sufficiently to transfer water vapour to the vessel containing pure water. The evaporator of the heat pump extracts the heat of condensation from the water vapour as it condenses in the vessel containing pure water. At the same time the external working fluid is vaporised at a temperature T_{EV} before being compressed and condensed at a higher temperature T_{CO} in the condenser of the heat pump located in the vessel containing impure water. The condensation temperature of the heat pump working fluid T_{CO} must be sufficiently higher than the temperature T_D in the vessel containing impure water to provide the necessary temperature difference driving force $(T_{CO} - T_D)$ for the required heat transfer rate. The heat pump working fluid is then condensed and the pressure of the condensate reduced by an expansion valve prior to being returned to the evaporator in the vessel containing pure water. The independent operating temperatures, the temperature difference driving force and the amount of superheat in the working fluid vapour at the suction side of the compressor have optimum values for a specific working fluid. (Refs.3,4).

2. Conventional absorption and reversed absorption (heat transformer) heat pump assisted water purification.

Absorption heat pumps have a number of advantages over the mechanical vapour compression systems. They are able to deal with large heat loads and do not require compressors which for the large heat loads in distillation processes, are large, expensive and not always readily available.

A heat transformer is uniquely attractive since it can make use of medium or relatively low grade heat energy such as solar energy, low grade geothermal brines or waste heat to drive the generator.

Figures 4 and 5 are schematic diagrams for a conventional absorption and a heat transformer assisted water purification system respectively.

In absorption heat pump assisted water purification, the evaporator is used to recycle the heat of condensation produced from the condensation of water vapour. The heat of condensation becomes the heat source to the heat pump evaporator. The absc and the condenser are used to heat the water solution in the water/salt container. A relatively high grade heat source is used to drive the generator. The gross temperature lift in the heat pump is

$$(T_{CO}-T_{EV}) = (T_D-T_S) + \Delta T_{EV} + \Delta T_{CO} \qquad (1)$$

where ΔT_{EV} and ΔT_{CO} are the temperature difference driving forces in the evaporator and the condenser.

In heat transformer assisted water purification, the evaporator of the heat transformer is located in the pure water container. The heat of condensation of the water vapour is then recycled. The absorber of the heat transformer is used to heat the water solution in the water/salt container. A medium or relatively low grade heat source is used to drive the generator. The gross temperature lift in the heat transformer is

$$(T_{AB}-T_{EV}) = (T_D-T_S) + \Delta T_{EV} + \Delta T_{AB} \qquad (2)$$

ENERGY REQUIREMENTS IN HEAT PUMP ASSISTED WATER PURIFICATION SYSTEMS

The evaporation process is entirely controlled by the rate of heat transfer within the system. The amount of energy required to evaporate water from water/salt solutions is a function of the thermodynamic properties of the water/salt solution and the coefficient of performance of the heat pump system. The main thermodynamic properties on which the energy calculations should be based are the boiling point elevation due to salt concentration, the enthalpy per unit mass for the water salt solution and the latent heat of vaporisation per unit mass at different temperatures.

In order to establish a basis for the calculation of the energy requirements for heat pump assisted water purification, basic cycles were considered without heat integration since this depends on the type of application. For example purified water produced from geothermal brine may be required at the highest possible temperature. In addition the following assumptions were also made :

1) Sufficiently good mixing takes place in the vessel containing impure water to ensure that the composition of the impure water outlet stream is the same as in the vessel.

2) The vaporised water is a saturated vapour at its boiling temperature.

3) The outlet pure water is saturated water at its condensing temperature.

It is interesting to compare the energy requirements for the closed cycle mechanical vapour compression heat pump (CMVC) with the open cycle mechanical vapour compression system which is already in common use.

A comparative energy analysis has been made between the two systems for the purification of sea water of a typical concentration of 3.4% salt to pure water and a concentrated outlet solution of double this typical concentration. The isentropic compression by the compressor in the open cycle system was considered to produce superheated steam with an enthalpy per unit mass of H_{SS} which was calculated from the following equation (Ref.5).

$$H_{SS} = H_{RCO} + (\phi_{SEV} - \phi_{RCO})\, T_{CO} \qquad (3)$$

where H_{RCO} is the enthalpy per unit mass of saturated steam at the condenser temperature,

ϕ_{RCO} is the entropy per unit mass of saturated steam at the condenser temperature,

and ϕ_{SEV} is the entropy per unit mass of saturated steam at the evaporator temperature.

By an overall heat and mass balance around the system using the thermodynamic properties of a typical sea water and the previously specified operating conditions, the energy requirements were calculated for different temperature lifts ΔT_{LF} and feed temperatures T_F.

$$(\Delta T_{LF})_{OMVC} = (T_D - T_F) \qquad (4)$$

$$(\Delta T_{LF})_{CMVC} = (T_D - T_F) = (T_D - T_S) \qquad (5)$$

$$(T_C)_{OMVC} = T_F + \Delta T_{LF} + \Delta T_{HT} \qquad (6)$$

and

$$(T_C)_{CMVC} = T_S = T_F \qquad (7)$$

The energy loads were plotted against temperature lift and feed temperature for both the open and closed heat pump assisted water purification systems as shown in Figures 6 and 7.

The calculations have shown that the work which is given by the compressor in the open cycle mechanical vapour compression system is not enough and additional heat is required. In both systems as the temperature lift or feed temperature for the same lift increases, the total energy required increases.

Figures 6 and 7 show that in contrast, for the OMVC system, work required from the compressor decreases with both temperature lift or feed temperature due to the lower (VPH) values of water vapour at higher temperatures. At about $T_F = 90^oC$ the work required from the compressor in the OMVC system is the same as for the CMVC system.

Figure 7 shows that if the work required from the compressor is the only factor to be considered, then it would appear that the CMVC system is preferable up to 90^oC and the OMVC system above 90^oC.

In absorption and heat transformer assisted water purification systems, the compressor is replaced by a secondary circuit in which a liquid absorbent is circulated by a pump. The mechanical energy required to pump the liquid is usually negligible compared with the input heat load. The energy requirements are a direct function of the coefficient of performance of the heat pump system. The rate of the distillation process depends on the pressure difference ΔP_{DF} between the two vessels containing the water/salt solution and the pure water. The pressure difference is determined by the temperature difference $(T_D - T_S)$ by which the temperature lifts $(T_{CO} - T_{EV})$ and $(T_{AB} - T_{EV})$ in the absorption heat pump or the heat transformer respectively are fixed. Figures 8 and 9 show the effect of the temperature difference $(T_D - T_S)$ on the coefficients of performance of an absorption heat pump and a heat transformer operating on water as the working fluid and aqueous lithium bromide solution as the absorbent. The coefficients of performance were calculated using the thermodynamic properties of the water-lithium bromide system (Refs.6,7). These enthalpy based coefficients of performance were estimated for a basic cycle with 10 K as a temperature difference driving force for the heat exchangers used in both the pure water and water/salt containers.

The equations used for the calculations of the enthalpy based coefficients of performance are as follows :

$$(COP)_{EHW} = \frac{H_{EW} + [(FR)-1]H_{GS} - (FR)H_{BS} + H_{GW} - H_{CW}}{H_{GW} + [(FR)-1]H_{GS} - (FR)H_{BS}} \qquad (8)$$

for the absorption heat pump, and

$$(COP)_{ETW} = \frac{H_{EW} + [(FR)-1]H_{GS} - (FR)H_{BS}}{H_{GW} + [(FR)-1]H_{GS} - (FR)H_{BS} + H_{EW} - H_{CW}} \qquad (9)$$

for the heat transformer. These equations can be drawn by individual equipment and overall heat and mass balances around the system.

It can be seen from Figures 8 and 9 that as the temperature difference between the two vessels increases, the coefficient of performance decreases. The rate of decrease in the enthalpy based coefficient of performance is higher at higher temperature differences.

The flow ratio (FR) appears in both Equations (8) and (9) and has a significant effect on the coefficient of performance. The flow ratio is essentially the mass flow rate of solution circulating in the secondary circuit to the mass flow rate of the working fluid circulating in the primary circuit. In a refined cycle a solution heat exchanger is normally placed between the generator and the absorber in the secondary circuit. The flow ratio has a major effect on the effectiveness of the economiser heat exchanger and the overall performance of the system and it has an optimum value (Refs.8,9,10).

WORKING FLUIDS FOR HEAT PUMPS USED FOR WATER PURIFICATION

The choice of a suitable working fluid in a mechanical vapour compression system or a working solution in absorption heat pumps and transformers is mainly dependent on the working temperatures in the purification process. This means that working fluids or solutions which may be suitable in the purification of sea water may not be suitable for use in the purification of geothermal brines or for other industrial applications.

1. Working fluids for mechanical vapour compression systems

Figure 10 shows the variation of vapour pressure with temperature for the most common working fluids. Figure 10 shows the effect of the increase of working temperatures on vapour pressures and hence the compression ratios.

Figure 11 is a plot of the required compression ratio (CR) = P_{CO}/P_{EV} against the temperature lift ΔT_{LF} for various working fluids for temperatures $T_D = 60^oC$ and $T_D = 100^oC$ in the vessel containing impure water. Figure 11 shows that the required compression ratios decrease with increasing evaporation temperature T_D. The rate of increase in compression ratio increases with temperature.

Figure 12 is a plot of the theoretical Rankine coefficient of performance $(COP)_R$ against the condensing temperature T_{CO} for the working fluids in Figure 11 for a gross temperature lift $(T_{CO} - T_{EV})$ of 30 K. Although R114 has lower $(COP)_R$ values than R11 and R113, it has excellent stability characteristics compared with R11 and R113 (Ref.11).

2. Working solutions for absorption heat pump assisted water purification

The thermodynamic properties of the working fluid/absorbent solution determines the maximum possible efficiency of an absorption system. The absorber is used in both the heat pump or the heat transformer to deliver the heat in the container of impure water. Depending on the evaporation temperature for the impure water, the solubility of the working fluid in the absorbent should be as high as possible at the temperature and pressure of the absorber. The absorber temperature is a function of the amount of heat of solution produced from the absorption process. In general, solutions with high heats of solution usually have lower vapour pressures which means lower circulation rates for the absorbent solution. Figure 13 shows the amount of lowering in vapour pressure against the heat of solution for some aqueous salt systems for similar conditions.

The purification of sea water at lower temperatures requires the absorbent to be sufficiently soluble and the viscosity of the solution to be sufficiently low. For the purification of geothermal brines at higher temperatures, an absorbent of moderate solubility but which possesses other desirable properties can be used.

In the ammonia/water system, ammonia is the working fluid in the primary circuit and water is the absorbent in the secondary circuit. In order to obtain a delivery temperature of 160^oC from a heat source of about 100^oC, the absorption process must take place at a pressure of about 70 bar, which requires the use of relatively high pressure equipment. Water/salt systems such as water/lithium bromide have the advantage that they can operate under the same temperature conditions at one sixtieth of the pressure.

CONCLUSIONS

A heat pump can be used to recycle heat energy in water purification systems. A closed cycle mechanical vapour compression system can be used to replace an open cycle system. A closed cycle system is smaller, requiring less expensive compressors and lower energy requirements.

An absorption cycle system has the advantage over mechanical vapour compression systems that it does not need a compressor and is suitable for the large heat loads needed for large scale water purification plants.

From an economic and energy saving point of view, the heat transformer looks very promising. Heat transformer assisted water purification is based on the use of low cost waste heat and the utilisation of the heat of condensation.

REFERENCES

1. Spiegler, K.S. and Laird, A.D., "Principles of desalination", 2nd Ed., Part B, Academic Press, New York, 1980.

2. White, D., "Characteristics of geothermal resources", Geothermal energy resources, production stimulation, Ed. by Kruger and Otte, Stanford University Press, 1973, pp 69-94.

3. Eisa, M.A.R., Supranto, S. and Devotta, S., "Operating characteristics of a water-to-water mechanical vapour compression heat pump using R114: Part I: Interaction between dependent and independent parameters", J. Heat Recovery Systems, 5(1), 1985, pp 57-64.

4. Eisa, M.A.R. and Devotta, S., "Operating characteristics of a water-to-water mechanical vapour compression heat pump using R114: Part II: Statistical analysis of the effect of lifetime and the individual operator of the heat pump", J. Heat Recovery Systems, 5(1), 1985, pp 66-69.

5. Holland, F.A., Watson, F.A. and Devotta, S., "Thermodynamic design data for heat pump systems", Pergamon Press, Oxford, 1982.

6. McNeely, L.A., "Thermodynamic properties of aqueous solutions of lithium bromide", ASHRAE Trans., 85(1), 1979, pp 413-434.

7. Keenan, J.H., Keyes, F.G., Hill, P.G. and Moore, J.G., "Thermodynamic properties of water including vapour, liquid and solid phases, John Wiley and Sons, New York, 1969.

8. Eisa, M.A.R., Devotta, S. and Holland, F.A., "A study of the economiser performance in a water-lithium bromide absorption cooler", Int.J. of Heat and Mass Transfer, 28(12), 1985, pp 2323-2329.

9. Smith, I.E., Carey, C.O.B. and Smith, G.F., "Absorption heat pump research", Proceedings of a Workshop in Berlin, 1982, pp 149-151.

10. Eisa, M.A.R., Sane, M.G., Devotta, S. and Holland, F.A., "Experimental studies to determine the optimum flow ratio in a water-lithium bromide absorption cooler for high absorber temperature", Chem.Eng.Res.Des., 63(4), 1985, pp 267-270.

11. Srinivasan, P., Devotta, S. and Holland, F.A., "Thermal stability of R11, R12B1, R113 and R114 and compatability with some lubricating oils, Chem.Eng.Res.Des. 63, 1985, pp 230-234.

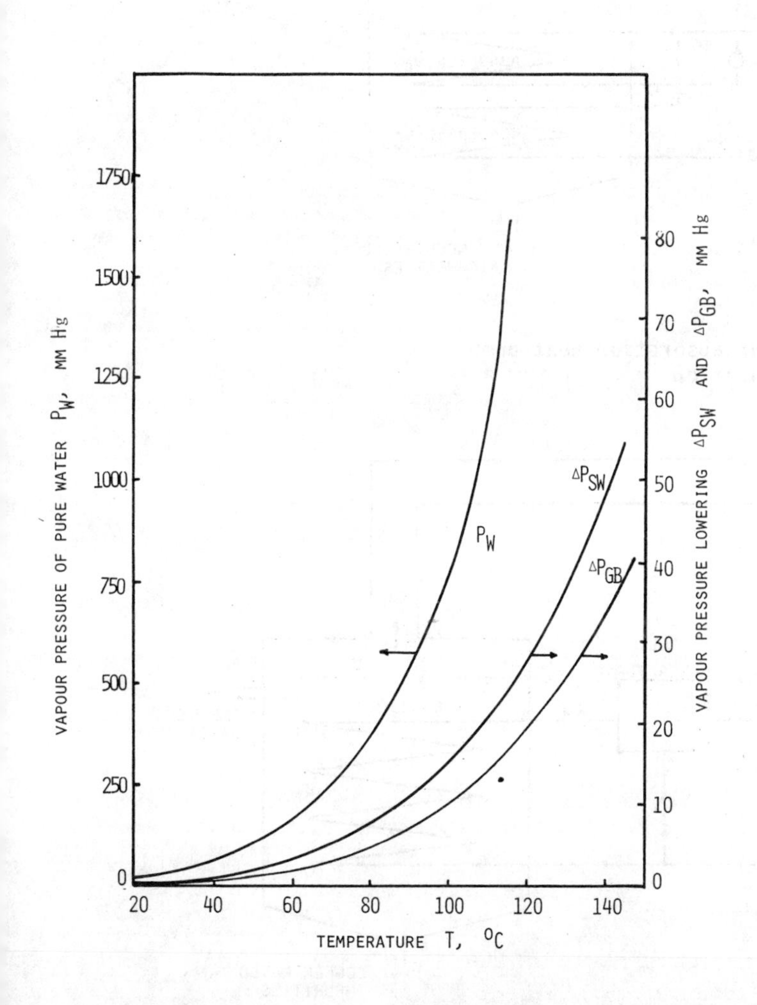

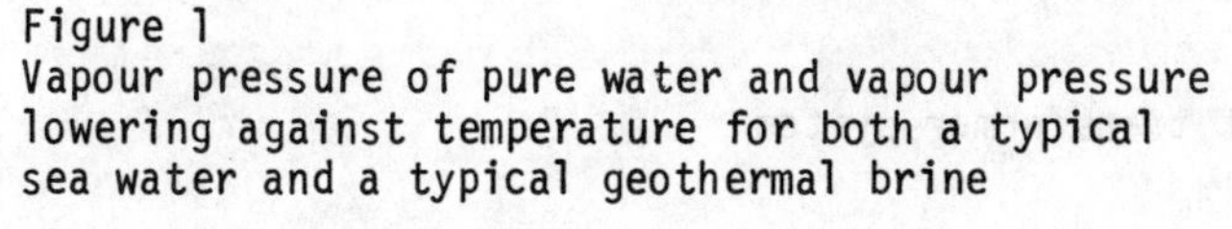
Figure 1
Vapour pressure of pure water and vapour pressure lowering against temperature for both a typical sea water and a typical geothermal brine

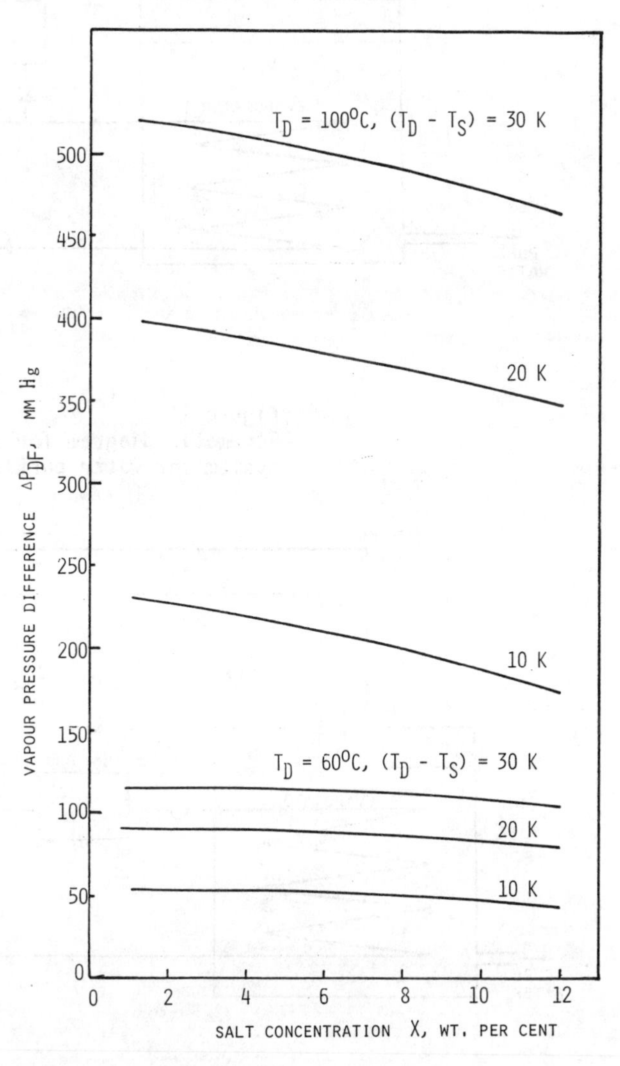

Figure 2
Vapour pressure difference against salt concentration at different levels of temperature difference (T_D - T_S) and evaporation temperature

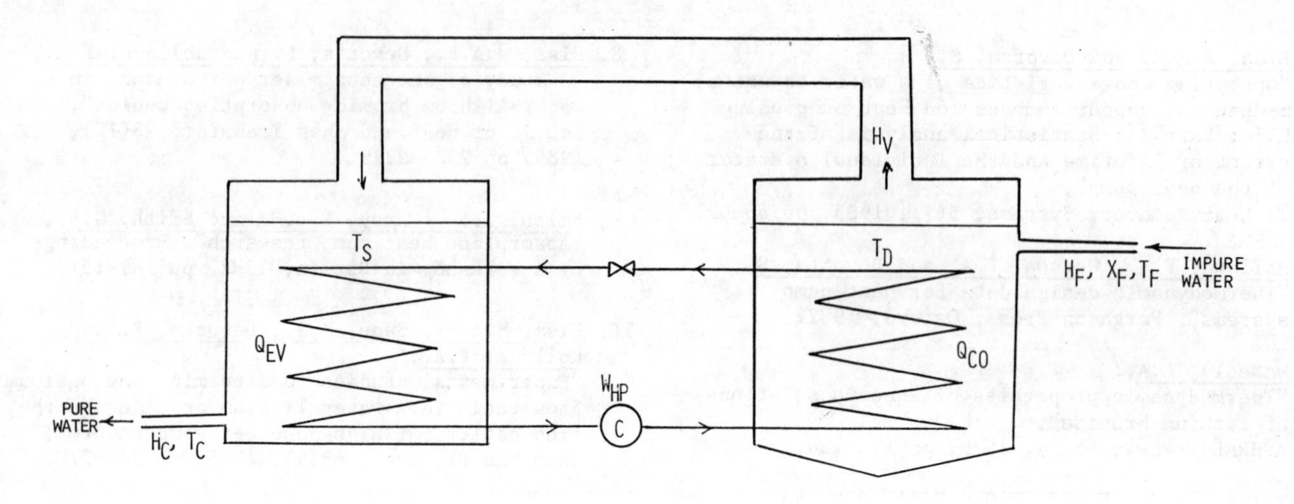

Figure 3
Schematic diagram for a closed cycle mechanical vapour compression system for water purification

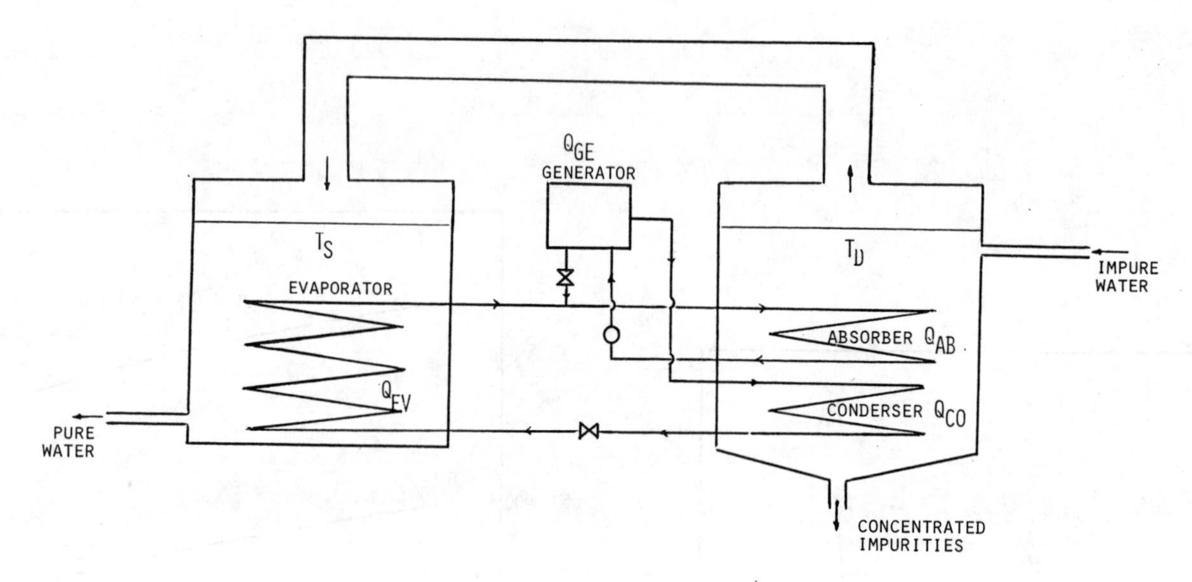

Figure 4
Schematic diagram for an absorption heat pump system for water purification

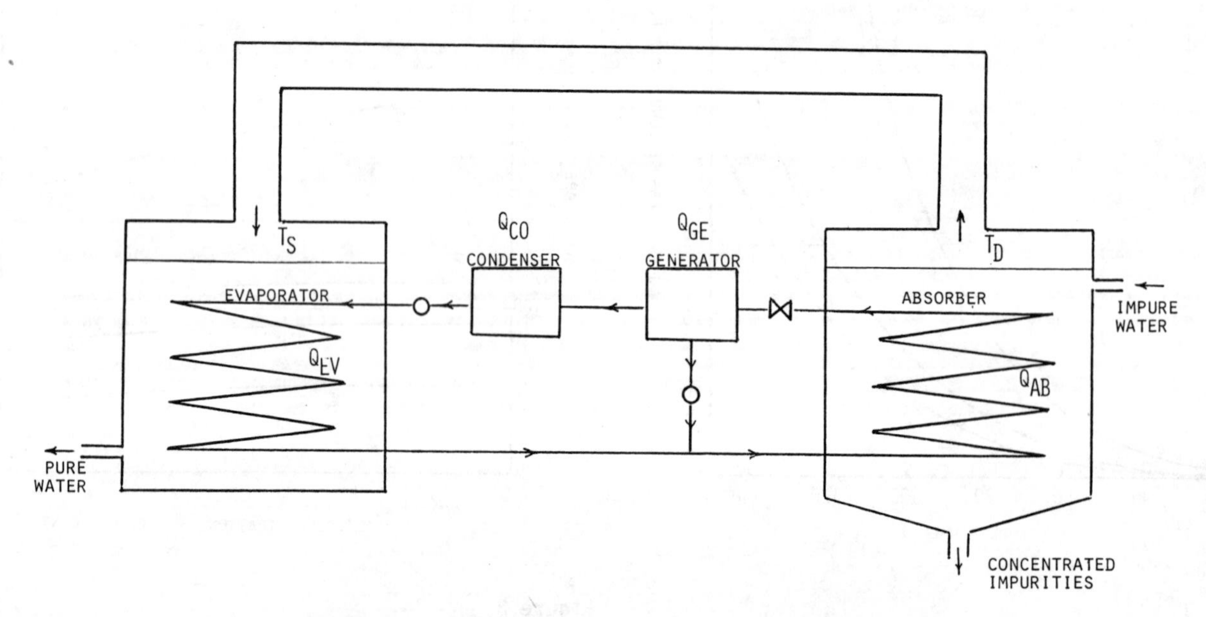

Figure 5
Schematic diagram for a heat transformer system for water purification

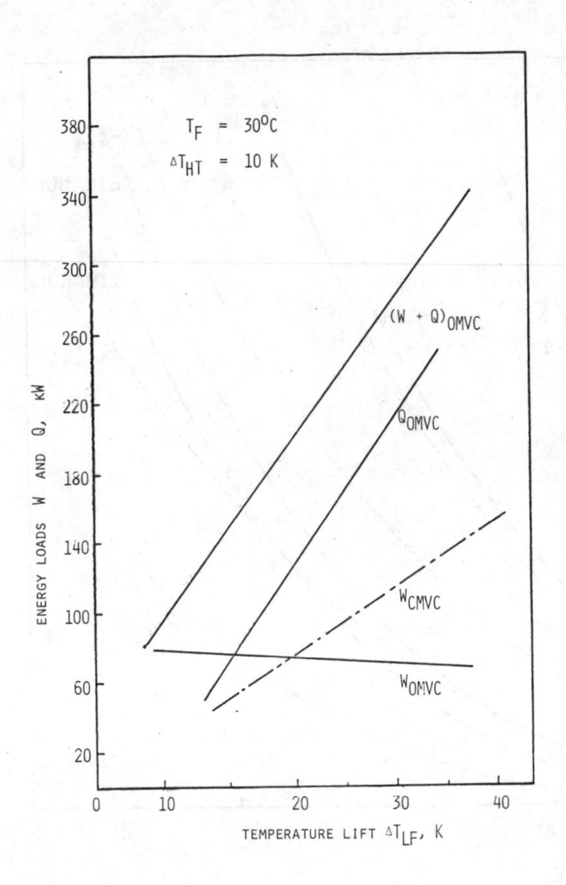

Figure 6
Energy loads against temperature lift for (OMVC) and (CMVC) heat pump assisted water purification systems

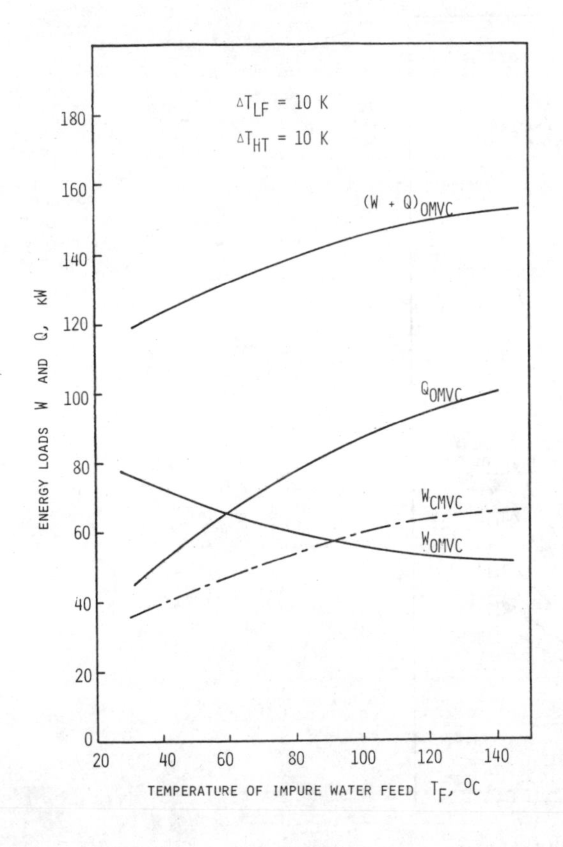

Figure 7
Energy loads against temperature of impure water feed for (OMVC) and (CMVC) heat pump assisted water purification systems

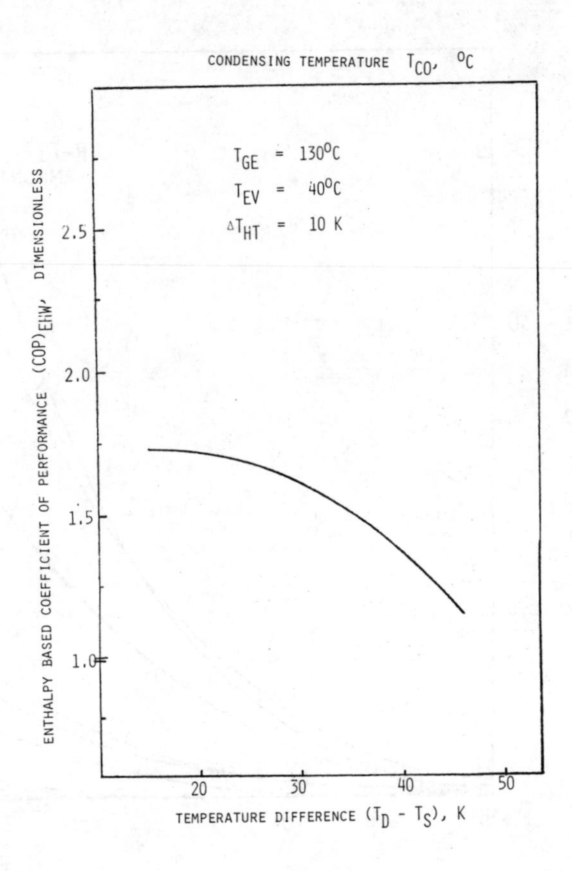

Figure 8
Enthalpy based coefficient of performance against temperature difference between the pure and impure water containers for an absorption heat pump assisted water purification system

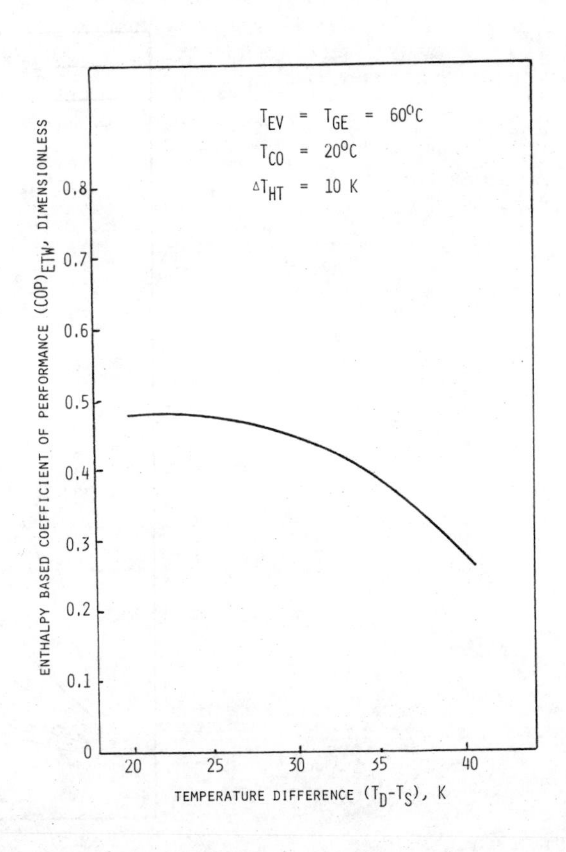

Figure 9
Enthalpy based coefficient of performance against temperature difference for a heat transformer assised water purification system

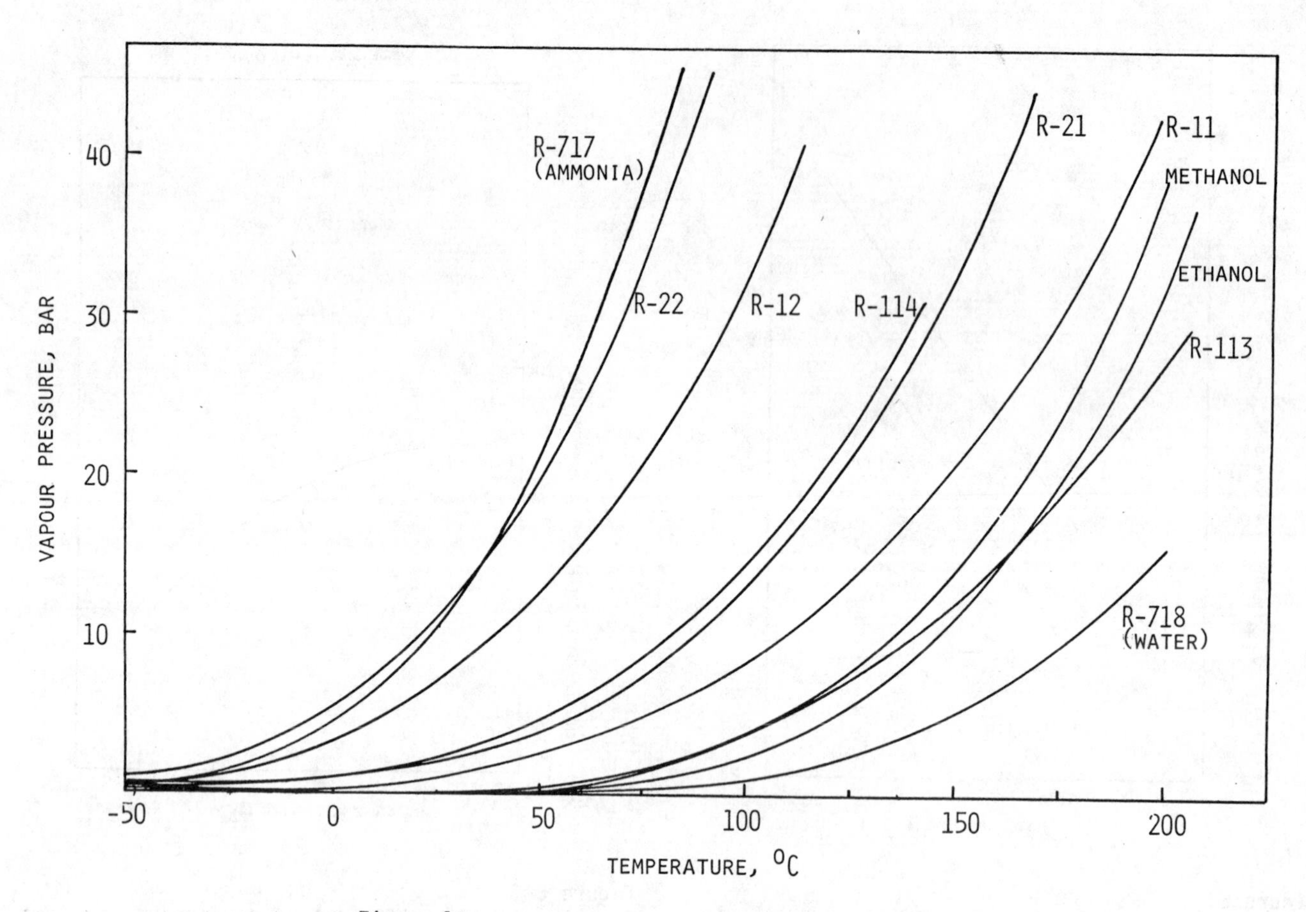

Figure 10
Vapour pressure against temperature for various working fluids

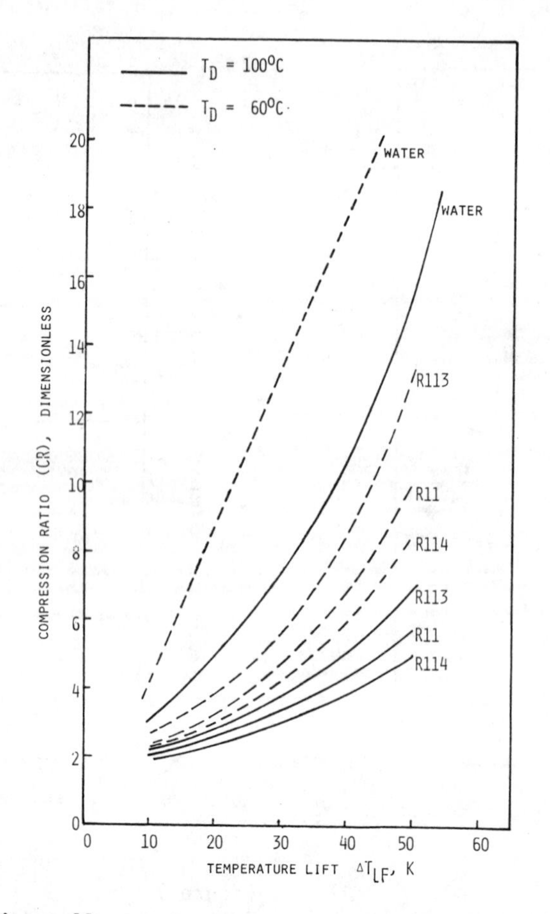

Figure 11
Compression ratio against net temperature lift for R114 R11 R113 and water

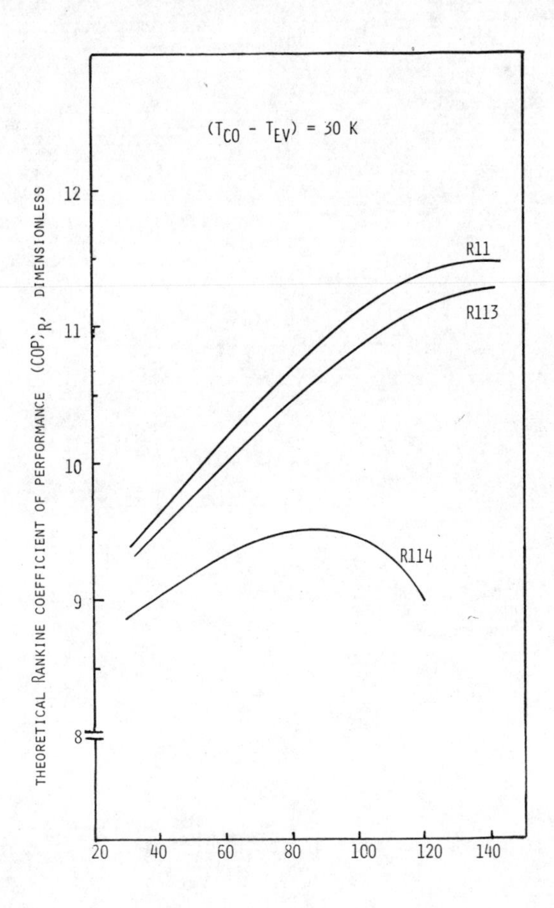

Figure 12
Theoretical rankine coefficient of performance against condensing temperature

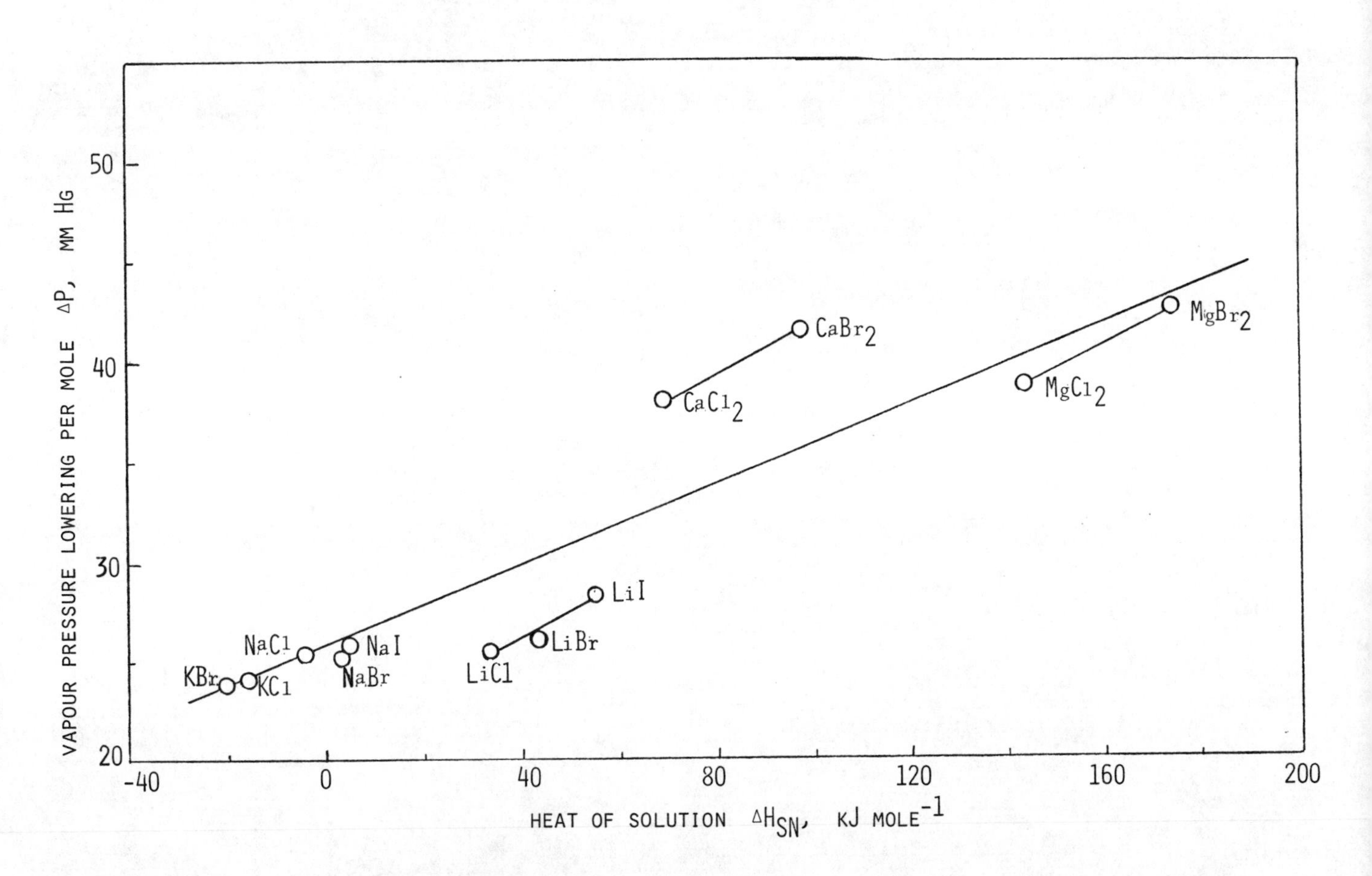

Figure 13
Vapour pressure lowering against heat of solution

3rd International Symposium on the

Large Scale Applications of Heat Pumps

Oxford, England : 25-27 March 1987

PAPER B3

PERFORMANCE OF AND OPERATIONAL EXPERIENCE WITH A LARGE SCALE HEATTRANSFORMER

P.F. Jansen, Delamine BV, Delfzijl, The Netherlands
J.W. Wormgoor, TNO, Apeldoorn, The Netherlands

SUMMARY

Mid October 1985, a large heat transformer has been taken into operation at the Delamine chemical plant. This special type of absorption heatpump, operating with LiBr-H_2O, utilizes 22 tons of 100 °C atmospheric water vapour, to produce 11 tons of saturated steam (150 °C), which means that the coëfficient of performance is over 45%.

Because of the energy saving potential of this innovative technology this installation has been selected by NEOM to be a demonstration project for other companies interested in this technology. The first year of operation an evaluation program, including a performance test, has been carried out.

The performance of the heatpump, measured one month after start-up, is in accordance with the specification. The operational experience in the first half year has been very satisfactory, later a number of corrosion problems have decreased the availability of the heatpump.

1. INTRODUCTION

With the production of ethylene amines at Delamine's plant at Delfzijl in the Netherlands, saturated watervapour of 100° C comes available from one of the destillation columns. This vapour has previously been condensed through aircoolers and discharged.

The application of a heatpump - in this case a heat transformer on the basis of the absorption principle - enables the recovery of a part of this heat as saturated steam of about 150° C, which can be fully used again in Delamine's production process.

Based on an attractive return on investment, due to the characteristic situation of generation and application of the steam in its own facilities, the decision was made in 1984 to build a heat transformer.

The project - being the first heatpump of this kind in the Netherlands - also received a subsidy from the Nederlandse Energie Ontwikkelings Maatschappij (NEOM), being a demonstration for other companies which might be interested in this type of energy savings.

Mid October 1985 the heatpump has been taken into operation. During the first 5 months the operation was quite successful. Thereafther serious corrosionproblems occured, considerably affecting the performance of the heatpump.

Being a demonstration project, the performance and operational experience of the heatpump have to be tested and evaluated during the first year of operation. The performancetest was carried out by TNO. In this paper the results of this test and the experiences during the first half year after the start of commercial operation are presented.

2. PRINCIPLE OF HEATTRANSFORMER

The heattransformer, a special type of absorption heatpump, is able to transform medium level heat to a higher level by means of a cooling water flow of low temperature level. The advantage of this system is that only little primary energy, actually only the energy to drive the pumps, is required to drive the process, producing heat at a usable level from waste heat.
Potentially it is a system that will need very little maintenance because the main parts are static heat exchangers.

The working media in this heattransformer are Libr, the absorbent, and water, acting as the refrigerant.
The main components of the heatpump are the regenerator and the condenser at the low pressure side, and the evaporator and the absorber at the high pressure side (see figure 1).
The waste heat feeding the installation is available as saturated water vapour at a temperature of 100° C. This flow is divided over the regenerator, to evaporate the water from the weak solution and the evaporator to evaporate the same water, which in the meantime has been condensed in the condenser, at a higher pressure level. In the absorber, the "high" pressure vapour is condensed and absorbed by the strong solution arriving from the regenerator, producing heat at a temperaturelevel of approximately 150° C. The diluted (weak) Libr solution is pumped back to the regenerator, closing the cycle.
In the recoverer, which is built as a plateheatexchanger, heat from the weak solution returning to the regenerator is transfered to the strong solution flowing to the absorber.

3. THEORETICAL BACKGROUND

The process in a heatpump can be shown using a log P, I/T-diagram, in which points of constant concentration of the Libr in water form straight lines (see figure 2).
The characteristics of the cycle applied in the Delamine heatpump are shown schematically in figure 3. In this figure point (1) represents the weak solution entering the regenerator, to be split up into water vapour, condensing in the condenser (5) and a strong solution going to the absorber (2). This part of the process takes place at the low pressure level p_c (approx. 80 mbar).
At the high pressure level p_e (approx. 950 mbar)

Held at St. Catherine's College Oxford, England. Organised and sponsored by BHRA, The Fluid Engineering Centre, Cranfield, Bedfordshire, MK43 0AJ England.

the strong solution (3) absorbs watervapour (6) arriving from the evaporator. The Libr concentration decreases from 64% to 60%, being the level of the weak solution (4). The heat produced during this phase is released into the steam cycle.

The efficiency of the cycle, neglecting the required pump-energy, can be defined as:

$$\epsilon = \frac{\text{heat released for steam production}}{\text{heat extracted from vapour}}$$

Theoretically, ϵ of an ideal cycle is described by the formula:

$$\epsilon = \frac{T_o - T_c}{T_1 - T_c} \cdot \frac{T_1}{T_o} \qquad (1)$$

4. DESIGN

The main components of the Delamine heatpump, the condenser, regenerator, evaporator, and absorber are constructed as vertically mounted shell and tube heatexchangers, connected two by two. The recoverer is a plate heatexchanger, mounted on groundlevel, together with all pumps. Figure 4 shows the complete installation. Figure 6 is a simplified PID of the installation.

According to the design data, the vapour input is 16 MW (22 ton/hour, 100° C), the coolingwater discharge temperature is 40° C and the steam production is 8.5 MW (10.5 ton/hour, 150° C, 4.7 bar).
Figure 5 is an enthalpy-flow diagram, based on the design data, showing the internal flows.

Because of the corrosive nature of Libr, specially at high temperature, the selection of the construction materials for the components is of prime importance.
Based on experience gained in an installation built earlier, the selected material for the regenerator tubes is titanium, this material has also been used for the plates of the recoverer. This design-aspect will be discussed further in connection with the operational experience.

5. PERFORMANCE TEST

The heatpump is instrumented rather extensively, among other reasons because the installation is monitored and controlled automatically by a computerized system.
Temperatures are measured with PT 100 RTD's, the main flows are measured with orifices equiped with Δp transducers. System-pressures are measured with pressure transducers. All signals are scanned and entered into the procescomputer and can be displayed in the controlroom as numerical values or graphically as a function of time.

After a complete calibration procedure the measuring system, as installed, is suitable for obtaining reliable and accurate performance test data. During the 48 hours full load performace-test, all appropriate data, scanned every minute, were averaged every 15 minutes. These averaged data were then stored and printed. A number of periods, each with a length of approx. two hours has been selected for the calculation of the heatpump performance.

There is some deviation between the actual operating conditions and the design conditions. The reason is that after careful analysis of the steam-consuming systems, it was concluded that the max. required steam temperature should be restricted to 145° C, which yields an improved heatpump efficiency in comparison with the design temperature of 150° C.

The system has operated very stable during the testperiod. The results that can be calculated for the various selected intervals are almost equal. So as a result of the test, one single set of operating data can be presented in table 1. Table 2 shows the calculated energy flows in the process.

Process variable	Dimension	Measured value
Vapour temperature	°C	103.3
Coolingwater in	°C	10.7
Coolingwater out	°C	36.3
Steam temperature	°C	144.9
Feedwater temperature	°C	130.8
Vapour flow	kg/s	6.08
Coolingwater flow	kg/s	64.6
Feedwater flow	kg/s	3.11
Steam pressure	bar	4.1
Condensate temperature	°C	103
Condensate flow	$m^3/s.10^3$	2.59
Weak solution flow	$m^3/s.10^3$	28.4

Table 1 Measured process variables

The test-data lead to a calculated efficiency of 49.4% (excluding electricity consumption). To make a comparison of this efficiency to the design-efficiency of 46.6%, this number has to be corrected for the deviation of the operating conditions, which are more favourable than the design conditions. This can be achieved by calculating the efficiency of the ideal process, according to Carnot, for both conditions with formula (1). For this comparison the measured external (steam and water) parameters instead of the internal (solution) parameters have been used, this yields only a minor error.
The testconditions lead to ϵ = 0.68, the design conditions to ϵ = 0.62, so the difference is 9%. Correcting the measured efficiency to a comparable efficiency at design-conditions gives ϵ = 0.45 which is within the tolerances specified by the supplier of the heatpump.

Flow	Energy kW
Vapour	+ 16370
Condensate	− 2630
Feedwater	+ 1670
Steam	− 8450
Coolingwater	− 6890
Electricity	+ 53

Table 2 Main energyflows calculated from measured data.

6. PLANT EXPERIENCES

6.1. Performance

After the mechanical completion of the total installation in September 1985, much attention has been paid to the internal cleaning and chemical treatment of the system before the actual start up, in view of the potential corrosion problems.

The leaktesting has also been carried out very carefully.
After introduction of the Libr into the system in the beginning of October 1985, the heatpump could be taken into operation quite succesfully. In the first period of operation, the heat pump had to be stopped a few times during a short period for repair and adjustments of some instruments.
In the same starting period the emergency stop sequences have been tested to be sure that the plantoperations could be continued without the heatpump, and the latter in a safe standstill position.

The heatpump operated quite well from the beginning and in such a way that the steam production raised to full capacity, which is about 11 ton per hour at full production of the plant.

Apart from a few stops of the heatpump due to instrumental failures (partly also introduced by the cold winter conditions), the whole operated satisfactorily until annual shut-down of the Delamine plant in April 1986.
Worth mentioning is the remarkably easy way of operation of the heatpump, following every single fluctuation in the supply of the overhead vapour into the system smoothly.
Unfortunately, the restart of the heatpump after the plant shut-down in April went very differently. Due to corrosion problems in the plate cooler the installation could only be operated up to half the capacity. More-over one of the circulation pumps became mechanically defect - also due to severe corrosion. And finally some leaking tubes in the evaporator were found to be the reason for the pressure increase in the system, very much affecting the efficiency of the heatpump.

6.2. Corrosion problems

- Plate heatexchanger

In the circulation of the refrigerant between the regenerator and the absorber a plate heatexchanger serves as heat recoverer, with a total capacity of 3.85 MW. The plate material is titanium (B 265 Gr 1) and the gaskets are made of FEM; in total 258 plates with a surface area of 195 m^2.
The maximum operating temperature is 152.8° C.

Almost right after the start the first leakage was discovered through one of the gaskets. Tightening of the bolts did not prevent further leakages. As the loss of working fluid was not too serious, we decided to continue the heatpump operation.
In the course of time the leakages became so large that we decided to replace this plate cooler by a new one with Ti-Pd plates, manufactured in Europe.
Checking the conditions of the platecooler during the plant shut-down it was found that almost half of the number of plates were heavily corroded. (see Figure 7). Regarding the delivery time for the new platecooler, the existing heat exchanger had to be reassambled with the remaining usable plates, giving only half of the original heat-transfer-capacity.

- Condensor

After the restart of the heatpump on April 24, we soon discovered a problem with an increasing waterlevel in the condenser. Despite regular discharge of water from the condenser, it was not possible to operate the condensersystem properly.
It was then decided to open the condenser for inspection. All tubes were checked with ultrasone leak detection, whereby leakage was found in two tubes. These tubes have been plugged with copper plugs and as precaution measurement another 5 tubes around these two. In total 7 tubes have been plugged. (see Figure 8).
It was supposed that these leakages have been introduced by thermal stress corrosion, because the position of the affected tubes in the condenser is opposite to the nozzle for the connection of the evacuation system.

- Weak solution pump

This pump became defect after about 5 months operation by shortcircuiting. The pump has an electromotor, encased in the pump. After dismounting it appeared that the inducer was broken off and an adjusting ring was cracked. These parts are made of AISI 316 L. Moreover, the impeller showed cracks.
To find the reason for shortcircuiting the cooling jacket was pressurized, but no leakage could be found. Also the stator can on the inside has been thoroughly investigated by penetration, but no cracks have been found. This can is made of Hastelloy - C 276.
Probably the inducer showed fatigue damages and also stress corrosion, whilst the impeller had been seriously attacked by stress corrosion. It is not impossible that the inducer has cavitated during operation, so that fatigue load could occur.
The manufacturer of the pump has been advised to apply better corrosion resistant materials.

- Evaporator (see Figure 9)

At the end of May 1986 leakages were detected in the vacuum system of the AHP. All flange connections were investigated by means of an ultrasone leak detector, but no leakage could be found. After a certain part of the pipingsystem of the heatpump was blocked off, the problem could be localised in the evaporator. After dismounting of the bottom cover, the system around the tubes was put under pressure by nitrogen. All tubes were checked by us-leakdetection and 4 were found to be leaking. Then the topcover was removed and the tubes were investigated visually. A number of tubes were thus found with internal corrosion patterns. It was decided to do some vortex flow measurements on the tubes with the leaks and those which were really visually attacked. During this investigation a number of tubes were found with pit corrosion. Based on the results of this measurement 55 tubes were plugged. The 4 leaking tubes have been replaced by new spare tubes.
Based on a materialanalysis of one tube the composition corresponds with SS 304, except for the Cr-percentage.

7. CONCLUSIONS

Considerable corrosion problems have affected the heatpump operations at the Delamine plant during the first year.
Investigations for better corrosion resistant materials in several parts of this type of heatpump are important and have to be carried out to obtain a reliable heat recovery system. An extensive research program with exchange of information between manufacturers and users of heatpumps, should lead to definite solutions for

these corrosion problems.
This is the more interesting, because the performance test and the first half year of operation have shown, that the heatpump can be operated quite easily, and meets the attractive performance specifications.

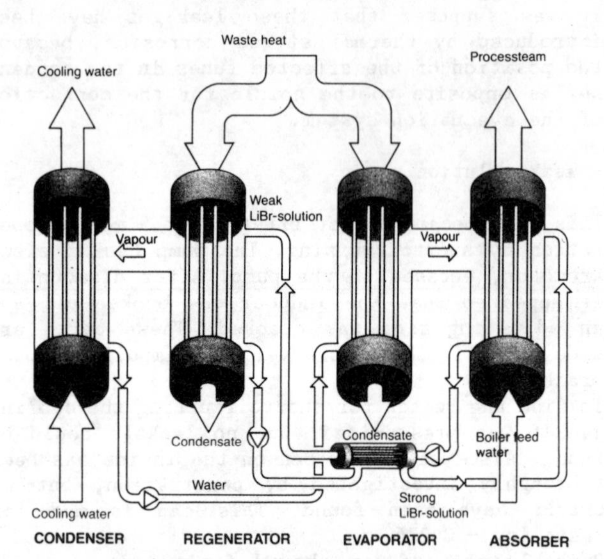

Figure 1. Simplified flowscheme of the heattransformer.

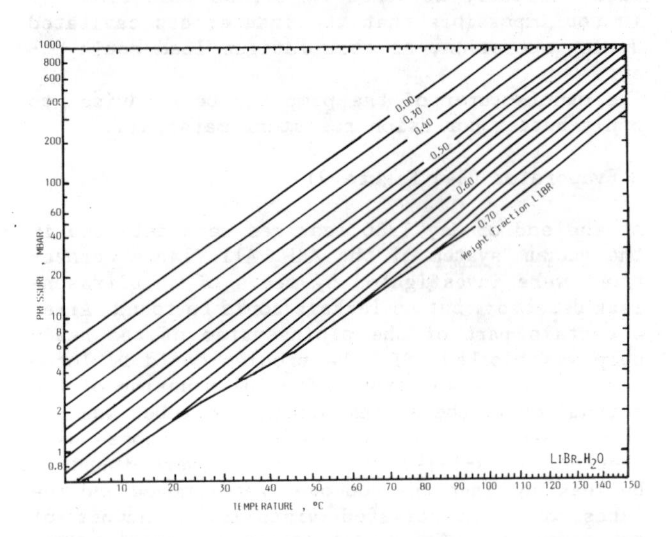

Figure 2. Log P-1/T diagram of LiBr - H_2O.

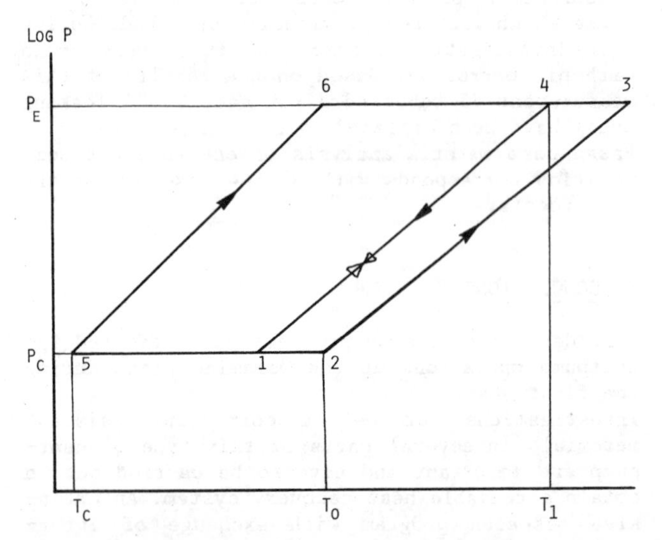

Figure 3. Heattransformer process in LogP-1/T diagram.

Figure 4. The Delamine heattransformer.

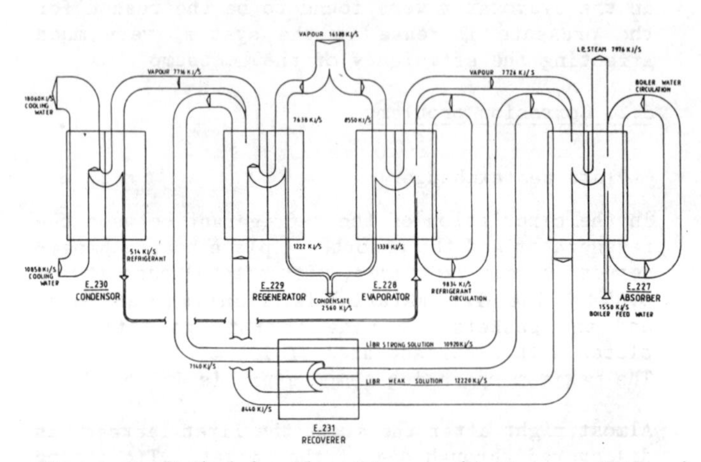

Figure 5. Enthalphy flow diagram (design conditions).

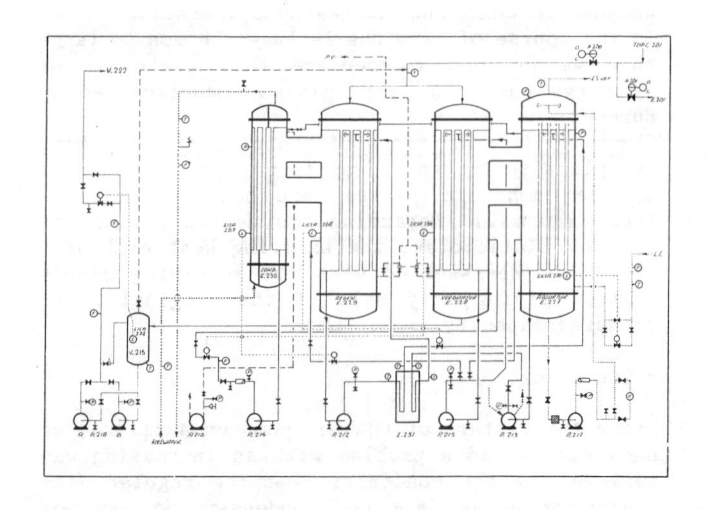

Figure 6. Simplified PID of heatpump.

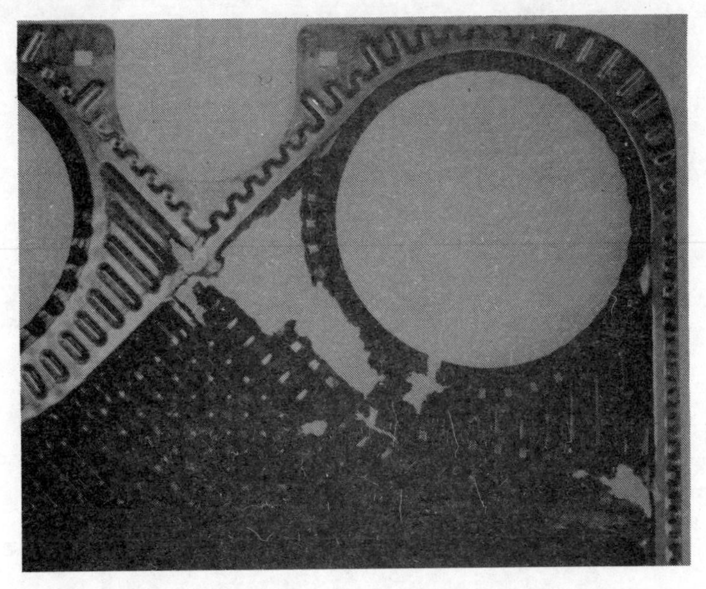

Figure 7. Condenser section with plugged tubes.

Figure 8. Corroded heatexchanger plate.

Figure 9. Evaporator section with plugged tubes.

3rd International Symposium on the

Large Scale Applications of Heat Pumps

Oxford, England : 25-27 March 1987

PAPER B4

PERFORMANCE CHARACTERISTICS OF A WATER-WATER HEAT PUMP USING NONAZEOTROPIC MIXTURES

V.R. Patwardhan, S. Devotta and V.S. Patwardhan

Chemical Engineering Division, National Chemical Laboratory, Pune -411 008, INDIA.

(NCL Communication Number: 4142)

Thermodynamic evaluations of nonazeotropic mixtures of R12, R11 and R114 are presented based on a modified Rankine cycle. Significant improvement on the coefficient of performance over that for pure fluids is predicted based on the theoretical calculations. The improvement is found to be more pronounced for the R12/R11 mixture than for the R12/R114 mixture.

Experimental performance data obtained using the mixture R12 (43 wt.%)/R11 in a water to water heat pump are presented along with data obtained using R11 under comparable conditions. $(COP)_A$ values obtained with the R12/R11 mixture are significantly higher than those with R11.

NOMENCLATURE

$(COP)_L$	Lorenz coefficient of performance	dimensionless
$(COP)_R$	Rankine coefficient of performance	dimensionless
$(COP)_{RM}$	modified Rankine coefficient of performance for mixtures	dimensionless
(CR)	compression ratio, P_{CO}/P_{EV}	dimensionless
H_X	enthalpy per unit mass	kJ kg^{-1}
P_X	pressure at state condition X	bar
T_X	temperature at state condition X	K or °C
$(\Delta T)_{MIN}$	minimum approach temperature in heat exchangers	°C
X	weight per cent of the first component in a mixture	per cent
λ_{CO}	latent heat of vaporisation at condensing temperature	kJ kg^{-1}
ρ_{EV}	vapour density at evaporating temperature	kg m^{-3}
ϕ	entropy per unit mass	kJ $kg^{-1}K^{-1}$

INTRODUCTION

It is well known that there is no single heat pump working fluid possessing all the desirable characteristics for a given application. The important criteria for the selection of a working fluid are reasonable (COP), moderate pressures and good chemical and thermal stability The ultimate choice is usually a compromise on the desirable characteristics. The use of a mixture of working fluids could eliminate drawbacks of a single working fluid. This has been already practised in low temperature processes such as airconditioning, refrigeration and cryogenics. The number of reports on the use of nonazeotropic mixtures in refrigeration and heat pump systems has increased recently. Thus there is a clear trend towards the use of nonazeotropic mixtures of working fluids in heat pump systems (Berntsson, 1982; Moser and Schnitzer, 1985).

The obvious advantages of nonazeotropic mixtures are the following:

The operating pressures can be adjusted by the choice of the composition of the mixture. The compressor discharge temperature can be reduced. The heat load capacity modulation can be achieved. The disadvantages are increased heat exchanger area owing to lower mean temperature difference and lack of proven control strategies and experience.

THEORETICAL ANALYSIS

When mixtures are used as working fluids, the classical Carnot cycle is no longer the thermodynamic reference cycle. As the absorption and the release of heat are not performed isothermally, the Lorenz cycle shown in Figure 1 is more appropriate as the theoretical reference cycle. The Lorenz coefficient of performance $(COP)_L$ can be defined similar to $(COP)_C$, based on the logarithmic mean temperature of the dew and bubble points of the mixture in the condenser and the dew and bubble points in the evaporator, as

$$(COP)_L = \frac{T_{MC}}{T_{MC} - T_{ME}} \tag{1}$$

T_{MC} is the logarithmic mean temperature in the condenser, defined as

$$T_{MC} = \frac{T_{DC} - T_{BC}}{\ln (T_{DC}/T_{BC})} \tag{2}$$

where T_{DC} is the dew point temperature in the condenser and T_{BC} is the bubble point temperature in the condenser.

Held at St. Catherine's College Oxford, England. Organised and sponsored by BHRA, The Fluid Engineering Centre, Cranfield, Bedfordshire, MK43 0AJ England.

Similarly, T_{ME} is the logarithmic mean temperature in the evaporator, defined as,

$$T_{ME} = \frac{T_{DE} - T_{BE}}{\ln (T_{DE}/T_{BE})} \qquad (3)$$

where T_{DE} is the dew point temperature in the evaporator and T_{BE} is the bubble point temperature in the evaporator. As with the Carnot cycle, the Lorenz cycle is defined entirely by the temperature levels in the condenser and the evaporator.

As a mixture consists of two working fluids or more, the Rankine cycle cannot be represented on a two-dimensional pressure enthalpy diagram. In addition to pressure and enthalpy, there is a third parameter, namely, the composition of the fluid mixture. Figure 2 is a three dimensional representation of a P-H diagram for a mixture of working fluids. The cycle shown in front is the Rankine cycle for component A (at the composition x=o), while the cycle at the rear is the Rankine cycle for component B (at the composition x=1). The Rankine cycle for the mixture M is shown by dotted line at a mixture composition X. However, one has to ignore the temperature scales on this plane.

The Rankine coefficient of performance for component A and B can be expressed as

$$(COP)_{RA} = \frac{H_{D1A} - H_{D3A}}{H_{D1A} - H_{S2A}} \qquad (4)$$

and

$$(COP)_{RB} = \frac{H_{D1B} - H_{D3B}}{H_{D1B} - H_{S2B}} \qquad (5)$$

where H_{D1A}, H_{D1B} are the enthalpies of superheated vapours of component A and component B respectively after isentropic compression to pressure P_{CO}.

H_{S2A}, H_{S2B} are the enthalpies of saturated vapours of component A and component B respectively before the isentropic compression from pressure P_{EV}.

H_{D3A}, H_{D3B} are the enthalpies of saturated liquids or component A and component B respectively.

P_{CO}, P_{EV} are the pressures in the condenser and evaporator respectively.

The Rankine coefficient of performance $(COP)_R$ can be calculated from the saturation thermodynamic properties, when the condensing temperature T_{CO}, and the temperature lift ($T_{CO} - T_{EV}$) are known (Holland et al, 1982). When a mixture of working fluids is used, condensation occurs over a range of temperature.

The enthalpy of a mixture at saturation for a given composition and pressure can be found from the properties of pure components at saturation, if the system behaves ideally. The working fluids (usually halogenated hydrocarbons of similar type), when mixed can be assumed to give an ideal mixture (Berntsson, 1982).

An attempt is made here to evolve a thermodynamic cycle (similar to the Rankine cycle) for a working fluid mixture, using simple mixing rules. The ideal coefficient of performance for a mixture is denoted by $(COP)_{RM}$. It is true that the Rankine coefficient of performance cannot be expressed for mixtures. However, it is denoted here as $(COP)_{RM}$, just to indicate its similarity to $(COP)_R$.

The $(COP)_{RM}$ can be defined as

$$(COP)_{RM} = \frac{H_{D1M} - H_{D3M}}{H_{D1M} - H_{S2M}} \qquad (6)$$

where H_{D1M} is the enthalpy of the superheated vapour mixture at system pressure P_{CO}, after the isentropic compression.

H_{S2M} is the enthalpy of the saturated vapour mixture at system pressure P_{EV}, before the isentropic compression.

H_{D3M} is the enthalpy of the saturated liquid mixture at the system pressure P_{CO}.

The superheated vapour enthalpy of the mixture H_{D1M} can be calculated from the entropies of the saturated vapour of the mixture at P_{EV} and P_{CO}.

Equation (6) can be used to calculate the coefficient of performance $(COP)_{RM}$, for the working fluid mixture.

The typical thermodynamic cycle for the mixture is shown in Figure 3. The highest temperature T_{DC} is the dew point temperature of condensing vapour at pressure P_{CO}. The minimum temperature T_E is the equilibrium flash temperature at pressure P_{EV}. Actually the bubble point temperature T_{BE} at P_{EV} is less than T_E but it is this temperature at which the mixture enters the evaporator at pressure P_{EV} as a two phase system. T_E lies between T_{BE} and T_{DE} (dew point temperature at P_{EV}). The temperature lift in the cycle is defined as the difference between the highest temperature T_{DC} and the lowest temperature T_E.

For the purpose of theoretical calculations three sets of temperatures were chosen for the R12/R11 and R12/R114 mixtures. These conditions are:

(1) $T_{DC} = 60°C$, $T_E = 10°C$

(2) $T_{DC} = 80°C$, $T_E = 30°C$

(3) $T_{DC} = 100°C$, $T_E = 50°C$

The calculations were carried out for various compositions of mixture (10-90 per cent by weight of R12).

Consider R12 condensing at a temperature of $T_{CO} = 80°C$ with a minimum temperature drop available in the heat exchanger T=5°C. Also suppose that a mass flow rate is fixed at some arbitrary value and the inlet temperature of sink is 70°C. The maximum temperature attainable by sink will be 75°C. The condenser pressure will be 23 bar. Any increase in the delivery temperature beyond this will not be feasible due to the pressure limitation. However, when a 50 per cent by weight of R12 in the R12/R11 mixture is used the sink can be heated from 70°C even up to 95°C as in case 3. This is also possible with R11 but with a lower heating capacity. The corresponding condenser pressure will be 13.6 bar for the mixture. The addition of R12 to fluids such as R11 and R114 will increase the heating capacity under identical conditions due to its significantly high ($\lambda_{CO}\ \rho_{EV}$) values.

It should be noted that the equivalent condensing temperature for the mixture is taken to be T_{DC}. However it does not mean that the entire quantity of heat can be delivered at a temperature lower by (Δ T)$_{MIN}$ in the heat exchanger as in the case of pure fluids. This is possible only when the inlet temperature of the sink in a countercurrent heat exchanger is lower than the bubble point T_{BC} in the condenser by (Δ T)$_{MIN}$ in the condenser heat exchanger. Any violation of this would lead a temperature crossover hence a reduction in the heating capacity. Similarly the entire cooling capacity can be realised only if the inlet temperature of the source is higher than the dew point T_{DE} in the evaporator by (Δ T)$_{MIN}$ in the evaporator heat exchanger.

Heat pumps are economically more attractive when it performs a dual function of heating and cooling. If it is required to cool a process stream (source) from 25°C to 15°C and heating simultaneously another process stream from 45°C to 55°C, it can be done either by using R12, R114 or a mixture of R12/R114. When R114 is used P_{CO} is 5.78 bar which is a reasonable value, but the compression ratio (CR) is 4.55 which is rather high. On the other hand, when R12 is used, (CR) = 3.6 which is reasonable, but P_{CO} is 15.25 bar, which is rather high. It is quite evident that a 50 per cent by weight of R12 in the R12/R114 mixture has both a reasonable condensing pressure P_{CO} of 9.12 bar and a compression ratio (CR) of 3.25. The working fluid R11 or a mixture of R12/R11 can also be used for the same duty. In the case of R11, the compression ratio (CR) of 5.12 is too high and also the evaporator pressure P_{EV} (= 0.60 bar) is subatmospheric. It is possible that air may ingress into the system. The 50 per cent by weight of R12 in the R12/R11 mixture offers both, relatively low P_{CO} (=5.37 bar) and low compression ratio (CR) (=2.36). The R12/R11 pair seems to be better suited than the R12/R114 pair, but the thermal stability favours the latter.

Figure 4 shows the variation of (COP)$_{RM}$ and compression rate (CR) for the R12/R11 mixture for three sets of temperatures. The compression ratio goes through a minimum. This effect is more pronounced for the R12/R11 mixture, than for the R12/R114 mixture as can be seen in Figure 5. This is because the latent heat values are widely different for R12 and R11, while they are quite comparable for R12 and R114. The effect of increasing the composition of low boiling component is to increase the pressure of system in a nonlinear manner (this effect is enhanced if latent heat values of the components are widely different). The line becomes more curved as the temperature is increased. Therefore, it is observed that the P_{CO} line is more curved than P_{EV} line.

The coefficient of performance is the ratio of heat delivered to the work done. Most of the heat delivered is due to the latent heat of condensation of working fluid mixture. The latent heat usually changes linearly with the composition of low boiling component in mixture. It has been stated earlier that the compression ratio (CR) goes through a minimum as composition is increased. The work of compression is directly proportional to the (CR), therefore it also goes through a minimum. As a result of such variation (COP)$_{RM}$ goes through a maximum with change in composition.

The volumetric latent heat ($\lambda_{CO}\ \rho_{EV}$) is the product of latent heat delivered at the condensing pressure P_{CO} and the vapour density at P_{EV}. For the R12/R114 mixture, ($\lambda_{CO}\ \rho_{EV}$) decreases up to 10 per cent by weight of R12 and then increases. On the contrary, in case of the R12/R11 mixture, no such reduction is noticed but the volumetric latent heat remains constant up to 10 per cent by weight of R12, and then increases. A maximum occurs at X=80 per cent of R12. The decrease after this composition is due to drastic reduction in latent heat value of R12 after 100°C and at higher X this effect is predominant. The difference in latent heat values of R12 and R11 are large as compared to the difference between R12 and R114, therefore this effect is observed to a greater extent in case of R12/R114.

EXPERIMENTAL

To assess the various benefits derived by using a mixture of working fluids over single working fluids, experiments have been carried out using R11 and R12 (43 wt.%)/R11 mixture.

A water to water heat pump, specially designed and well instrumented for research purposes, has been used for this study.

The compressor used in the heat pump was an open type Poladaire PLE4 model designed for use with R12. This was matched by a 2.2 kW motor. The condenser used in the heat pump was a Dunham-Bush CSTC-B548 model and the evaporator a Dunham-Bush CH-324A both with a safe working pressure of about 20 bar. The superheat of the vapour entering the compressor was controlled manually by a high precision throttle valve. The equipment also included various auxiliary units and safety devices standard in the refrigeration industry. In the case of the mixtures, while charging, the fluid with lower vapour pressure was charged first and the other afterwards.

RESULTS AND DISCUSSION

The performance experimental results have been presented in a characteristic plot of (COP)$_{A}$ against temperature lift (T_{CO} -T_{EV}) in Figure 6. In the case of the mixture, the temperature equivalent to the condensing temperature is the dew point temperature of the vapour mixture at the condensing pressure P_{CO} and the temperature equivalent to the evaporating temperature is the temperature of the working fluid at the inlet of the evaporator. It can be seen from Figure 6 that the actual coefficient of performance for the mixture is consistently higher for the mixture than for R11. This is consistent with the theoretical results presented earlier.

It is to be noted here that the mean condensing temperature for the mixture is 85°C and for R11 it is 73°C. It is generally known that the coefficient of performance is more sensitive to temperature lift than to condensing temperature (Holland et al, 1982). Therefore the difference in the mean condensing temperature does not significantly alter the superior performance of the mixture.

Figure 7 shows the variation of the heat delivered Q_D and the shaft power to the compressor W with temperature lift ($T_{CO}-T_{EV}$). The shaft power for the mixture as well as for R11 does not appear to vary significantly However, the heat delivered in the condenser is consistently higher for the mixture. The heat pump capacity for R11 has been improved by the addition of R12 while the shaft power has not been affected. This explains the increase in $(COP)_A$ for the mixture over R11.

In Figure 8 the actual coefficient of performance data have been presented along with the theoretical modified Rankine coefficient of performance and the Lorenz coefficient of performance for the mixture R12/R11. It is to be noted that there is not significant difference between $(COP)_{RM}$ and $(COP)_L$ for a specified temperature lift. However the actual coefficient of performance is much lower than the ideal cycle performance. For a typical lift of 42°C, the modified Rankine heat pump effectiveness factor, which is the ratio of $(COP)_A$ to $(COP)_{RM}$, was 0.509 while the corresponding Lorenz heat pump effectiveness factor was 0.453.

For the purpose of comparison a few data from the experiments have been selected for R11 and for the mixture R12/R11. The data were chosen for comparable temperature lifts. For the first pair of data sets, for a temperature lift of about 50°C, the coefficient of performance for the mixture R12/R11 was 3.92, a significant improvement over the value of 2.94 for R11. Similarly the Q_D values were 4.5 kW and 2.89 kW respectively. The condensing pressure for the mixture was 6.94 bar which is significantly lower than the corresponding saturation pressure of 20.87 bar for R12.

In the next pair of data sets, for a temperature lift of about 42°C the coefficients of performance for the mixture and for R11 were 4.32 and 3.90 respectively. The corresponding heat delivered Q_D values were 4.40 and 3.90 kW. The equivalent condensing temperature for the mixture was 91.0°C which is impossible, in a practical system, to be obtained with R12 as its corresponding saturation pressure would be about 28.0 bar, much beyond the capacity of the compressor used as well as the design pressure of the condenser used.

CONCLUSION

A modified Rankine cycle has been evolved for the theoretical analysis of the performance characteristics of nonazeotropic mixtures as heat pump working fluids. Some of the advantages of the nonazeotropic mixtures have been experimentally demonstrated using R11 and a R12/R11 mixture with a conventionally designed water to water heat pump.

REFERERNCES

Berntsson, T., 'Measurements on a water to water heat pump using mixtures of R22 and R114 as working fluids' Report No. 8, Chalmers University of Technology, Goteborg, Sweden (June 1982).

Holland, F.A., Watson, F.A., and Devotta, S. 'Thermodynamic design data for heat pump systems', Pergamon Press, Oxford, U.K. (1982).

Moser, F., and Schnitzer, H., 'Heat Pumps in Industry', Elsevier Science Publishers B.V. Amsterdam, The Netherlands (1985).

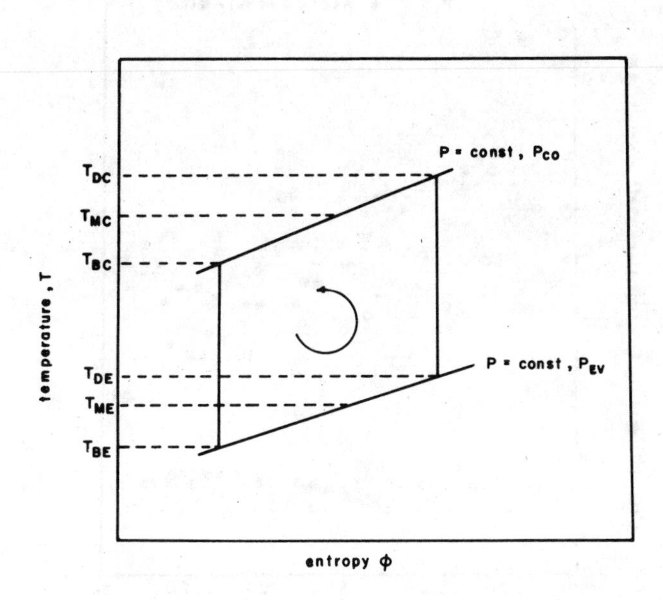

Figure 1 Lorenz Cycle

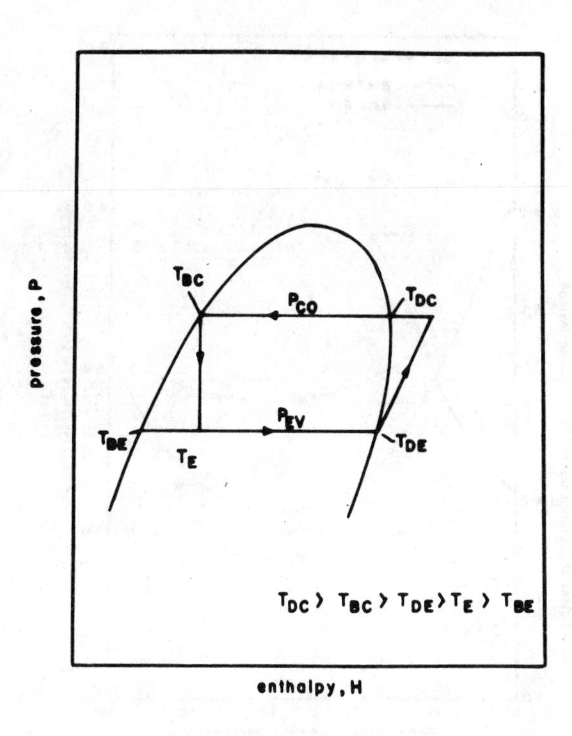

Figure 3 Pressure – Enthalpy diagram for Nonazeotropic mixture of working fluids

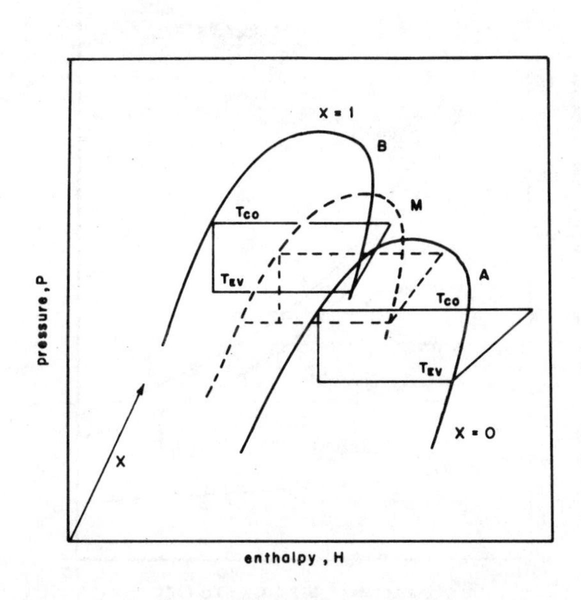

Figure 2 Pressure (P) – Enthalpy (H) – Composition (X) Diagram for mixture of working fluids

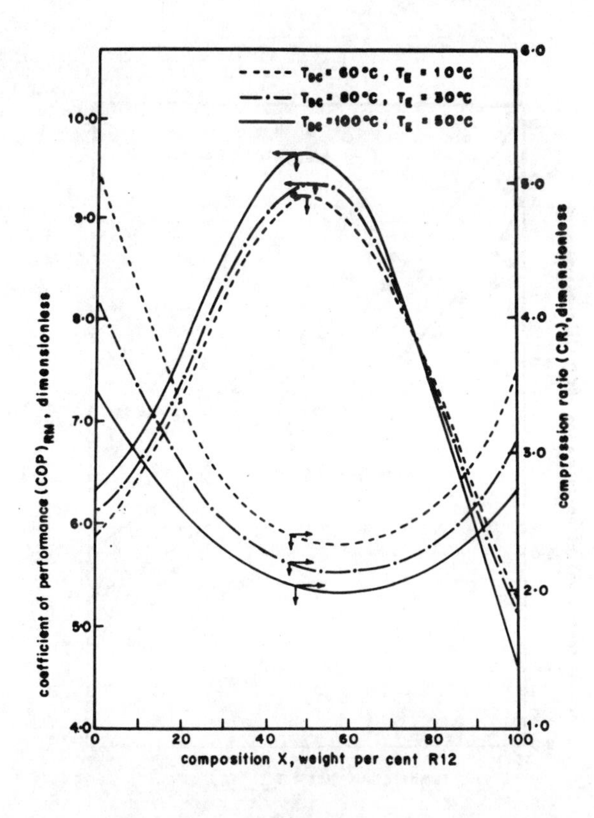

Figure 4 Coefficient of preformance $(COP)_{RM}$ and compression ratio (CR) against composition (X) for the mixture R12 / R114

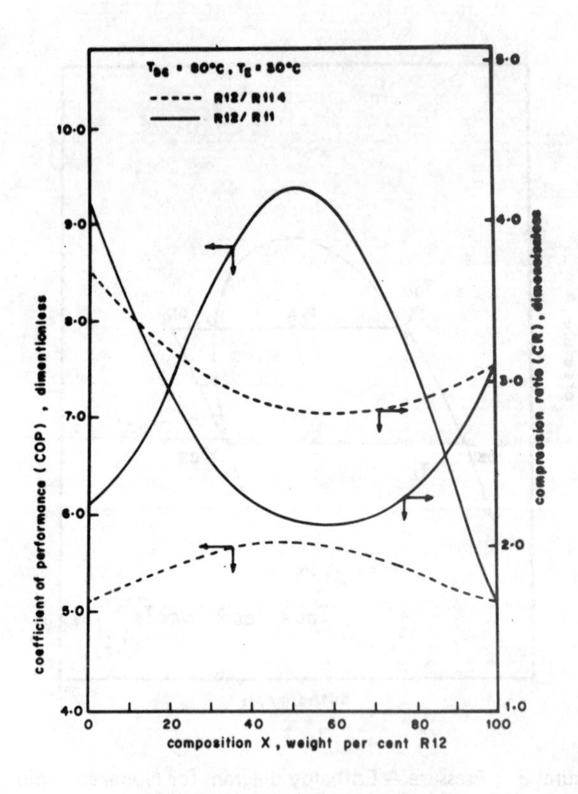

Figure 5 Comparison of coefficient of performance and compression ratio for the mixture R12 / R114 and the mixture R12 / R11 against composition

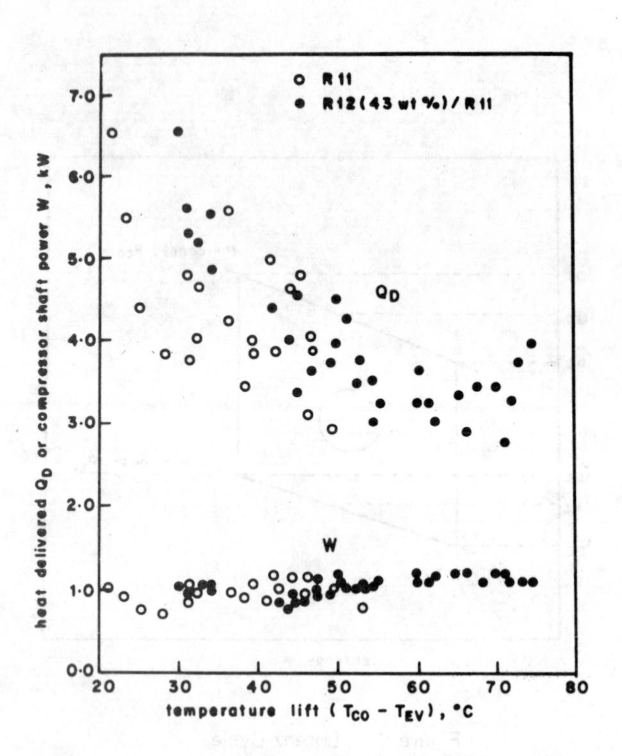

Figure 7 Heat delivered and shaft power against temperature lift for R11 and R12 (43 wt %) /R11

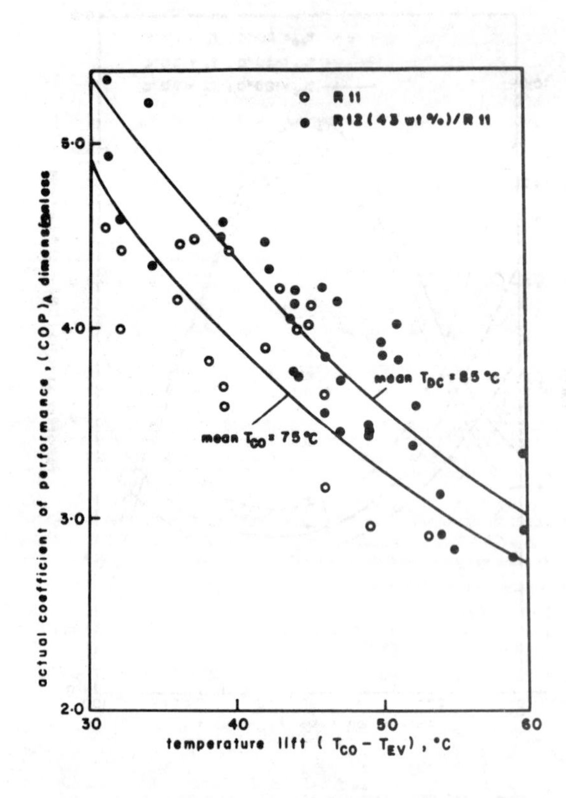

Figure 6 Actual coefficient of performance against temperature lift for R11 and R12 (43 wt %) /R11

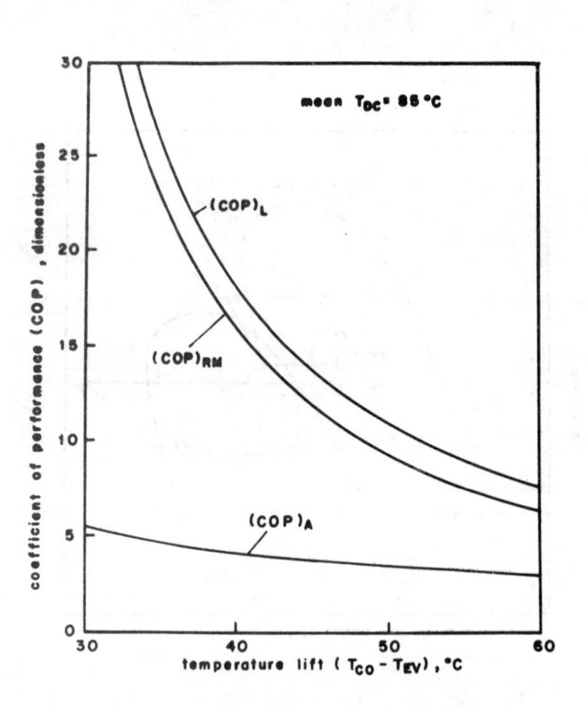

Figure 8 Coefficient of performance against temperature lift for R12 (43 wt %) /R11

3rd International Symposium on the

Large Scale Applications of Heat Pumps

Oxford, England : 25-27 March 1987

PAPER C1

THE PRELIMINARY DESIGN OF AN INDUSTRIAL HEAT PUMP WASTE HEAT RECOVERY SYSTEM BY AN INTERACTIVE COMPUTER PROGRAM

F.R. Steward

The author, Dr. F.R. Steward, is Chairman of the Department of Chemical Engineering, University of New Brunswick, Fredericton, NB, Canada, E3A 5A3. The Co-authors, Dr. D. Karman and S. Sunderarajan, are also affiliated with the University of New Brunswick.

SYNOPSIS

An interactive computer program has been developed to make a preliminary design of an industrial heat pump waste heat recovery system. The program was prepared for an Apple IIe computer.

The program contains a number of modules for the design of various pieces of equipment such as heat exchangers, heat pumps, coolers and heaters. These modules can be arranged among the two or more streams exchanging heat.

The results from the program for two case studies are presented. These studies make an economic comparison of the arrangement of the equipment in the waste heat recovery scheme.

INTRODUCTION

It has become well established practise to use interactive design programs to make preliminary designs of certain common processes in the chemical industry (1).

The recovery of waste heat within an industrial process has certain general characteristics which lends itself to a design analysis of this type. Two previous studies have used the batch form in Fortran (2,3). While the type of computer program presented in both of these references can be used to give a preliminary design of certain waste heat recovery systems, a considerable amount of effort is required to learn how to use and operate such a program. This study has developed an easy to use interactive computer program which can be used to give a preliminary design of certain basic types of waste heat recovery systems. The program has been formulated so that additional sophistications can be readily added in the future as desired.

DEFINITION OF THE PROBLEM

A common problem for waste heat recovery is to have available one or more streams leaving a process or within a process with potential for heat recovery. This heat can be added to streams entering the process or within the process. The flow rates and temperatures of both sets of streams would normally be known. The desired temperatures of the streams to receive the heat would also be known. In some instances the temperature to which the streams are to be cooled is a requirement.

The program assumes that the equipment available to recover the waste heat consists of four types, heat exchangers involving the direct exchange of heat, heat pumps, heaters using an external heat source and coolers using an external cooling source.

It should be possible to arrange any number of these pieces of equipment in any pattern among the streams available for the interchange of heat.

THE EQUIPMENT MODELS

The Heat Exchanger

The heat exchanger model is based on the following assumptions:

1. The process streams contain only sensible heat.
2. The heat capacities of the fluids are independent of temperature.
3. The overall heat transfer coefficient is uniform throughout the exchanger.

The major design equations for the heat exchanger are those commonly used

$$Q_{HE} = m_h C_{Ph} (T_{h\,in} - T_{h\,out}) \quad (1)$$

$$Q_{HE} = m_c C_{pc} (T_{c\,out} - T_{c\,in}) \quad (2)$$

$$Q_{HE} = F_t UA\Delta T\ell m \quad (3)$$

where

$$\Delta T\ell m = \frac{(T\,h\,in - T\,c\,out) - (T\,h\,out - T\,c\,in)}{\ln \dfrac{(T\,h\,in - T\,c\,out)}{(T\,h\,out - T\,c\,in)}} \quad (4)$$

m is the mass flowrate, kg s^{-1}
c_p is the heat capacity of the fluid J $kg^{-1}\,{}^oK^{-1}$
T is the temperature, ^oK
U is the overall heat transfer coefficient, J $s^{-1} m^{-2}\,{}^oK^{-1}$
A is the area of the heat exchanger, m^2
subscripts
h high temperature stream
c low temperature stream

Held at St. Catherine's College Oxford, England. Organised and sponsored by BHRA, The Fluid Engineering Centre, Cranfield, Bedfordshire, MK43 0AJ England.

$$F_t = f(T_{h\ in}, T_{h\ out}, T_{c\ in}, T_{c\ out}, \text{number of shell and tube passes}) \quad (5)$$

F_t accounts for the multiple passes within the exchanger.

The basic output of the heat exchanger program is the area of the exchanger and the amount of heat exchanged. The latter gives the outlet temperatures of the streams knowing the inlet temperatures.

The capital cost of the heat exchanger was assumed to be given by an equation of the form

$$Cost_{HE} = K\ (Area)^a \quad (6)$$

where K and a were determined from standard references for specific construction materials (4,5). The values used in this case studies are given in table 2.

The operating cost for the heat exchanger is the cost of pumping the fluids which is determined from the equation

$$Cost_{oper} = \frac{v \Delta P\ t_o\ C_e}{n} \quad (7)$$

where
- v is the volumetric flowrate of the fluid, $m^3 s^{-1}$
- ΔP is the pressure drop across the exchanger, $N\ m^{-2}$
- n is the efficiency of the pump
- t_o is operating time in s per year
- C_e is cost of electricity, $\$\ J^{-1}$

Reasonable values of ΔP and n are assumed for a preliminary analysis.

THE HEAT PUMP

The heat pump models are based on the following assumptions:

1. The heat pump is the vapor compression type and follows a Rankine cycle.
2. All inefficiencies within the heat pump can be accounted for by an overall efficiency factor.
3. The pressure drop across the evaporator and condenser on the refrigerant side can be given a set value.
4. The process streams contain only sensible heat.
5. The heat capacity of the process streams are independent of temperature.
6. The overall heat transfer coefficients for the evaporator and condenser are uniform throughout.

The major design equations for the heat pump are listed below.

The coefficient of performance of the heat pump is based on an analysis of the Rankine cycle which has been shown to be well represented by the relation (6).

$$COP\ (Rankine) = A\left(\frac{T\ cond}{T\ cond - T\ evap}\right) + B \quad (8)$$

where

$$A = A_0 + A_1\ T\ cond + A_2 T cond^2 \quad (9)$$

$$B = B_0 + B_1\ T\ cond \quad (10)$$

All temperatures in the above equations are in K. The constants in the above equation have been determined for the commonly used refrigerants and are given in Table 1.

The actual COP of a heat pump can be determined from the Rankine COP by the relation.

$$COP\ (actual) = n_r COP\ (Rankine) \quad (11)$$

where n_r has been found to be between .6 and .7 based on operating data for several units (7).

The condenser and evaporator refrigerant temperature is set at fixed values above and below the process stream exit temperature

$$T\ cond = T_{c\ out} + \alpha \quad (12)$$

$$T\ evap = T_{h\ out} - \beta \quad (13)$$

The areas and heat duties of the condenser and evaporator are determined from relations identical to those given for the heat exchanger model.

Equations 8-13 assume condensation and evaporation occur at a uniform temperature in the condenser and evaporator respectively. For a more detailed analysis of the refrigerant cycle, the small pressure drops in these exchanges as well as subcooling in the condenser and superheating in the evaporator can be considered. In this case, temperature (and saturation pressure) at one end of the exchanger is fixed by eqn's 12 or 13. Conditions at the other end are then calculated from the specified pressure drops, superheating and subcooling, using physical properties for the refrigerant. The flowrate of the refrigerant can then be calculated using enthalpy data and the calculated duty.

The compressor capacity is determined directly from the condenser and evaporator heat duties

$$W\ comp = Q\ cond - Q\ evap$$
$$= Q\ cond / COP\ (actual) \quad (14)$$

The capital cost of the heat pump was determined from the relations

$$Cost_{HP} = F_z\ (\text{cost of major components}) \quad (15)$$

where F_z accounts for minor components and is approximately 1.9 - 2.0

The major components of the heat pump are the evaporator, the condenser and the compressor.

The capital cost of the evaporator and condenser were determined in the same manner as the heat exchangers for direct transfer between the process streams, equation (6).

The capital cost of the compressor was determined

from the relation

$$\text{Cost comp} = K\,(\text{shaft power})^a \qquad (16)$$

where K and a were obtained, from standard references (4,5). The values used are presented in Table 2.

The operating cost for the heat pump is the cost of pumping the process fluids and the cost of driving the compressor. This can be determined from the equation

$$\text{Cost}_{\text{oper}} = \left[\frac{v\Delta P}{n} + W\ \text{comp}\right] t_o\, C_e \qquad (17)$$

where the quantities in the equation are the same as for equation (7) giving the heat exchanger operating costs.

Heat pump models have been developed for single stage and two stage units. The two stage unit can be either a two stage compression with intercooling by combining with interstage liquid or two single stage units operated through an evaporator condenser.

Heaters and Coolers

The equations used to design and determine the costs of heaters and coolers are the same as those used for heat exchangers except were a boiler is used as the heater. In the case when a boiler is used as the heater the capital cost of the boiler is determined from the capacity of the boiler required. The operating cost of the boiler is determined from the firing rate, hours of operation and cost of fuel.

Economic Evaluation

Once the capital costs of the installation and the yearly operating costs have been determined for the equipment certain economic evaluations can be made.

Installation and maintenance costs are determined from the following relations

$$\text{Cost}_{\text{install}} = Ki\ (\text{cost cap})$$

where Ki can vary from .5 to 2 (4).

$$\text{Cost (main)} = Km\ (\text{Cost}_{\text{install}} + \text{Cost}_{\text{Cap}})$$

where Km can vary from .02 to .20 (4).

Any of several techniques for economic evaluation can be made. The ones presently available in the program are payback time, net present worth, return on investment and life cycle cost. Each of these are determined from the normal definition of the terms (4,8).

The Interactive Program

The models discussed above are used in a general program named WASHET which controls the calculation procedure. The program, WASHET, consists of six major programs each performing a specific task. Each program possesses a menu which allows the user to perform parts of the program or bypass it if desired. The program was written for and operated on an Apple II e. A flow of logic diagram for the program WASHET is shown in Figure 1.

A short description of the flow of logic is as follows:

The main menu allows for data input from the OLD SYSTEM CONFIGURATION where the data has been stored on a diskette or NEW SYSTEM CONFIGURATION through the INPUT ROUTINE. The EXIT from the program is also at this point.

The DISPLAY CHANGE ROUTINE allows the data to be verified and some changes to be made should an input error be detected.

An energy or energy and economic analysis of the system as defined is performed by the EXECUTE routine.

Any error detected during the analysis will be displayed as a message on the screen to allow for its correction.

If a change of parameter study is being made the DISPLAY CHANGE ROUTINE is called.

Once a desired result has been achieved a summary of input data and calculation results is displayed on the screen. A hard copy of this display can be obtained at this point and the program returned to the MAIN MENU ROUTINE for further case analysis as desired.

Additional details and operating instructions for this program can be obtained from the original thesis (7).

Case Studies

The program described above, WASHET, was used to give a preliminary design of a waste heat recovery system for two cases. Various types of optimizations were performed in each case study depending on the circumstances.

Case 1

This case study was based on an installation made at Monarch Fine Foods, an edible oils plant in Rexdale, Ontario. Details on this installation and operating experience of the system were reported previously (9).

The problem is defined as follows:

A stream of warm water is available at 44 C and a flowrate by 1364.9 kg/min from the plant cooling water system. The necessary heat removal from this stream was accomplished by two plant cooling towers. The waste heat in this stream is to be used to preheat a boiler feedwater make up stream of 500.4 kg/min at a temperature of 1.4 C. It is desired to preheat the boiler feedwater make up to 52 C.

Two schemes were analyzed to recover heat from the cooling water stream.

1. A heat exchanger with the remainder of the heat to be supplied by the boiler.

2. A heat exchanger and heat pump operated in series. A flow diagram for this scheme is given in Figure 2.

Each scheme was optimized with respect to the heat exchanger boiler feedwater make up exit temperature. The heat pump was also optimized with respect to the temperature differences between the exit process streams and the refrigerant temperature in the condenser and evaporator.

The input data required for the two schemes is given in Table 3. The values used for such quantities as the heat transfer coefficients, equipment pressure drops, pump efficiencies and pump cycle efficiency are considered to be approximate estimates for the type of equipment to be used. The economic factors were those in use or best estimates for Canada in late 1985 when the calculations were performed.

The results of the preliminary design are presented in Figures 2-4 and Table 4.

Figure 3 gives the present worth of only the heat exchanger installed vs. the temperature difference between the exit sink stream and the inlet source stream of the heat exchanger. The diagram indicates that the optimum installed heat exchanger on this basis would heat the boiler feedwater make up to within 1.4 C of the cooling water inlet temperature.

Figure 4 presents the discounted payback period vs. the temperature difference between the exit sink stream and the inlet source stream for the combined heat pump and heat exchanger recovery system and each individual unit. The heat pump was optimized for internal parameters such as the condenser and evaporator exit stream temperature differences (α and β).

The optimum for the combined system is quite insensitive to the heat exchanger temperature difference and was found to be at 1.4 C. The discounted payback time for the combined system was found to be 2.33 years with a present worth of $556,739. A complete run down of the total waste heat recovery system including both the heat exchanger and heat pump is given in Table 4.

Case 2

This case study was based on an installation made at Agrinove, a dairy plant is St. Claire, Quebec. Details on this installation and operating experience of the system were reported previously (9).

The problem is defined as follows:

A stream of warm water is available at 35 C and a flowrate of 1162 kg/min from condensate collected from several sources throughout the plant. This stream was previously going to sewer. The waste heat in this stream is to be used to preheat a boiler feedwater make up stream of 136 kg/min at a temperature of 1 C. It is desired to preheat the boiler feedwater make up to 71 C.

Two schemes were analyzed to recover heat from the combined condensate stream.

1. A heat exchanger and heat pump operated in series.

2. A heat exchanger and heat pump operated in parallel with respect to the source stream as shown in Figure 5.

The first scheme was optimized with respect to the heat exchanger boiler feedwater make up exit temperature. The second scheme was optimized with respect to the split in flow of condensate between the heat exchanger and the heat pump. The heat pumps were optimized with respect to internal parameters.

The input data required for the two schemes is given in Table 5. Those quantities which are the same as Case 1 have been omitted.

The results of the preliminary design are presented in Figures 5 and 6 and Table 6.

Figure 6 gives the discounted payback period vs. the fraction of total condensate flow fed to the heat pump. The diagram indicates that the optimum split sends 60-70% of the flow to the heat pump.

A complete run down of the two schemes for optimized operating conditions is presented in Table 6. The results indicate that scheme 2 gives slightly better economics than scheme 1. When the flow rate of the source stream is large relative to the sink stream parallel operation of the heat exchanger-heat pump recovery system is normally somewhat better than series operation.

CONCLUSIONS

An interactive computer program has been developed that can give a preliminary design for a waste heat recovery system consisting of various combinations of heat exchangers and heat pumps. A systematic variation of certain important internal parameters allows for the optimization of the system.

It was found that heat exchangers should be used to bring the temperature of the cold stream up to a temperature within 2 C of the source stream before the cold stream enters the heat pump.

In those cases where the source stream flow rate is significantly greater than the sink stream flow rate a division of the source stream between the heat exchanger and the heat pump evaporator for parallel operation gives superior operating conditions to a series arrangement of the heat exchanger and heat pump.

REFERENCES

(1) Weber, J.H., "Basic Programs for Chemical Engineering Design", Marcell Dekker: New York, (1984).

(2) Urdaneta-Bohorquez, A.H., "Energy and Economic Analysis of Industrial Process Heat Recovery with Heat Pumps", University of Texas, PhD Dissertation, (1978).

(3) MacKay, C.H., "Optimization of Waste Heat Recovery in the Process Industry", University of New Brunswick, BSc Thesis, (1982).

(4) Peters, M.S. and Timmerhaus, K.D., "Plant Design and Economics for Chemical Engineers", McGraw-Hill Publishing Co., pp160, 273 (1980).

(5) Hall, R.S., Matley, J. and McNaughton, K.J., "Current Cost of Process Equipment", Chemical Engineering, 89, 80-116 (1982).

(6) Omidey, T.O., Kasprzycki, J. and Watson, F.A., "The Economics of Heat Pump Assisted Distillation Systems - I: A Design and Economic Model", Journal of Heat Recovery Systems, 187-200, (1984).

(7) Sunderarajan, S., "Optimization of Waste Heat Recovery Systems", University of New Brunswick MSc Thesis, (1986).

(8) Sauer, H.J. Jr. and Howell, R.H., "Heat Pump Systems", John Wiley and Sons, Chapter 9, (1983).

(9) Wright, J.R. and Steward, F.R., "Three Industrial Heat Pump Installations Operating in Canada", 2nd BHRA International Symposium on the Large Scale Applications of Heat Pumps, York, England, 245-54, September 1984 also published Journal of Heat Recovery Systems, 5, 81-8, (1985).

Table 1

Constants for Common Refrigerants in Determining the Rankine COP from Equations 8-10

Refrigerant	Temp Range K	A_o	$A_1 x10^{-3}$ K^{-1}	$A_2 x10^{-5}$ K^{-2}	B_o	$B_1 x10^{-3}$ K^{-1}
R-12	274-380	.9497	.346	-2.26	-.6224	-4.04
R-22	274-365	.9367	.223	-5.29	-.5696	-.181
R-11	274-343	.9944	.0685	-.920	-.5185	-3.49
	343-339	.9559	.999	-.920	-.5185	-3.49
R-114	274-343	.9805	.158	-.862	-.3388	-13.76
	343-400	.8706	2.84	-2.53	-.3388	-13.76

Table 2

Capital Cost Coefficients for Various Types of Equipment

Type of Equipment	Capacity and Range	K	a
	Area m^2		
Plate and Frame Heat Exchange	1 - 500	610	.778
	Area m^2		
Shell and Tube Heat Exchange	5 - 10	1030	.536
	10 - 20	890	.607
	20 - 800	1185	.665
	Shaft Power kw		
Centrifugal Compressor	50 - 4000	490	.925
	Shaft Power kw		
Reciprocating Compressor	50 - 4000	831	.873
	Heat duty kw		
Boiler	600 - 6000	270	.807
	6000 - 60,000	220	.841

Cost = K $(capacity)^a$
Cost in 1984 Canadian $

Table 3

Input Data for Case #1, Monarch Fine Foods

STREAM DATA FOR UNIT HEAT EXCHANGER

DATA FOR THE SINK STREAM

1) STREAM NUMBER OF INCOMING STREAM = 1
2) TEMPERATURE OF STREAM #1 (Deg K) = 274.4
3) STREAM NUMBER OF OUTGOING STREAM = 2
4) TEMPERATURE OF STREAM #2 (Deg K) = 315.6
5) FLOW RATE OF STREAM (Kg/min) = 500.4
6) SPECIFIC HEAT OF STREAM (kJ/Kg-K) = 4.19

DATA FOR THE SOURCE STREAM

1) STREAM NUMBER OF INCOMING STREAM = 4
2) TEMPERATURE OF STREAM #4 (Deg K) = 317.0
3) STREAM NUMBER OF OUTGOING STREAM = 5
4) TEMPERATURE OF STREAM #5 (Deg K) = 0.0
5) FLOW RATE OF STREAM (Kg/min) = 1364.9
6) SPECIFIC HEAT OF STREAM (kJ/Kg-K) = 4.19

DATA FOR UNIT HEAT EXCHANGER

i) NUMBER OF PASSES ON SOURCE SIDE (1-9 ONLY) = 4
ii) TYPE OF HEAT EXCHANGER (ENTER 1/2)
 1) SHELL AND TUBE
 2) PLATE AND FRAME = 2
ii) HEAT TRANSFER COEFFICIENT (kW/sq m-K) = 1.136
iv) SOURCE SIDE PRESSURE DROP (kPa) = 44.84
v) SINK SIDE PRESSURE DROP (kPa) = 14.95
vi) PUMP EFFICIENCY (%) = 65

STREAM DATA FOR UNIT HEAT PUMP

DATA FOR THE SINK STREAM

1) STREAM NUMBER OF INCOMING STREAM = 2
2) TEMPERATURE OF STREAM #2 (Deg K) = 315.6
3) STREAM NUMBER OF OUTGOING STREAM = 3
4) TEMPERATURE OF STREAM #3 (Deg K) = 325.0
5) FLOW RATE OF STREAM (Kg/min) = 500.4
6) SPECIFIC HEAT OF STREAM (kJ/Kg-K) = 4.19

DATA FOR THE SOURCE STREAM

1) STREAM NUMBER OF INCOMING STREAM = 5
2) TEMPERATURE OF STREAM #5 (Deg K) = 0.0
3) STREAM NUMBER OF OUTGOING STREAM = 6
4) TEMPERATURE OF STREAM #6 (Deg K) = 0.0
5) FLOW RATE OF STREAM (Kg/min) = 1364.9
6) SPECIFIC HEAT OF STREAM (kJ/Kg-K) = 4.19

A value of 0.0 indicates that quantity must be calculated.

DATA FOR UNIT HEAT PUMP

i) HEAT PUMP CYCLE EFFICIENCY (%) = 65
ii) WORKING FLUID (R12, R22, R114, R11) = R114
ii) PUMP EFFICIENCY (PUMP SOURCE/SINK STREAM, %) = 65

DATA FOR HEAT PUMP CONDENSER

i) APPROACH TEMP DIFFERENCE (Deg K) = 2.8
ii) WATER SIDE PRESSURE DROP (kPa) = 14.95
iii) REFRIGERANT PRESSURE DROP (kPa) = 14.95
iv) HEAT TRANSFER COEFFICIENT (kW/sq m-K) = 2.271

DATA FOR HEAT PUMP EVAPORATOR

i) APPROACH TEMP DIFFERENCE (Deg K) = 2.8
ii) WATER SIDE PRESSURE DROP (kPa) = 44.84
iii) REFRIGERANT PRESSURE DROP (kPa) = 14.95
iv) HEAT TRANSFER COEFFICIENT (kW/sq m - K) = 1.703

DATA FOR HEAT PUMP COMPRESSOR

i) MAXIMUM COMPRESSION RATIO = 4.0
ii) TYPE OF COMPRESSOR (ENTER 1/2):
 1) RECIPROCATING
 2) CENTRIFUGAL = 2

ECONOMIC INPUT DATA

i) ANNUAL RATE OF INTEREST (%) = 12
ii) RATE OF TAXATION (%) = 35
iii) RATE OF FUEL COST INFLATION (%) = 7
iv) RATE OF ELECTRICITY COST INFLATION (%) = 7
v) RATE OF MAINTENANCE COST INFLATION (%) = 7
vi) LIFE TIME OF EQUIPMENT (Years) = 10
vii) OPERATING TIME (Hours/Year) = 6000
viii) COST OF FUEL (INCLUDING EFFICIENCY, $/1E9 J) = 4.49
ix) COST OF ELECTRICITY ($/KWhr) = .0322
x) PERCENTAGE MAINTENANCE COSTS (% of equipment costs) = 151
xi) PERCENTAGE MAINTENANCE COSTS (% of equipment costs) = 10
xii) COST OF COOLING UTILITY ($/1E9 J) = 0

TABLE 4

Summary of the Results of an Analysis
of the Case # 1 for both
Heat Exchanger and Heat Pump
In Waste Heat Recovery System

UNIT NO	PURCHASE COST ($)	OPERATING COST($/YEAR)	HEAT DUTY 1E9 J/Hr	CHARACTERISTIC
1	34,415	340	5.19	177.14 sq m
2	56,075	11,496	0.97	5.69 (COP)

SUMMARY OF UNIT DATA

DATA FOR UNIT 1, HEAT EXCHANGER

1) TEMPERATURE OF STREAM # 5 = 301.5 Deg K
2) HEAT EXCHANGER EFFICIENCY = 97%
3) TEMPERATURE CORRECTION FACTOR = 0.89

DATA FOR UNIT 2, HEAT PUMP

1) TEMPERATURE OF STREAM # 6 = 299.7 Deg K

SUMMARY OF ECONOMIC RESULTS

1) PAYBACK PERIOD = 2.33 Years
2) PRESENT WORTH = 556,739 $
3) TOTAL EQUIPMENT COST = 90,490 $
4) TOTAL HEAT RECOVERED = 6.16 x 1E9 J/Hr
5) TOTAL FUEL COSTS FOR NO HEAT-RECOVERY = 171,081 $/Year
6) BOILER/HEATER COSTS = 0 $

A 0 in line (6) indicates a Retro-Fit analysis.

DATA FOR HEAT PUMP CONDENSER

1) CONDENSER HEAT DUTY (TOTAL) = 1.183 x1E9J/Hr
2) CONDENSING TEMPERATURE = 54.48 Deg C
3) AVERAGE CONDENSING TEMPERATURE = 53.91 Deg C
4) DEGREE OF SUPERHEAT = 2.71 % of total condenser duty
5) MASS FLOW RATE OF REFRIGERANT = 10,314.37 Kg/Hr
6) DEGREE OF SUBCOOLING = 1.37% of total condenser duty

DATA FOR HEAT PUMP EVAPORATOR

1) EVAPORATOR HEAT DUTY (TOTAL) = 0.983 x1E9 J/Hr
2) EVAPORATING TEMPERATURE = 22.57 Deg C.
3) AVERAGE EVAPORATING TEMPERATURE = 23.66 Deg. C.
4) FLOWRATE OF LIQUID REFRIGERANT ENTERING EVAPORATOR = 7,943.59 Kg/Hr
5) DEGREE OF SUPERHEAT = 1.44 % of total evaporator duty

DATA FOR HEAT PUMP VALVE

1) COMPRESSION RATIO ACROSS VALVE = 2.30
2) VAPOUR FRACTION AT VALVE EXIT = 22.0%

DATA FOR HEAT PUMP COMPRESSOR

1) ACTUAL POWER INPUT = 57.78 kW
2) ISENTROPIC POWER INPUT = 41.18 kW
3) ISENTROPIC EFFICIENCY = 71 %
4) COMPRESSION RATIO = 2.55
5) VOLUMETRIC FLOW RATE AT INLET = 700.69 cu m/hr
6) VOLUMETRIC FLOW RATE AT EXIT = 288.00 cu m/hr

HEAT PUMP CYCLE DATA

LOCATION	PRESSURE (kPa)	TEMPERATURE (DEG C)	PHASE
COMP EXIT	505.59	61.00	SUP HT VAP
COND INLET	505.59	54.58	SAT VAP
COND EXIT	490.70	53.34	SAT LIQ
VALVE INLET	490.70	50.57	SUB COOL LIQ
EVAP INLET	213.39	24.75	VAP/LIQ MIX
EVAP EXIT	198.57	22.57	SAT VAP
COMP INLET	198.57	25.34	SUP HT VAP

Note that a superheat and subcooling of 2.8 C has been specified by the user.

Table 5

Input Data for Case # 2, Agrinove

Stream conditions given in Figure 5

Other conditions same as Case 1 Table 3 except the following

Source side Pressure Drop for Scheme 1 = 59.78 kPa

Source Side Pressure Drop for Scheme 2 = 44.84 kPa

Number of source side passes for heat exchanger for scheme 1 = 3

Number of source side passes for heat exchanger for scheme 2 =4

Cost of fuel including boiler efficiency = 6.13 $\$/10^9 J$

Cost of Electricity = .028 $/kwh

Percentage installation costs = 75%

Table 6

Results of the Analysis for Case # 2

PARAMETERS	SCHEME 1 Series	SCHEME 2 Parallel
1) Heat supplied to sink (1E9 J/Hr)		
a) Heat Exchanger	1.101	1.101
b) Heat pump	1.290	1.290
2) Heat Recovered (1E9 J/Hr)		
a) Heat exchanger	1.101	1.101
b) Heat pump	0.921	0.943
3) Power required (kW)		
a) Heat exchanger (H.E)	1.88	0.24
b) Heat pump	104.44	97.61
4) Characteristics of unit		
a) Area of Heat exchanger (sq m)	29.21	37.90
b) Efficiency of HE(%)	95	95
c) COP of Heat pump	3.49	3.71

PARAMETERS	SCHEME 1 Series	SCHEME 2 Parallel
1) Discounted Payback period (Years)		
a) Heat exchanger	0.57	0.70
b) Heat pump	8.41	7.35
c) Overall	3.39	3.22
2) Simple Payback period (Yrs)	2.21	2.11
3) Net discounted present worth ($)	194,999	206,047
4) Total Installed cost ($)	137,658	134,449
a) Heat exchanger	14,816	18,144
b) Heat pump	122,842	116,305
5) Total operating costs ($/Yr)		
a) Heat exchanger	1,163	1,077
b) Heat pump	24,565	24,121
6) Net fuel savings ($/Yr)	62,180	63,787

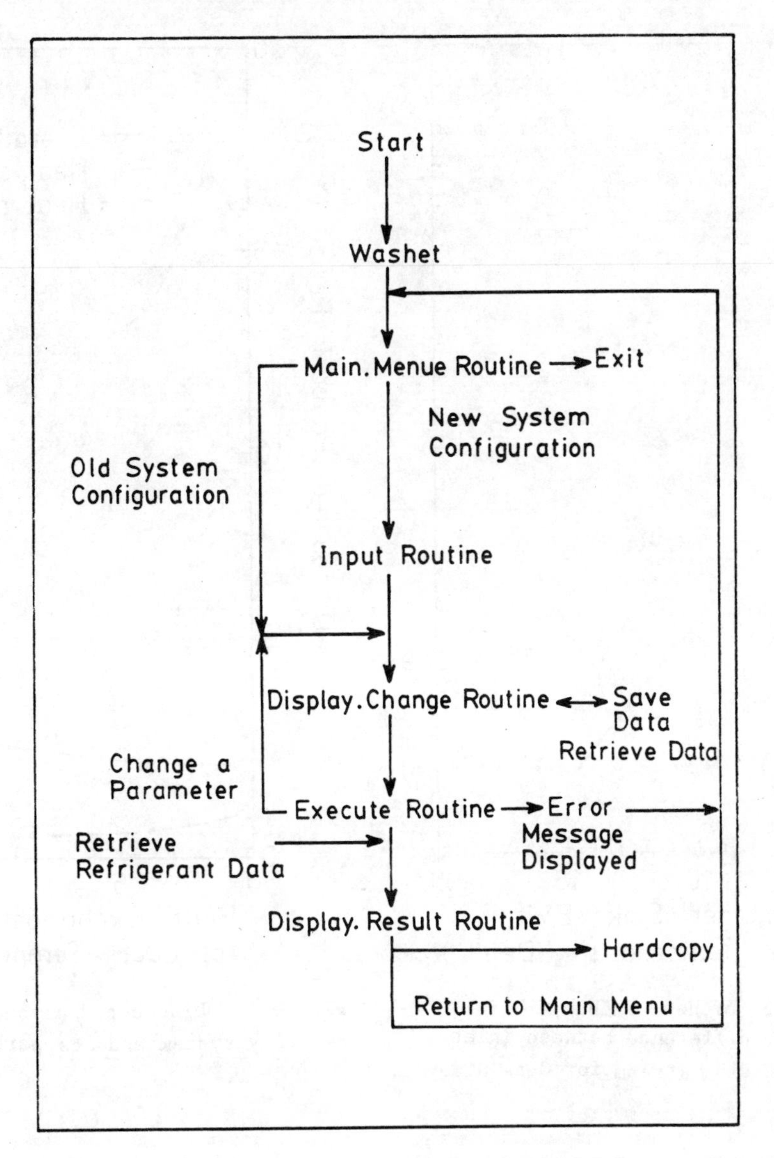

Figure 1: Flow of logic for interactive computer program

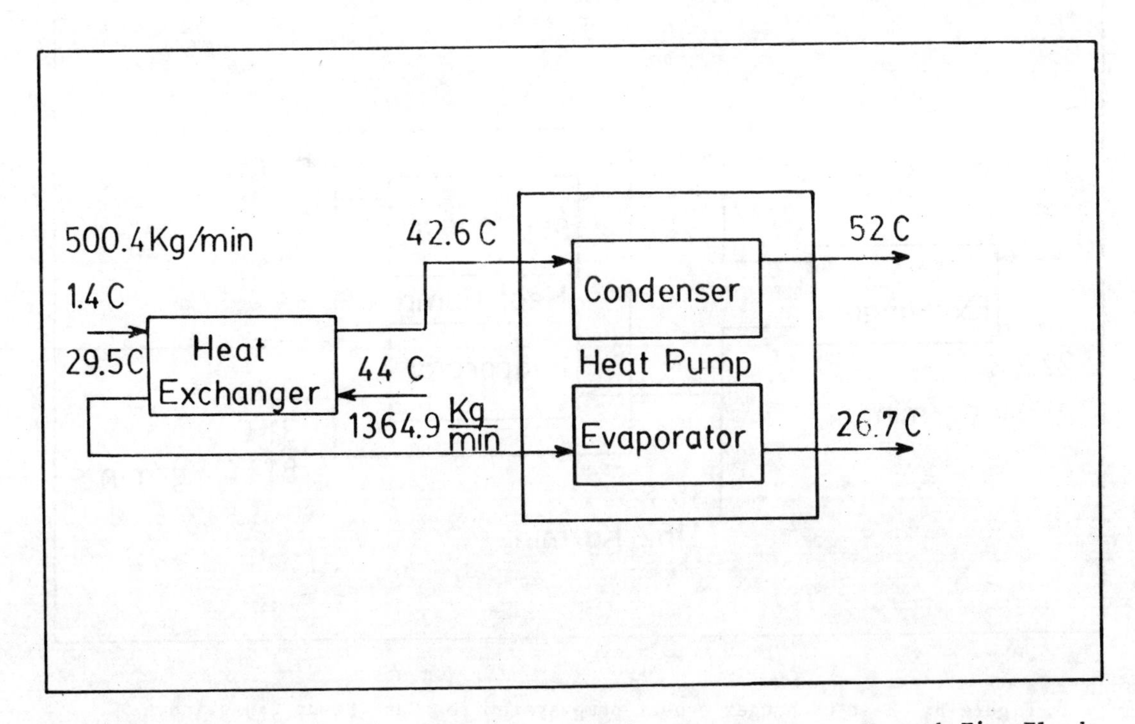

Figure 2: Heat exchanger - heat pump system for Case Study 1: Monarch Fine Floods

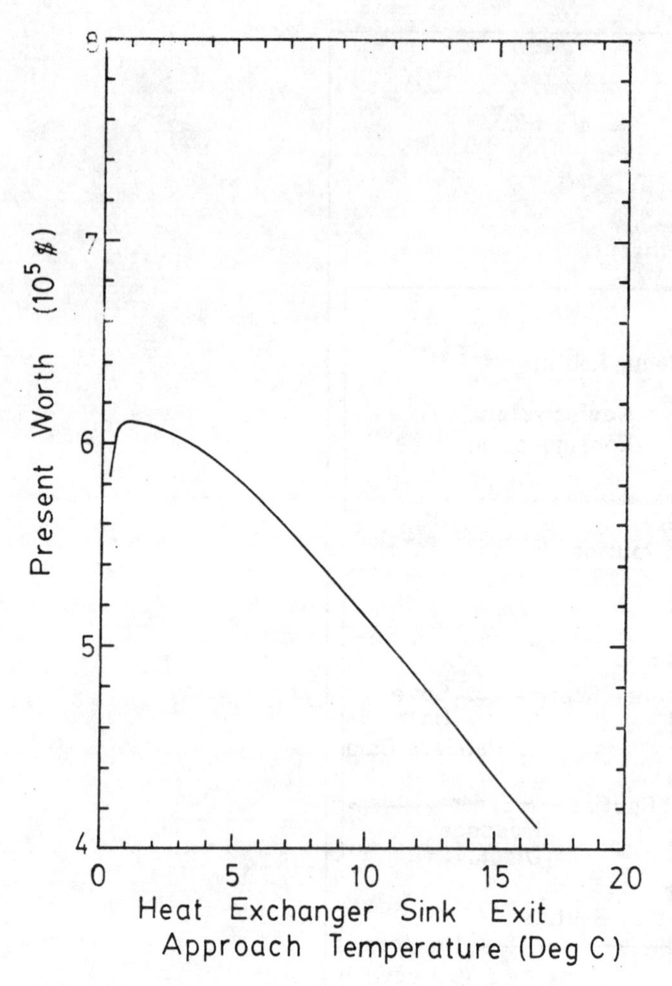

Figure 3: Optimization of heat exchanger with respect to temperature difference between inlet source stream and exit sink stream for Case Study 1.

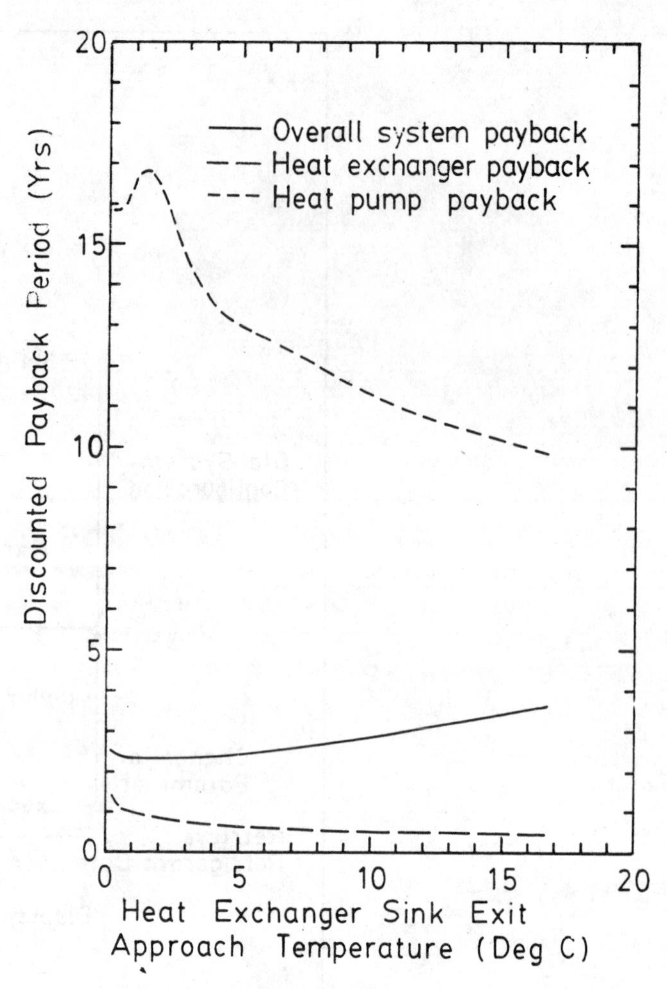

Figure 4: Discounted payback period for waste heat recovery system and its parts for Case Study 1.

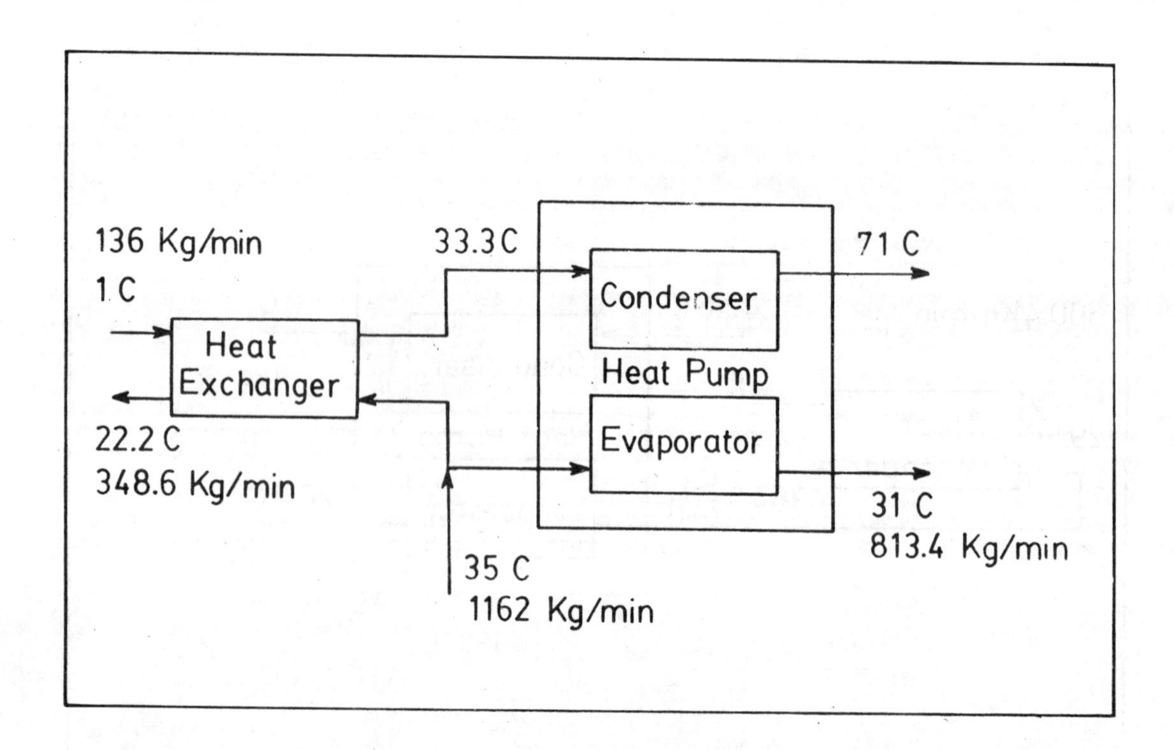

Figure 5: Heat exchanger - heat pump system for Case Study 2: Agrinove

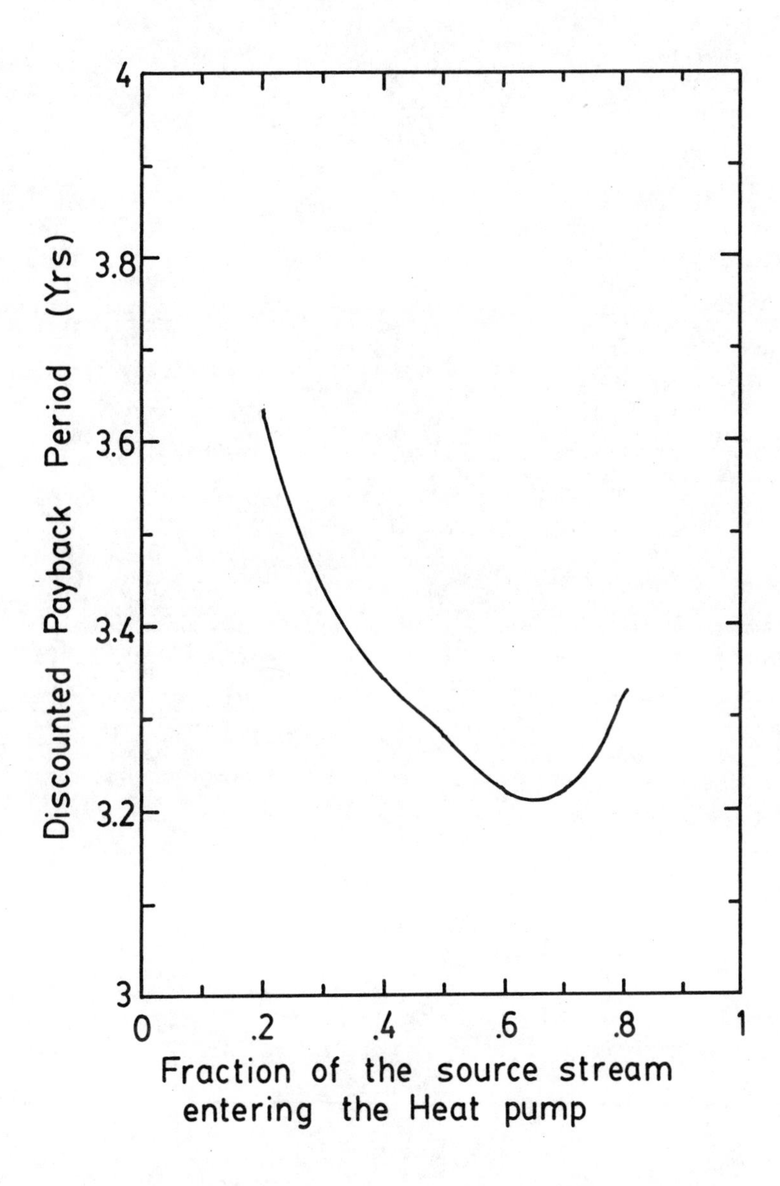

Figure 6: Optimization of division of source stream flow between heat exchanger and heat pump for Case Study 2.

3rd International Symposium on the

Large Scale Applications of Heat Pumps

Oxford, England: 25-27 March 1987

PAPER C2

AN ABSORPTION CYCLE HEAT PUMP FOR DRYING APPLICATIONS IN CANADIAN INDUSTRY

W.K. Snelson
Division of Mechanical Engineering, National Research Council of Canada, Ottawa, Ontario

J.B. Codrington
Acres International Limited, Niagara Falls, Ontario

SYNOPSIS

A concept has been developed for using an open cycle absorption heat pump in industrial drying applications. The study is based on an actual paper drying process using the final drying stage of a 400 tonne/day paper machine, but the underlying principle could be equally applied to other industrial drying processes.

Moist air from the drying system is fed into the absorber and contacted directly by the liquid absorbent solution in a contacting tower. Heat and mass transfer effects take place as some of the water vapour in the air is absorbed by the solution. The net result is a reduction in moisture content and an increase in temperature of the moist air returning to the dryer. The excess water content in the solution is evaporated away in the low pressure regenerator using a source of low grade waste heat.

Low cost sodium hydroxide/water was chosen as the heat transport medium. This working fluid pair is well suited to an open cycle system where losses could be an economic burden. Available thermodynamic fluid properties were reviewed and refined to provide formulae suitable for computation purposes.

Component design criteria were established, and a computer-aided design technique was developed for estimation of operating conditions within a contacting tower absorber containing open type packing. The computer program models the absorption effects and performs stage-by-stage heat and mass transfer calculations using fluid thermodynamic data previously formulated.

Preliminary cost estimates and an economic analysis established attractive energy benefits and payback terms. Materials considerations were evaluated in detail, and the scope of a supporting research and development program was identified.

NOMENCLATURE

- a = surface area of liquid per unit volume of tower, $m^2\ m^{-3}$
- c = specific heat of gas, $kJ\ kg^{-1}/^{\circ}C$
- G = gas loading, $kg\ s^{-1}m^{-2}$
- h = overall coefficient of heat transfer, $kJ\ s^{-1}\ m^{-2}/^{\circ}C$
- h_{ℓ} = specific enthalpy of solution, $kJ\ kg^{-1}$
- h_v = specific enthalpy of superheated water vapour, $kJ\ kg^{-1}$
- K_x = overall coefficient of mass transfer, $kg\ s^{-1}\ m^{-2}$/unit concentration
- L = liquid loading, $kg\ s^{-1}\ m^{-2}$
- Le = Lewis number, dimensionless
- M_a = molecular weight of air, dimensionless
- M_w = molecular weight of water, dimensionless
- m = mass of water vapour diffused, $kg\ s^{-1}$
- NTU = number of transfer units, dimensionless
- P = water vapour pressure in gas stream, bar
- P' = water vapour pressure at gas/liquid interface, bar
- q_c = heat transferred by convection, $kJ\ s^{-1}$
- q_d = heat transferred by diffusion, $kJ\ s^{-1}$
- T = gas temperature, °C
- T' = liquid temperature, °C
- V = tower volume, m^3
- X = moisture content of gas body (mass of water vapour per unit mass of dry gas), $kg\ kg^{-1}$
- X' = moisture content of gas at liquid interface (mass of water vapour per unit mass of dry gas), $kg\ kg^{-1}$
- λ = weight concentration of absorbent per unit mass of solution, $kg\ kg^{-1}$

Subscripts

- 1 = first stage conditions
- n = n^{th} stage conditions
- n+1 = $(n+1)^{th}$ stage conditions

1. INTRODUCTION

The coordinated interdepartmental program of energy research and development within the Canadian federal government was established in 1974. The National Research Council (NRC) acted as the lead agency in many of the work areas, including heat pumps, and the Division of Energy of NRC provided a central focus for this activity. In addition to in-house energy research and development carried on in the laboratory divisions, NRC also funded a substantial amount of contracted out R&D work.

The national heat pumps program was aimed broadly at exploiting both near term opportunities for energy conservation in industry and investigating the practical feasibility of future advanced systems. This interest in longer term technologies led to a review of the potential for introduction of absorption heat pumps into suitable areas of Canadian industry. As a result of

Held at St. Catherine's College Oxford, England. Organised and sponsored by BHRA, The Fluid Engineering Centre, Cranfield, Bedfordshire, MK43 0AJ England.

this study, certain specific areas of particular industrial opportunity were identified, and it was considered worthwhile to explore some of these applications in more detail. This paper summarizes the work done in one of those feasibility studies completed in 1985. More detailed information is contained in the full report (Ref. 1).

2. CONCEPT

The application selected for the study was a paper drying operation based on an actual 400 tonne per day paper machine at the Ontario Paper Company plant in Thorold, Ontario. The aim is to reduce the amount of plant steam used at the drying rolls for extraction of moisture from the wet paper. A simplified schematic diagram (Fig. 1) illustrates the principal features of the process.

The absorption heat pump operates in this case as an open cycle device and acts as a dehumidification dryer. Warm moist air at 77°C and absolute humidity 15% leaves the paper drying machine and enters the absorber section of the heat pump. The moist air is contacted by a concentrated sodium hydroxide solution which enters a contacting tower at 80°C and 60% NaOH weight concentration. Mixing of the two streams passing downwards through the tower causes mass transfer of some of the water vapour from the moist air stream to the solution. This in turn releases the heat of dilution within the NaOH solution. Convection heat transfer occurs simultaneously in the reverse direction. The net effect is to extract the required moisture from the air stream which leaves the tower dryer and warmer at 11% humidity and 106°C. The diluted NaOH solution also leaves the tower at an increased temperature of 107°C.

A demister is used to extract any remaining NaOH solution from the air stream before it returns to the dryer. The dilute 57.6% solution flows from the bottom of the tower through a control valve which lowers the pressure before the solution enters the regenerator. A source of low pressure waste steam is used to evaporate the excess water out of the solution. The extracted water vapour is condensed in a desuperheater/condenser and then dumped to drains. The concentrated NaOH is pumped from the regenerator back to the top of the absorption tower.

The open cycle system is especially applicable to the drying process, and direct contact in the absorber between sodium hydroxide solution and water vapour in the drying air eliminates the need for any evaporator.

3. WORKING FLUID

3.1 Selection

A solution of sodium hydroxide-water was selected as the working fluid pair. Most industrial absorption heat pumps built to date have used lithium bromide-water as the working fluid. However, crystallization problems at low temperatures have led to a search for alternatives where higher temperature boosts are required. In this regard other workers (Ref. 2, 3, 4) have identified sodium hydroxide-water as one promising combination. Although high temperature lift is not a requirement for the paper drying application considered here sodium hydroxide is preferred over lithim bromide in this case mainly because of the wide cost difference between the two fluids. Sodium hydroxide is relatively low cost which is an important consideration for a system operating with large inventory in an open cycle where any losses would need to be made up. Sodium hydroxide is also commonly used in pulp and paper plants for other purposes and operating staff are familiar with handling it. Any spillage causes no corrosion problems since atmospheric carbon dioxide acts to neutralize the sodium hydroxide.

3.2 Thermodynamic Data

To design an absorption heat pump and determine the operating characteristics it is necessary to know the thermodynamic properties of the working fluid combination. In particular it is important to obtain the specific enthalpy of the solution as a function of concentration and temperature, and also the vapour pressure of the refrigerant (in this case water) over the solution as a function of concentration and temperature. The data are required in a suitable form and consistent units such that they can be handled conveniently in a computer model. The literature was searched extensively to obtain the most accurate data available to serve as a basis for calculations.

3.2.1 Enthalpy/concentration/temperature

Original experimental data obtained by Wilson and McCabe (Ref. 5) tabulate the heat content of sodium hydroxide-water as a function of concentration and temperature over the range of conditions of interest. Converting to SI units and using linear regression techniques the equation

$$h_\ell = (4.418-2.491\ \lambda)\ T' + 1704.63\ \lambda - 635.86$$

is obtained. The expression is valid within ±0.2% over the range of λ from 0.54 to 0.64, and a solution temperature range between 76.7°C and 115.6°C.

3.2.2 Vapour pressure/concentration/temperature

Data most frequently referred to for water vapour pressure as a function of sodium hydroxide temperature and concentration are contained in the International Critical Tables (Ref. 6). However, the water vapour pressure is expressed relative to the vapour pressure over pure water, and the resulting structure of complex polynomials is inconvenient for use in engineering calculations. More recent detailed experimental data were obtained (Ref. 7), and these were used for calculation purposes in this application. The data were reduced to obtain the usable numerical relationship

$$P'=10^{[(6.437-0.309\lambda)-(2143.31+952.2\lambda)/(273+T')]}$$

which is valid over the range $0.5 \le \lambda \le 0.6$ and $80 \le T' \le 120$. This equation establishes the vapour pressure within ±0.5% over the above ranges.

4. DESIGN APPROACH

4.1 Basis

The design basis used for the proposed absorption heat pump was a drying capacity equivalent to that of a typical industrial paper machine. In this case the Ontario Paper Company machines produce 400 tonnes of oven dried paper per day from wet paper entering the machine with 60% moisture by weight and leaving with 8% moisture content. This is equivalent to a water removal rate of 6.58 kg/s which forms the basic design condition for the heat pump. From this basis and other given paper dryer operating criteria the required heat pump system parameters were developed. Results of the overall heat balance calculations are indicated in the system schematic (Fig. 2). Only the major equipment items are shown in this diagram. Smaller sub-systems such as chemistry and inventory control are omitted for clarity purposes.

Plant design was carried out to the level of detail necessary to size equipment and components and to prepare specifications suitable for accurate cost estimating purposes. The following sub-sections briefly summarize the criteria used in design of the main components. The Appendix describes in more detail the design technique and computerized calculation procedure developed for sizing the absorber.

4.2 Absorber

For drying purposes some of the water vapour contained in the moist air must be extracted by contacting the absorption working fluid. The most suitable equipment for this dehumidification process is a contacting tower in which direct contact is made between liquid and gas streams as they pass over contact surfaces arranged in stages within the tower. This is similar to cooling towers commonly used in industry.

As water vapour is absorbed by the sodium hydroxide heat and mass transfer effects occur and the heat of dilution is released by the solution. In this application it is desirable to transfer heat back to the air stream, and therefore a parallel flow configuration of moist air and solution downwards through the tower seems the most suitable. The fluid is sprayed into the top of the tower and runs down the packing material in parallel with the air flow. For calculation purposes assumptions were made about the performance of typical open type fill used for packing in the tower stages. This open type of packing allows high air flows with low pressure drops.

The transfer unit concept commonly used in tower design (e.g. Ref. 8) was adapted to suit the particular application, and a computer program was developed to allow sequential stage conditions to be estimated from temperature and mass concentration data. Heat and mass transfer calculations were based on tower design techniques used by Kern (Ref. 9). Typical values of gas and liquid loading were used which in practice have been found to give satisfactory flow velocities.

In calculating heat transfer from the solution to the gas resulting from absorption of the water vapour the overall heat transfer coefficient was assumed to be identical to the gas-side coefficient, i.e. the liquid film resistance was assumed to be small. This may not be completely true for an aqueous solution, and it is recognized that more involved treatments of the problem may be adaptable (Ref. 10). However, the degree of added complexity involved was not considered justifiable for the current aims of the investigation. The simplified approach is consistent with assumptions made by other workers in this field (Ref. 11).

The program was able to predict the number of transfer units required, and hence the overall tower dimensions were established. Two separate tower units connected in parallel were required due to vessel sizing constraints.

4.3 Regenerator

The absorbed water vapour must be extracted and the solution reconcentrated in the regenerator. A forced circulation type of evaporator was selected for this purpose. This type of evaporator is commonly used for sodium hydroxide concentration purposes in the chemical industry. It consists of a vapour head vessel which acts as a flash chamber connected to a separate tube-in-shell heating element. A pump withdraws liquid from the bottom of the vapour head and forces it through the heating element back to the flash chamber. The heating element is located far enough below the liquid level in the flash chamber to prevent boiling in the tubes.

Incoming dilute solution from the absorber is mixed with a larger recirculating flow of concentrated solution taken from the liquid collection section at the bottom of the vapour head. The combined flow passes through the heating element where heat is added by condensing steam outside of the tubes carrying the solution. After leaving the tubes the solution discharges into the vapour head which is maintained under vacuum thereby causing excess moisture to flash from the dilute solution and be driven off as a vapour. The pressure and temperature of the water vapour removed from the regenerator are controlled by cooling water conditions at the inlet to the desuperheater/condenser.

Heat contained in the weak solution provides some of the reconcentration energy required, with the additional necessary heat being supplied by a source of waste steam supply at close to atmospheric pressure. Heat transfer from the heating steam to the sodium hydroxide was calculated using conventional heat transfer methods, allowing for the characteristics of the fluid and the boiling point rise resulting from concentration of the solution. The calculated film coefficients, tube wall metal correction factor, and appropriate fouling factors were applied, and the resulting coefficient was compared and verified against reported chemical industry practices.

All information required for design of the heating element and vapour head was derived from a heat balance on the regenerator, and by using differential heat of dilution and water vapour pressure data developed in Section 3. A nominal recirculation ratio was used, and vapour velocities in the vapour head were limited to values which have been proven in practice to minimize liquid entrainment problems.

4.4 Condenser

The water vapour discharged from the regenerator vapour head is superheated at the required vacuum operating conditions. Desuperheating and condensing can be achieved in the same vessel using either a surface or jet spray type condenser. A tube-in-shell surface type was chosen for this design, and established overall coefficients and fouling factors were used for sizing purposes.

Cooling water is in abundant supply in this particular location either from in-plant sources or from a large canal which flows directly by the plant. The direct contact type jet condenser or closed circuit cooling system may be required for other potential sites with less favourable cooling water supplies.

4.5 Miscellaneous

Piping and ducting sizes were based on suitable established velocities for sodium hydroxide, water, water vapour, and air to provide reasonable pressure drop values.

Wall thickness of vacuum vessels were calculated using the ASME Section VIII pressure vessel code.

4.6 Plant Layout

The basic philosophy adopted for the physical layout was to locate the new equipment as a package in its own independent building. The heat pump is connected to the existing equipment by two large air ducts at inlet and outlet from the dryer. Separation from the existing plant layout has some advantages for the heat pump particularly during the construction and startup phases. However, this approach adds to the capital cost.

An equipment layout was developed to produce a suitably compact arrangement of the vessels, allowing space for lifting apparatus, maintenance access, services, etc. The resulting overall building size is 25m long × 18m wide × 12m high.

5. MATERIALS

Material properties were investigated in detail for the temperatures and sodium hydroxide concentration levels anticipated in the proposed system. Usual chemical industry practice is to use nickel as the wetted surface for those components exposed to high concentrations and higher than ambient temperatures. Other alternative potentially less expensive materials were considered for this application.

The most promising of these alternatives was a fibreglass reinforced plastic (FRP) lined with a higher performance plastic on the wetted surface. There are a number of fluorinated polymer plastics which meet the high performance requirements of the absorption heat pump operating environment. The most widely known and commercially available is polytetrafluorethylene (PTFE). Fluorinated polymers have excellent corrosion resistance and high temperature stability.

The fluorinated polymer liners combined with FRP as a base material form dual-polymer laminates (known simply as dual laminates). Investigations and discussions with suppliers indicated that these materials would be practical for this type of application. However, reliable prices suitable for estimating purposes proved difficult to obtain from the few exclusively Canadian sources contacted. There were some indications from these sources that costs could be as high as those for pure nickel. Such information was not considered to be necessarily representative, and it is possible that further investigation into U.S. or overseas suppliers could produce competitively-priced equipment. Also, as their properties become more widely known it is probable that the future cost of these materials will be reduced if they are introduced more widely into industrial process applications.

Another possible material for pipe sizes up to 305mm (12") diameter is carbon steel with a PTFE liner, and pumps are also available in this material. Again, pricing information was difficult to obtain for this material combination.

For the purposes of this study therefore, costs for those components exposed to the sodium hydroxide solution were based on nickel-clad carbon steel plate. It was considered that such costs would provide an upper level criterion which could ultimately be optimized. Clad plate is used extensively throughout the chemical industry and fabrication techniques are well established.

6. COSTS AND ECONOMICS

6.1 Cost Estimates

The capital cost estimates were based on prices obtained from equipment manufacturers using component specifications developed in the design phase. In the absence of complete detailed design the prices obtained were essentially budget costs, but the equipment involved is considered conventional enough that the estimates provided are reasonably accurate.

Estimates were broken down into different sections for the heat pump unit itself and support facilities, including ductwork, piping, controls, crane, civil/structural work, and miscellaneous mechanical and electrical items. Installation costs were included and indirect cost allowances were made for interest during construction, engineering design and construction management fees, commissioning, and contingency costs. The total of these direct and indirect costs amounted to $2,992,000 in 1985 Canadian dollars.

Operating costs were derived from estimates of electric motor units required to drive the circulating pumps, using an annual operating capacity factor of 0.85. It was assumed that a part-time operator would be necessary, and allowances were made for labour, maintenance, and material costs. The total operating and maintenance cost was estimated at $69,000 per year.

6.2 Benefit

The real value derived from the heat pump is represented by the reduction in energy normally required to dry the paper. Under normal circumstances fresh air is brought into the dryer continuously and heated sufficiently to absorb the necessary amount of moisture before exhausting from the dryer hood. A portion of the sensible heat supplied to the air in this operation is usually recovered, but even allowing for this it was calculated that the heat pump effectively reduces the energy required by 6,750 kJ/s from a normal requirement of 17,000 kJ/s. (It is recognized that energy input from the atmospheric steam supplied to the regenerator is a genuine waste steam source which is otherwise unused, and therefore no value is assigned to it.) Based on a natural gas price in southern Ontario of 0.427¢/MJ ($4.50/MBtu) the value of the energy saved amounts to $1,197,000 per year. This assumes that the normal steam supply to the dryer is produced in a conventional fossil-fuel steam boiler, with an efficiency taken to be 70% and an allowance of 5% made for mechanical, piping, and miscellaneous losses. Allowing for the operating and maintenance costs gives a net value of energy saved amounting to $1,128,000 per year.

6.3 Economic Feasibility

Based on the benefits indicated above and the total estimated capital, operating, and maintenance costs the simple payback period is 2.65 years after construction of the plant is complete.

An economic analysis was also done using discounted cash flow techniques and allowing a 15 year operating life for the plant equipment. Using a real discount rate of 10% it was shown that a net present value (NPV) of $4.247M would be required to produce an annual net cash flow equivalent to the net energy savings value over the 15 years of operation. This compares favourably with the required initial capital investment of $2.99M.

A corresponding internal rate of return (IRR) was calculated based on the annual net cash flow over 15 years and the required capital investment. If an inflation rate of 5% is added to the chosen discount rate of 10% the implication is that a rate of return on investment greater than 15% is required for the project. The calculated IRR value was very attactive at 34.3%, and reflects the continuing energy benefits over the life of the plant after the initial cost has been paid off.

Sensitivity to effects of fuel price increases and discount rate choice were also checked. Using a 2% annual escalation in fuel prices improves the economics of the heat pump, since it displaces increasingly expensive fossil fuel. A slight reduction in payback occurs and the NPV improves significantly to $5.317M. The IRR also increases to 37.6%. A range of discount rates between 7 and 13% were tested. This had a small effect on payback periods, but showed a range of NPV varying between $5.94M and $3.038M.

7. CONCLUSIONS AND RECOMMENDATIONS

From technical and economic information developed in the previous sections it is clear that an absorption heat pump could be utilized for a drying/heating requirement such as the large pulp and paper machine studied here. The application of such systems is very site specific, and a suitable source of waste heat must be available in adequate quantity in order to satisfy economic feasibility. Alternative competing heat recovery technologies must be evaluated in each particular case, and some systems considered may prove to have a lower capital cost. However when capital and operating costs are considered over the life of a project the absorption heat pump will have little serious competition when the circumstances are favourable.

Although this study has indicated that such systems can be effectively applied to full size operating plants it is recognized that further investigation and demonstration work on a smaller scale is required before considering large scale implementation. Accordingly it has been proposed that a research and development program should be undertaken to verify the industrial potential of the absorption heat pump in this type of application. Successful results in this area would subsequently lead the way to pilot plant demonstration scale activity.

In particular it is planned to design and construct a laboratory scale test facility capable of establishing the thermodynamic cycle performance characteristics and operating concepts developed in this industrial application study. The major area for optimization and potential savings in plant cost is in the contacting tower, and testing work should concentrate on investigations of heat and mass transfer effects within the absorber aimed at verification and demonstration of the accuracy of the design methods and criteria developed in the study. The test rig should also be capable of demonstrating the versatility of the working fluid under a wide range of operating conditions. Exposure of the selected materials to various fluid temperature/concentration conditions should be examined and alternative material combinations evaluated.

Conventional equipment utilized in the concept such as the regeneration system and condenser also need further performance optimization study.

ACKNOWLEDGEMENT

The authors wish to acknowledge the contribution made by others to the original study (Ref. 1) on which this paper is based, and in particular the valuable assistance provided by Mr. A.T. Carter, Mr. D.C.K. Ngo, and Mr. B.A. Tulloch of Acres International, and by Dr. J.S. Wallace of the University of Toronto.

REFERENCES

1. Acres International Limited: "Chemical Heat Pump Phase 'A' Study," March 1985.
2. Isshiki, N.: "Study on the Concentration Difference Energy System," J. Non-Equilib. Thermodyn., Vol. 2, 1977, pp. 85-107.
3. Carey, C.O.B., Khahra, J.S., Smith, I.E.: "The Sodium-Hydroxide Water Absorption Heat Pump," Proceedings 16th International Congress of Refrigeration, Paris 1983, Commission B1, pp. 269-274.
4. Patterson, M.R., Perez-Blanco, H.: "Design of an Advanced Absorption Heat Pump for Minimum Payback Period," Oak Ridge National Laboratory, CONF-851125-7, 1985.
5. Wilson, H.R., McCabe, W.L.: "Specific Heats and Heats of Dilution of Concentrated Sodium Hydroxide Solutions," Industrial and Engineering Chemistry, Vol. 34, No. 5, May 1942.
6. International Critical Tables, prepared by the National Research Council of the USA, 1926.
7. Krey, J.: "Dampfdruck und Dichte des Systems H_2O-NaOH," Zeitschrift für Physikalische Chemie Neue Folge, Vol. 81, 1972, pp. 252-273.

8. ASHRAE Handbook - 1983 Equipment, published by American Society of Heating, Refrigerating and Air Conditioning Engineers.
9. Kern, D.Q.: "Process Heat Transfer," published by McGraw-Hill Book Co., Inc., 1950.
10. Grossman, G.: "Simultaneous Heat and Mass Transfer in Film Absorption under Laminar Flow," International Journal of Heat and Mass Transfer, Vol. 26, No. 3, 1983, pp. 357-371.
11. Webb, R.L.: "A Unified Theoretical Treatment for Thermal Analysis of Cooling Towers, Evaporative Condensers, and Fluid Coolers," ASHRAE Transactions, Vol. 90, Part 2, 1984.
12. Keenan, J.H., Keyes, F.G., et al.: "Steam Tables (International System of Units - S.I.)," published by John Wiley and Sons, 1978.

APPENDIX

Design Method for Direct Contact Drying Tower Absorber

Figure 3 shows a diagrammatic cross-section of the tower indicating the liquid and gas streams entering at Stage 1, and the relevant nomenclature used in successive stage calculations.

Initial conditions were established for L_1 and G based on typical tower loadings used in practice. T_1 and X_1 were derived from the dryer hood exhaust conditions, and values of T_1' and λ_1 were arrived at by trial and error solution.

Based on unit ground surface area, equations can be derived for the following parameters. Conditions apply to n^{th} stage unless indicated otherwise.

(i) Heat transfer by convection

$$(q_c)_n = (h\ a\ V)_n\ (T_n'-T_n)$$

The heat transfer coefficient is derived from the definition $\frac{h}{K_x c}$ = Le which can be rewritten as

$$[h\ a\ V]_n = [K_x\ a\ \frac{V}{L}]_n\ L_n\ \text{Le}\ c_n$$

$$[K_x\ a\ \frac{V}{L}]_n = \text{NTU/stage}$$

An average value of Le for air/water systems can be obtained from Fig. 17.4 in Ref. 9 for the temperature range anticipated.

Using data obtained from Keenan, Keyes et al. (Ref. 12)

$$c_n = 1.0111+1.92\ X_n$$

(ii) Water content of gas at liquid film surface

From Section 3.2.2

$$P_n' = 10^{\left[(6.437-0.309\ \lambda_n)-\frac{(2143.31+952.2\ \lambda_n)}{(T_n'+273)}\right]}$$

For a tower operating at atmospheric pressure

$$X_n' = P_n'/(1-P_n')\ M_w/M_a$$

(iii) Mass of water vapour diffused

$$m_n = [K_x\ a\ V]_n\ [X_n-X_n'] = [K_x\ a\ \frac{V}{L}]_n\ L_n\ [X_n-X_n']$$

(iv) Water content of gas stream in $(n+1)^{th}$ stage

$$X_{n+1} = (G\ X_n-m_n)/G$$

(v) Specific heat of gas stream in $(n+1)^{th}$ stage

$$c_{n+1} = 1.0111+1.92\ X_{n+1}$$

(vi) Temperature of gas stream in $(n+1)^{th}$ stage

$$T_{n+1} = T_n+(q_c)_n/(G\ c_{n+1})$$

(vii) Liquid loading at $(n+1)^{th}$ stage

$$L_{n+1} = L_n+m_n$$

(viii) Solution concentration in $(n+1)^{th}$ stage

$$\lambda_{n+1} = L_n\ \lambda_n/L_{n+1}$$

(ix) Partial pressure of water vapour in gas stream

$$P_n = M_a\ X_n/(M_a\ X_n+M_w)$$

(x) Specific enthalpy of water vapour in gas stream

From Keenan, Keyes et al. (Ref. 12) an expression can be obtained to show that

$$(h_v)_n = 2499.87-26.426\ P_n+(0.14\ P_n+1.89)\ T_n$$

(xi) Specific enthalpy of liquid solution (n = 1 only)

From Section 3.2.1

$$(h_\ell)_n = (4.418-2.491\ \lambda_n)\ T_n'+1704.63\ \lambda_n-635.86$$

(xii) Heat transfer by diffusion

$$(q_d)_n = m_n\ (h_v)_n$$

(xiii) Specific enthalpy of liquid solution in $(n+1)^{th}$ stage

$$(h_\ell)_{n+1} = [L_n\ (h_\ell)_n+(q_d)_n-(q_c)_n]/L_{n+1}$$

(xiv) Temperature of liquid solution in $(n+1)^{th}$ stage

From equation in Section 3.2.1

$$T_{n+1}' = [(h_\ell)_{n+1}+635.86-1704.63\ \lambda_{n+1}]/(4.418-2.491\ \lambda_{n+1})$$

Using a typical incremental value for NTU/stage the above steps can be repeated for successive stages until the liquid and gas stream temperatures converge to an acceptable approach condition. The corresponding outlet moisture content of the gas stream in relation to the

inlet value establishes the mass flow of dry gas necessary to remove the required amount of water vapour. Comparing this air mass flow to the assumed gas loading gives the required tower cross-sectional area.
For the chosen liquid and gas loadings corresponding values of K_xa can be located for specific types of tower fill. Using the appropriate K_xa value and the total NTU computed for all stages the required tower height can be calculated.

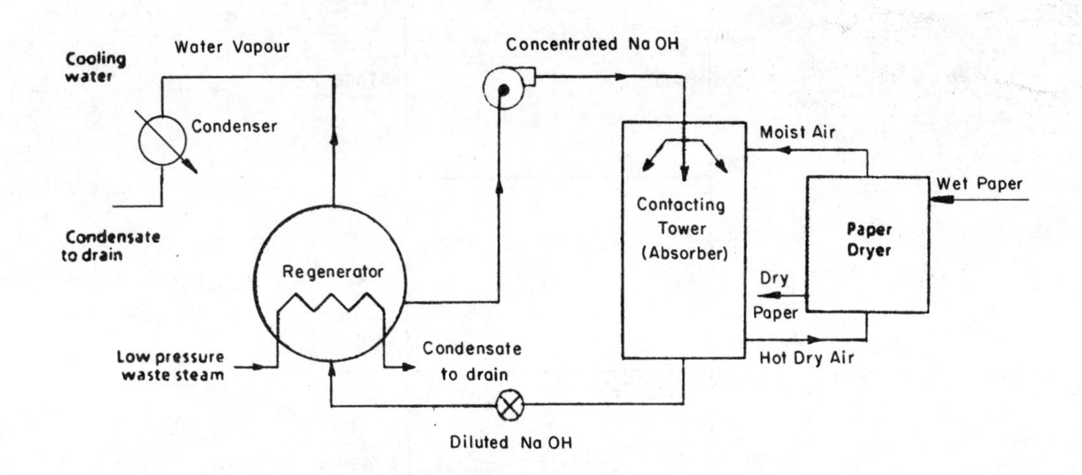

1. Absorption heat pump drying application concept

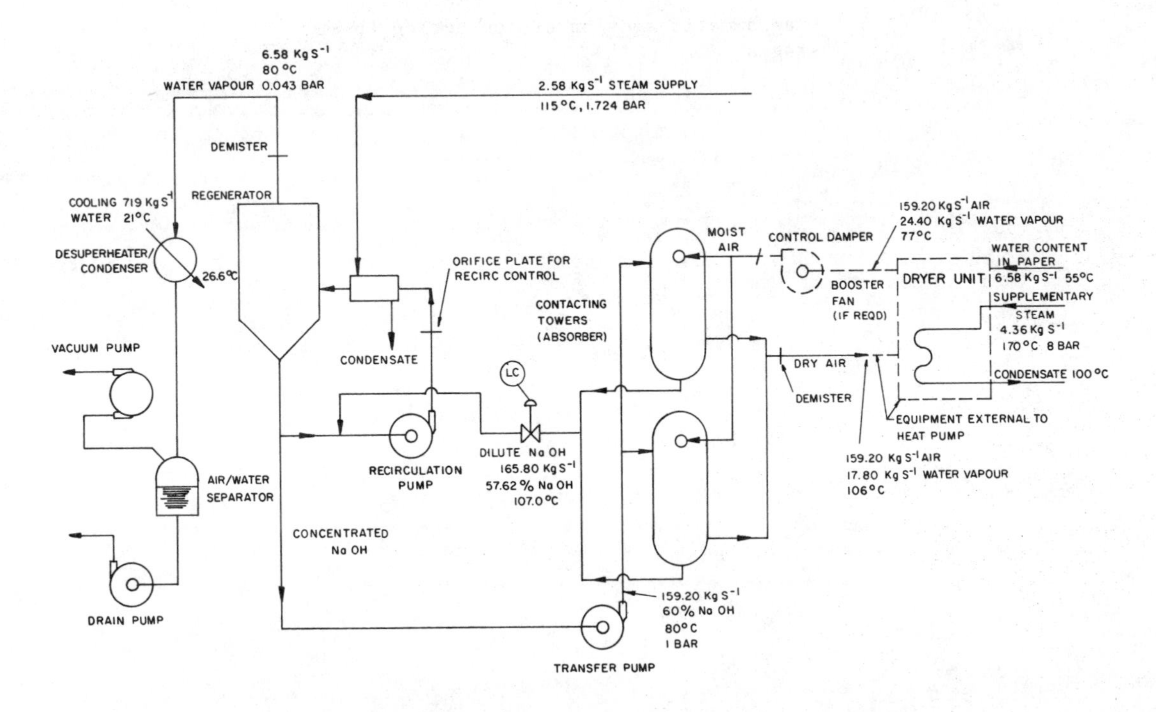

2. Simplified schematic diagram of absorption heat pump system for paper drying unit

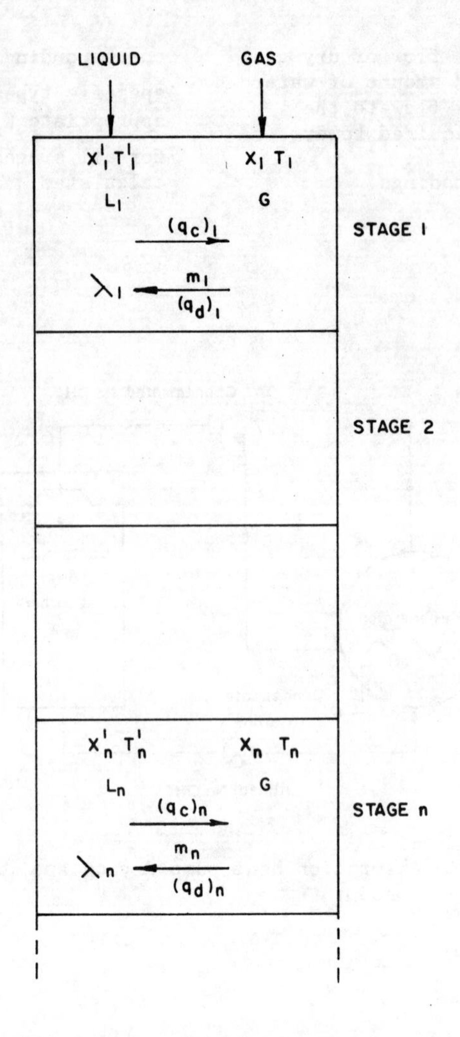

3. Diagrammatic section of contacting tower stages

3rd International Symposium on the

Large Scale Applications of Heat Pumps

Oxford, England : 25-27 March 1987

PAPER C3

INTEGRATION OF NATURAL GAS ENGINE DRIVEN VAPOR RECOMPRESSION HEAT PUMPS INTO INDUSTRIAL PROCESSES

D.S. Severson

The author is a project manager in the Industrial Utilization Department of the Gas Research Institute, Chicago, Illinois, U.S.A.

SYNOPSIS

Integration of a vapor compression heat pump into the process it serves is very important to insure that maximum benefits from the system are achieved. This paper describes three Gas Research Institute research and development projects involving process integrated industrial heat pumps. The results of these projects show that a thorough understanding of the process and proper matching of the heat pump to that process are prerequisites for a successful application.

INTRODUCTION

Integrating heat pumps into industrial processes is a concept that has generated considerable interest in the United States in recent years. The energy crisis of the 1970's brought increased emphasis to the use of heat pumps in industrial processes. A large amount of effort was put into developing heat pump hardware for applications identified during that era. The Gas Research Institute (GRI), the United States Department of Energy (DOE), and the Electric Power Research Institute (EPRI) participated in projects developing industrial heat pumps [1,2,3], and several private corporations developed heat pumps on their own. Mechanical vapor recompression, for example, has been considered for distillation and evaporation unit processes with increasing frequency in both new and retrofit applications. Even now, when lower cost energy is available in the U.S., manufacturers are looking for ways to control energy costs, including the use of heat pumps.

In addition to heat pump development, extensive work has been done in the area of process integration. Heat recovery through careful matching of hot and cold process streams for maximum energy efficiency has been heavily researched, leading to methodologies of heat exchanger networking and "pinch" technology[4]. Research into pinch technology indicates it can be used to quantify the minimum energy requirements, identify the best scheme for exchanging heat between hot and cold streams, and identify the best placement for heat pumps and heat engines in a process. Although it is not the subject of this paper, pinch technology may have a large impact on the installation of such equipment. This technique has wide application, especially in the complex plants of the chemical and petroleum industries, and several major manufacturers in these industries are using these techniques to reduce energy requirements in their plants. In smaller plants, the options for heat exchange are limited by the small number of streams available and complex analysis is not always necessary. Pinch technology techniques have not been used in the applications mentioned below but will be used in the future.

Integration of the heat pump into the process is vital to achieve maximum benefits from its installation. The heat pump must be appropriately placed in the process and, in addition, exhaust streams from the gas fueled prime mover must be used effectively to achieve economical operation. A discussion of three GRI projects and the benefits derived from their close integration into the process they serve will be the subject of this paper. Two projects have integrated heat pumps very closely into their respective process, and have shown that the energy efficiency and the productivity of that process can be increased. A third project will integrate an open-cycle steam heat pump into a cogeneration system, compressing low pressure steam produced by the cooling jacket and increasing the amount of process steam available from the system.

THE GRI HEAT PUMP PROGRAM

GRI initiated a research and development program in heat pump technology in 1979 to overcome some of the technical barriers that were present [5]. A major emphasis of this program is directed to Mechanical Vapor Recompression (MVR) because it is a well proven technology widely used in industrial applications, especially in evaporators. In order to address this segment of industrial applications, GRI has sponsored projects developing open-cycle steam recompression heat pumps. The program's current focus is on three MVR projects using steam as the working fluid. The first project, a current field test, developed a rotary screw vapor recompression heat pump for a rendering cooker. The system can operate on contaminated vapors and the engine exhaust is used to supply the remaining process heating requirements, achieving an overall system thermal efficiency of 57%. The second project, also in field test, developed a centrifugal vapor recompression heat pump for an evaporator in the textile industry. Engine exhaust is used to supply process heat to a dryer downstream in the same process. The third project, currently in the prototype development phase, is the novel application of vapor recompression to a cogeneration system

Held at St. Catherine's College Oxford, England. Organised and sponsored by BHRA, The Fluid Engineering Centre, Cranfield, Bedfordshire, MK43 0AJ England.

driven by a natural gas fueled piston engine. Steam produced by the cooling jacket is compressed for use as process steam. The system can then provide high pressure process steam from both the engine exhaust and the cooling jacket for applications where additional hot process water is not needed, resulting in greatly enhanced siting possibilities and overall system thermal efficiency.

ROTARY SCREW COMPRESSOR FOR MVR APPLICATIONS

The development goal for this project was to apply rotary screw compressor technology to MVR applications involving contaminated steam [6]. Many industrial processes produce waste steam or condensate that has been contaminated. Previous state-of-the-art compressors were not be used to recompress vapors from these sources because the contaminants would deposit on surfaces in the compressor or corrode the materials of construction. Thermo Electron Corporation (TECO) has developed an MVR system to recompress contaminated vapors produced in an animal rendering operation. The heat pump uses a rotary screw compressor fabricated from 316L stainless steel driven by a natural gas fueled reciprocating engine.

The animal rendering process reduces animal byproducts to bone meal and tallow. The raw feed is sized and sent to the cooker. The rendering cooker uses steam introduced into its shell and paddles to evaporate the water contained in the byproducts. After leaving the cooker the tallow is separated from the meal, and the two products are sent to final processing. The contaminated vapor, leaving the cooker at near-ambient conditions, is condensed and the contaminated condensate sewered.

The steam recompression heat pump is integrated into the system to reduce process energy requirements. Figure 1 compares a conventional rendering cooker with a vapor recompression cooker. A rotary screw compressor, constructed of 316L stainless steel for corrosion resistance, is driven by natural gas fueled reciprocating engine. Contaminated vapor from the process is recompressed for use as steam in the cooker, rather than being condensed and sewered as in a conventional system. Hot exhaust gases from the engine supply additional heat to the process through a hot water loop. A specially modified cooker was designed to operate with contaminated steam, and the system was tied together with a control system designed to require minimum operator attention. The stainless steel compressor has a design inlet flow of 57 m^3/min (2,000 cfm) at 1.03 bar (15 psia) and a built in pressure ratio of 4. The input horsepower required at design point is 400 kW (300 hp), supplying 4.14 bar (60 psia) steam to the process. A rotary screw configuration was chosen for two reasons. First, a screw machine is able to accept wet steam at the inlet. In fact, entrained moisture evaporates during compression, making the compression process more nearly isothermal, lowering compression energy requirements. Second, the compressor can accept small amounts of contamination in the inlet. These impurities can deposit on the rotors, reducing clearance and increasing volumetric efficiencies. These deposits build up to the point that rotor-to-rotor clearance goes to zero. Beyond that point, contact between the layers prohibits any further deposit build up.

The prime mover supplying shaft power to the compressor is a 440 kW (330 hp), naturally aspirated, natural gas fueled reciprocating engine. It operates at 1200 rpm and supplies 330°C (625°F) exhaust gas. The engine exhaust is used to heat a circulating hot water loop. The hot water heats a portion of the cooker shell. The inherent variable speed capability of the engine adds flexibility in controlling the capacity of the system. It was not practical to recover waste heat from the cooling jacket in this installation. A radiator is used to reject this heat.

The rendering cooker is a modified version of a standard production cooker. The basic design was modified to match the cooker capacity to that of the steam compressor under operating conditions. This included using the 4.14 bar recompressed contaminated steam in place of the 10 bar (145 psig) boiler generated steam used in the plant's current cookers. The steam flow paths examined to limit the amount of air and non-condensibles that could collect in the system, reducing heat transfer. As a result, the shaft and shell were connected in series, with steam flow purging the shaft. Any non-condensibles will collect in the shell and be vented with special air-venting traps. The shell of the cooker was also modified to be heated with hot water, recovering the heat from the engine exhaust.

The control system was designed to operate simply, reliably, and automatically, requiring a minimum of operator attention. If the steam available from the cooker is not sufficient for complete cooking, the control system will draw vapor from the adjacent cooker. If compressor discharge pressure increases beyond the process' needs, the control system will slow the engine to keep the steam pressure within its set point. If there is still high steam pressure when the engine has been slowed to idle, the control system will vent steam to the condenser until the pressure is reduced.

As the above description shows, the heat pump is very closely integrated into the system. The process has been optimized, with the components designed to operate together as a unit. The energy performance of the system at design and as measured during operation is shown in Table 1. During the one year field test, there has been no major operational breakdowns and no evidence of accelerated corrosion. As the results show, less than half the amount of steam needed to evaporate 1.0 kg of water in a conventional system is needed in the MVR system. Other operational savings include the fact that no boiler operator is needed, the system requires less attention because it is controlled automatically, and boiler water treatment costs are reduced because the system uses contaminated vapor to produce the steam in place of boiler feedwater.

HIGH LIFT CENTRIFUGAL COMPRESSOR FOR MVR APPLICATION

GRI's goal for this project is to transfer aerospace technology to the industrial sector

[7]. Industrial centrifugal compressors currently used for steam compression are limited to pressure ratios of 2:1 or less. This limitation is the result of low tips speeds required by materials and manufacturing methods and moderate shaft speeds necessary to avoid vibration and bearing problems. A higher compression ratio would open more opportunities for vapor recompression. Processes involving fluids with high boiling point elevations may require higher temperature differentials, and therefore higher compression ratios. Capacities can be increased in processes where higher evaporator temperature differentials will not affect product quality. Also, a high compression, single stage compressor will have lower first costs when compared to a multistage compressor or several low pressure single stage compressors in a retrofit of a multiple effect evaporator. AiResearch Manufacturing Co., Torrance, California, U.S.A., had previously designed high speed, centrifugal air compressor impellers for 3:1 compression ratio. The project applies this technology to an MVR evaporator in a mercerizing process for the textile industry.

The mercerizing process is applied only to cotton. The process causes chemical and physical changes within the fiber. Mercerization improves the luster, smoothness, and tensile strength of the fabric. Other benefits of mercerization include improvement in chemical reactivity, dye affinity, and stabilization. In the process, the fabric or yarn is held under tension and treated with 20% sodium hydroxide solution. The caustic solution must penetrate the fiber rapidly and uniformly, and after the reaction is complete, must be removed quickly and thoroughly. The dilute caustic solution rinsed from the fiber is collected and concentrated for reuse in the process by evaporation.

The steam recompression heat pump developed in this project is integrated into a retrofit for increasing process capacity. a schematic of the vapor recompression evaporator is shown in Figure 2. A centrifugal compressor is driven by a natural gas fueled gas turbine engine. Water vapor produced by the first effect is recompressed and used as the steam source for that effect. Turbine exhaust gases are used as a heat source in tenter-frame dryers downstream in the process.

The centrifugal compressor produces a pressure ratio of 2.94:1 with a flow of 1130 m^3/min (40,000 ACFM) with an adiabatic efficiency of 82.5%. The high lift is achieved by driving a 566 mm (22.3 in) diameter titanium impeller at a tip speed of approximately 610 m/sec (2000 ft/sec). The titanium construction of the impeller is necessary to achieve the mechanical strength required by these high tip speeds and corrosion resistance to any caustic carryover. Inlet guide vanes are supplied to increase part load performance of the compressor. Also, different diffuser/shroud assemblies with varying diffuser vane angle settings can be installed to adapt the compressor to a variety of flow ranges. The operating parameters are 6,855 kg/hr (15,100 lb/hr) of 0.19 bar (2.8 psia) and 59°C (139°F) water vapor at the inlet and 0.57 bar (8.2 psia) and 84°C (184°F) at the outlet. The prime mover is a natural gas fueled gas turbine engine, supplying 670 kW (500 hp) to the shaft of the compressor. The exhaust gas from the turbine provides a hot air source of approximately 12,260 kg/hr (27,000 lb/hr) at 480°C (900°F). This exhaust supplies the heat input to the tenter-frame dryers, drying the fabric after the mercerizing and dyeing processes. The air flow is diluted with make-up air, and the total flow is ducted to the dryers. The energy content of the stream is high enough to supply more than one dryer.

The evaporator is a retrofit application increasing the capacity of the site host's current equipment. Dilute feed of 15490 kg/hr (34,125 lb/hr) at 4.5% NaOH is concentrated to 8640 kg/hr (19,025 lb/hr) at 8.1%, reducing the amount of water that must be removed in the existing two effect evaporator. The use of MVR reduces the steam demand of the process from 16300 kg/hr (35,900 lb/hr) for the original equipment to a combined demand of 7280 kg/hr (16,032 lb/hr) for both the retrofit and original equipment.

The control system was designed to interface with the plant's distributed system, operating reliably with a minimum of operator attention. The engine controls are integrated with the process controls for the evaporator effect it serves, allowing capacity modulation. Inlet guide vanes can be closed to reduce flow to the compressor, reducing capacity. If capacity must be reduced further when inlet guide vanes reach their closed position, the engine can be slowed as needed.

As in the previous example, the heat pump supplies maximum benefits by being closely integrated to the process it serves. The compressor, engine, and evaporator have been designed to work as a unit. Steam needs for evaporation are halved, and fuel to the turbine is offset by replacing primary fuel in the dryers with hot turbine exhaust gases. Also, as in the previous example, no boiler operator is needed, reducing the associated operating costs.

MVR APPLICATIONS FOR COGENERATION

This project transfers MVR technology to reciprocating engine driven cogeneration [8]. State-of-the-art cogeneration systems powered by reciprocating engines provide two readily recoverable heat sources: exhaust gases and cooling water. Maximizing the energy recovery requires that hot water or low pressure steam can be utilized by the process. In many cases, this demand is just as easily supplied from several other low quality heat sources. A heat balance for current engine systems typically converts 28% of the fuel input to shaft power, 32% to hot exhaust gases, 30% to low pressure steam or hot water from the cooling jacket, and 10% to miscellaneous losses. Thermo Electron Corporation is currently developing a system that will integrate a rotary screw compressor into a cogeneration system, allowing the cooling jacket heat to be recovered as process pressure steam. The 30% of energy normally recovered as low grade heat can then be used to produce a higher form value heat source. This increases the percentage of fuel input energy recovered as process steam from 17% to 47%. This will greatly facilitate the siting of these reciprocating engine systems. The heat pump, a rotary screw steam compressor, is connected to the engine of a cogeneration system through the front power take-off. Figure 3 is a schematic of the integrated system. The inlet of the compressor receives low pressure steam produced by the engine cooling jacket, typically 2.05 bar (15 psig), and compresses it to a nominal pressure of 8.0 bar (100 psig). The compressor capacity is 15 m^3/min (500 cfm) at 1.79 bar (26 psia) inlet conditions, producing 955 kg/hr (2100 lb/hr) of 100 psig steam. The compressor requires 3 - 6% of the engine fuel input. Steam production of the cogeneration system is increased from 500 kg/hr (1100 lb/hr) to 1450 kg/hr (3200 lb/hr).

The control system is designed to modulate the steam and electric production to meet the demands of the process. If electric demand decreases during stable or increasing steam demand, the engine will produce the maximum amount of steam possible at that power setting. If steam demand decreases during stable or increasing electric demand, the compressor bypass will limit the steam production of the system. For maximum power output during high electric power demands, the compressor can be disengaged and the engine-generator can be operated as a conventional cogeneration system, producing the maximum continuous output of the generator.

The remainder of the system is a conventional cogeneration system, using components that are currently available. The prime mover is a turbocharged, natural gas fueled reciprocating engine, producing 619 kW at 1200 rpm. It can either be ebulliently cooled or flash steam cooled. The generator will produce 430 kW_e when the compressor is engaged, but is rated at 565 kW_e for maximum continuous operation. The exhaust heat recovery boiler is a commercially available unit, producing 500 kg/hr (1100 lb/hr) of 8.0 bar steam.

The performance of the MVR-Cogeneration system is compared to other cogeneration system configurations in Table 2. The MVR-Cogeneration system has a system efficiency of 74%, while recovering 50% of the thermal energy as high pressure steam. Currently available "total energy systems" have system efficiencies of 75%, but recover only 17% of the thermal energy as high pressure steam. The data show that carefully integrating the heat pump into the system provides maximum energy performance, but also provides the energy in the most productive form. The components are currently available production units, carefully chosen to match their performance characteristics. The components are tied together with a custom control system to insure they operate as a unit, not a collection of separate parts.

CONCLUSION

Natural gas fueled heat pumps can reduce energy costs when they are applied to an industrial process. Additional benefits may also include capacity and productivity improvements and reduced labor intensiveness, as well as energy savings. However, the heat pump must be closely integrated with the system to achieve the maximum impact of these benefits. As shown in the above examples, the performance characteristics of the system must be carefully matched to the overall process. The system must be designed to follow process demands. Each component must also be selected to match the performance characteristics of the others, and the overall design optimized so that all components work together as a single unit process.

MVR will have broad applications in processes utilizing evaporation or distillation. The rotary screw unit will apply to processes that have contaminated low pressure waste steam such as rendering, pulp and paper, and whey processing. The high lift compressor will provide processes with 3:1 compression ratios in a single stage compressor, whith lower capital costs than two stage units. MVR-Cogeneration system will make small cogeneration feasible in applications where there is no need for hot water from the cooling jacket, but there is a need for high pressure steam. This will include hospitals and small manufacturing locations.

GRI projects have shown that process integration of natural gas engine driven vapor compression heat pumps is vital to obtain maximum benefits.

ACKNOWLEDGEMENT

The author acknowledges Kathy Overby for typing and proof reading the manuscript.

REFERENCES

1. Becker, F.E. and Ruggles, A.E.: "Open-Cycle Vapor Compression Heat Pump System". Gas Research Institute Report, GRI 85/0084, March 1985.

2. "Brayton-Cycle Solvent Recovery System, Phase I: System Optimization", AiResearch Manufacturing Co. Project Report, United States Department of Energy Report, 82-19051, June 1981.

3. Harris, G.E.: "Heat Pumps in Distillation Processes". Electric Power Research Institute Report, EPRI EM-3656, Project 1202-23, August 1984.

4. Linnhoff, B. and Vredeveld, D.R.: "Pinch Technology has Come of Age". Chemical Engineering Progress, 80 (7), p. 33, 1984.

5. Kearney, D.W. and Insight West Associates: "An Assessment of Vapor Compression Heat Pump Technology and Applications for Industrial Processes". Gas Research Institute Report, GRI 80/0103, April 1982.

6. Ruggles, A.E.: "Contaminated Vapor Rotary Screw Heat Pump". Gas Research Institute Report, GRI 85/0272, November 1985.

7. Iles, T.L., Burgmeier, L.R., and Liu, A.Y.: "Open-Cycle Centrifugal Vapor-Compression Heat Pump". Gas Research Institute Report, GRI 85/0118, April 1985.

8. DiBella, F.A., Balsavich, J., and Becker, F.E.: "Integrated Natural Gas Engine Cooling Jacket Vapor Compressor Program". Gas Research Institute Report, GRI 86/0074, December 1985.

	Predicted	Measured (Over a 4-hr Period)
Raw Material Composition (Meat)		
Water (%)	45.0	45.0
Tallow (%)	27.5	-
Crax (%)	27.5	-
Tallow and Crax (%)	55.0	55.0
Raw Material Throughput: (kg/hr)	3,995	4,014
Product Discharge Temperature: (°C)	127	124-129
Compressor Discharge Steam Flow Rate: (kg/hr)	2,088	3,137
High Temperature Hot Water from Engine Exhaust: (10^6 J/hr)	485-570	591
Inlet Steam Pressure: ($10^3 Nm^{-2}$)	101.4	100.0
Discharge Steam Pressures: ($10^3 Nm^{-2}$)	344.8-413.7	365.4-399.9
Engine Fuel Input: (10^9 J/hr	2.72-3.17	3.22
Specific Evaporation Rate of Recompression Cooker		
$\frac{\text{(engine load)}}{H_2O \text{ evap}}$ (10^6 J/kg)	1.52-1.76	1.77
Specific Evaporation Rate of Conventional Cooker		
$\frac{\text{(boiler load)}}{H_2O \text{ evap}}$ (J/kg)	4.00	4.11
Engine Savings of Recompression Cooker Over Conventional Cooker (%)	61.2-56.0	56.9

Table 1. Energy Profile for Mechanical Vapor Recompression Rendering Cooker

Energy Conversion	Engine-Generator Set Heat Recovery Options			
	(1) No heat Recovery	(2) High-Pressure Steam From Exhaust Gas Recovery Boiler	(3) High Pressure Steam From Exhaust Gas Recovery Boiler Plus Low-Pressure Steam From Cooling Jacket	(4) IMVRS-COGEN System Delivering 4.14 bar Steam From Cooling Jacket and Exhaust Gas Recovery Boiler
Thermal				
$\frac{\text{J Steam}}{\text{J Fuel Input}}$ %				
4.14 bar Steam From Exhaust Gas Heat (%)	0	17	17	17
1.03 bar Steam From Cooling Jacket Heat (%)	0	0	30	0
4.14 bar Steam From Cooling Heat	0	0	0	33
TOTAL Steam (%)	0	17	47	50
Electrical				
$\frac{\text{J Elect (kWh)}}{\text{J Fuel Input}}$ %	28	28	28	24
Total Energy				
$\frac{\text{J Steam and Elect}}{\text{J Fuel Input}}$ %	28	45	75	74

Note: Engine Based on 30-Percent Power Conversion Efficiency
Generator Based on 93-Percent Electric Conversion Efficiency

Table 2 Energy Conversion Utilization from Engine-Generator Sets

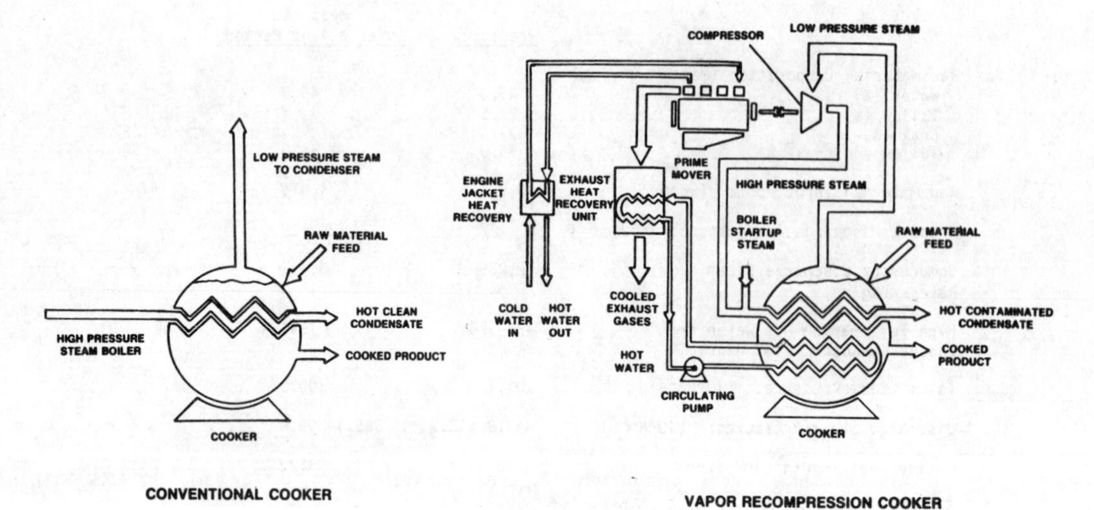

Figure 1. Integrated MVRS in a Meat Rendering Operation

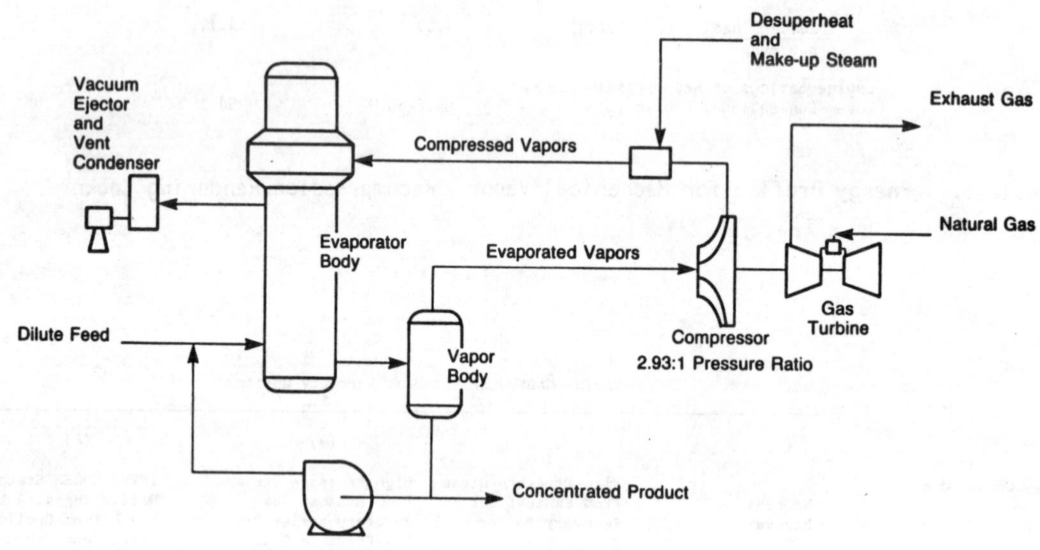

Figure 2 Mechanical Vapor Recompression Caustic Evaporator

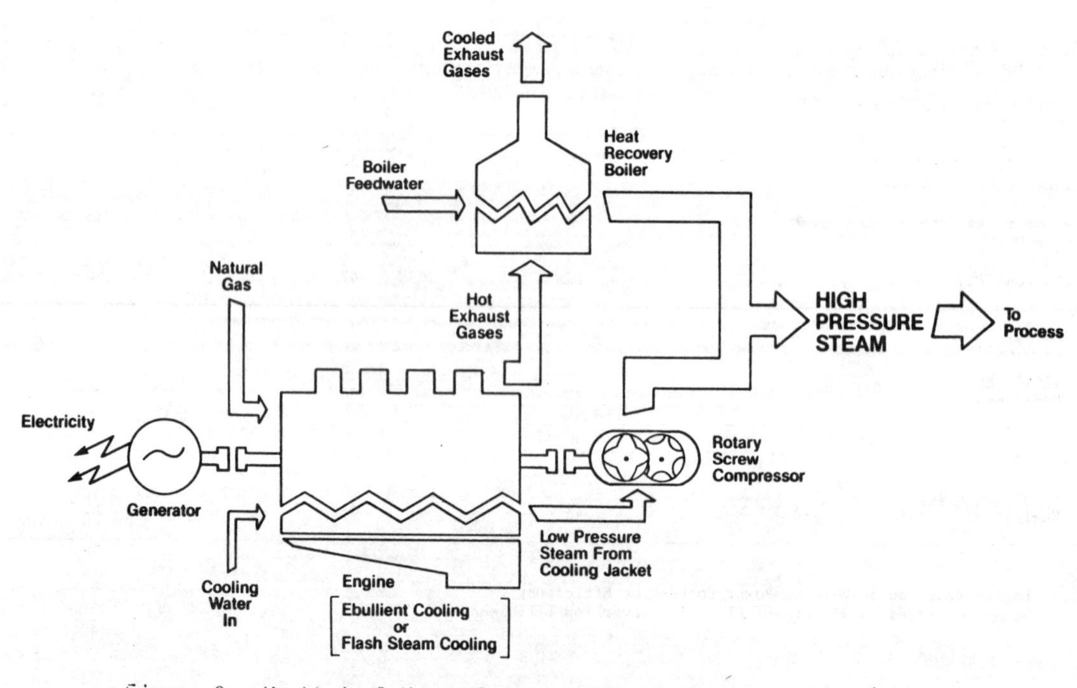

Figure 3 Mechanical Vapor Recompression for Cogeneration System

3rd International Symposium on the
Large Scale Applications of Heat Pumps
Oxford, England: 25-27 March 1987

PAPER D1

INDUSTRIAL HEAT PUMPS: NEW INSIGHTS ON THEIR INTEGRATION IN TOTAL SITES *

S.M. Ranade, E. Hindmarsh and D. Boland

All authors work for TENSA Services (a unit of ICI Americas) in Houston, Texas.

SYNOPSIS

Most modern day industrial complexes are total sites consisting of processes and the facilities that supply the process heat and power. Any meaningful heat pump integration procedure should, therefore, be based on a clear understanding of the site thermodynamics and economics, especially the utility system-process interactions. Pinch technology provides an excellent framework for systematic analysis of such interactions.

In this paper, options for appropriate placement of heat pumps in total sites are clearly delineated. A general equation for maximum economic lift is presented. Application of this equation to different heat pump types provides some novel insights on role and characteristics of heat pumps. An expression for maximum theoretical economic lift is also derived. This expression can be used to define, apriori, the region in temperature space for optimal integration of heat pumps.

Held at St. Catherine's College Oxford, England. Organised and sponsored by BHRA, The Fluid Engineering Centre, Cranfield, Bedfordshire, MK43 0AJ England.

NOMENCLATURE

Upper Case

A	Heat Transfer Surface Area
C	Capital Investment Including Installation
H	Hours of Operation Per Year
MV	Marginal Value ($/Energy)
N	Efficiency
Q	Heat Accepted or Delivered by the Heat Pump (Energy/Time)
T	Temperature (Deg. Absolute)
W	Work Input (Energy/Time)

Lower Case

n	Index Used in Summation
p	Payback Period

Greek

α	Fraction of Area Reduced (By Heat Pumping) that can be Reused.
Δ	Net Increase (e.g. ΔA)

Subscripts

C	Compressor (e.g. C_C), Temperature of Heat Acceptance (e.g. T_C)
CP	Source of Heat for the Heat Pump (e.g. Q_{CP})
CO	Cooler
d	Heat Pump Driver Utility
H	Sink for Heat Pump (e.g. Q_H), Temperature of Heat Rejection (e.g. T_H), Heater (e.g. A_H)
h	Hot Utility
HPC	Heat Pump Condenser
HPE	Heat Pump Evaporator
i	Index of Summation
I	Interchanger
j	Index of Summation
M	Mechanical Efficiency
p	Practical, does not Include Mechanical Losses
pi	Signifies Actual Includes Mechanical Losses
T	Thermodynamic Efficiency
T	Total Aréa
U	Utility
1,2..., 11,12	Subscripts Denoting Areas in Figure A-1

INTRODUCTION

Most modern day industrial complexes are total sites i.e. they consist of a utility system integrated with the processes on site. Early work concerning the evaluation and placement of heat pumps concentrated on integration with specific unit operations (e.g. distillation columns and evaporators). More recent works [1,2,3,4] placed the heat pumps in a total process context and showed the drawbacks of the unit operations based approach and the importance of understanding thermodynamics of process integration particularly the pinch concepts. These methods still lack a full capability to assess the impact of utility systems on placement of heat pumps in total sites.

Correct assessment of heat pumping opportunities necessitates a full understanding of the thermodynamics and economics of the process and the utility system and the associated interactions. Pinch technology together with an understanding of heat pump thermodynamics and utility system economics provides an excellent framework for systematic analysis of these interactions.

For the past year, TENSA Services, a unit of ICI Americas, has been conducting research funded by the Electric Power Research Institute to develop a procedure for appropriate integration of heat pumps in total sites. Some key findings of this research are presented in this paper. The specific objectives of this paper are:

1) To delineate opportunities for appropriate placement of heat pumps in total sites,

2) To present a general equation relating the site economics to heat pump thermodynamics,

3) To define "maximum theoretical economic lift" and illustrate its significance in the context of optimal integration of heat pumps, and

4) To characterize roles of heat pumps in total sites.

APPROPRIATE PLACEMENT IN A TOTAL SITE CONTEXT

The main aim of this section is to clearly identify the opportunities for appropriate placement of heat pumps in total sites.

Most modern day chemical complexes consist of several processes deriving energy form a single on-site utility (i.e. heat and power) system. Heat pumps can play different roles in such sites. A schematic of these roles is presented as Figure 1.

Besides accepting heat from and rejecting heat to the same process, heat pumps can, in principle, accept the waste heat form a process and deliver it at a higher temperature into some other process on the same site. With reference to Figure 1, this corresponds to the case: $i \neq j$. In fact, a heat pump may operate completely within a utility system e.g. it can be used to generate higher level steam from lower level steam, if there is a demand for that higher level of steam.

The next question is what are the available options for appropriate placement of these different types of heat pumps in the context of a process? To answer this question, one must take recourse to the grand composite curve - a powerful tool for process heat integration emerging from the field of pinch technology. Pinch technology and its applications have been discussed in many publications [e.g. 2, 4, 5-8].

The grand composite curve is a plot on temperature-heat load axes. It shows the heat surplus and deficit corresponding to different temperature levels in a process. It defines the heat acceptance and rejection profiles for a process. The pinch divides the problem into a heat sink above it and a heat source below the pinch. The key step in identifying opportunities for appropriate placement of heat pumps is to represent the process and the site utility system as shown in Figure 2. It should be stressed that for this analysis, a process is <u>defined</u> as consisting of all processes and unit operations on a given site that can be heat integrated. Processes that are physically far apart or unit operations (e.g. evaporator section of a pulp mill) that cannot be integrated due to some constraints (e.g. product quality, controllability) should be treated as separate processes.

The opportunities for heat pumping that are thermodynamically appropriate are identified in Figure 2. It can be seen from this figure that the utility levels divide the problem into temperature regions which in turn influence the placement of heat pumps. What is the influence of the marginal values [9] of these utility levels on heat pump lifts? Also, is it possible to define, apriori, the temperature levels between which the optimum heat pump lift must lie? These two questions are addressed in the next section.

MAXIMUM ECONOMIC LIFT

Computation of the maximum economic lift requires correct assessment of marginal values of utility levels in a site and evaluation of the heat transfer area penalty resulting from integration of heat pumps. These concepts are discussed first.

Marginal Values of Utilities [9]

In any site utility system, there is an interaction between fuel, steam and electricity. In an existing utility system, if there is any increase or decrease in a certain utility level requirement, then there will be a corresponding change in the fuel fired and/or a change in the import/export of steam and/or electricity. This change expressed in terms of monetary value (e.g. $/MM Btu) is defined as the marginal value (MV) of that utility level.

Note that the marginal value of cooling water or refrigeration cooling is negative, and only cooling which produces heat (in the form of steam or hot water) which can be used in another process has a positive marginal value.

Heat Transfer Area Penalty

Appropriately integrated heat pumps lift heat from below the process pinch where it is in surplus to above the process pinch where it is required, thereby saving both hot and cold utilities. However, this requires additional capital expenditure for increased heat transfer area and heat pump hardware (compressor, piping, etc.).

Integration of heat pump results in a decrease in utility heater and cooler areas and an increase in the heat interchanger area. It can be shown that the net increase in heat transfer area resulting from heat pump integration is given by:

$$\text{Net Increase in Area} = A_{HPE} + A_{HPC} + A_I - \alpha_H \sum (A_{Hi} - A'_{Hi}) - \alpha_{CO} \sum (A_{COj} - A'_{COj}) \quad (1)$$

A deriviation of this equation is presented in Appendix A. The symbols are defined in the nomenclature section.

General Equation for Maximum Economic Lift

For a heat pump operating in a temperature region bounded by the temperatures of low and high level utilities [i.e. $T_{Ul} \leq T_C < T_H \leq T_{Uh}$ (illustrated in Figure B-1)] with marginal values $(MV)_l$ and $(MV)_h$, respectively, one can derive a general equation to obtain the maximum economic lift. A derivation of this equation is presented in Appendix B and the final equation is as follows:

$$T_C/T_H = \{[C_{HPE} + C_{HPC} + C_I + C_C - \alpha_H(\sum C_H) - \alpha_{CO}(\sum C_{CO})] \times (N_T N_M / HQ_H p) + (MV)_d - N_M(MV)_l - N_T N_M \times [(MV)_h - (MV)_l]\}/[(MV)_d - N_M(MV)_l] \quad (2)$$

where, Q_H is the amount of energy delivered at T_H, p is the simple payback period, H is the hours of operation per year, C_i terms are capital investment terms, $(MV)_d$ is the marginal value of the driver utility and N_T and N_M are the thermodynamic and mechanical efficiencies of the heat pump.

Application of the Equation and Special Cases

Example Problem

Figure 3 [10] illustrates a heat pump operating on a site with two steam levels. It is desired to find out the minimum economic temperature of heat reception, T_{min}, for the heat pump under the following conditions:

a) LP steam can be exported from the process.

b) LP steam cannot be exported from the process.

The data for the problem is presented in Table 1. In terms of the variables in Equation (2),

Case (a): $(MV)_d$ = 51.0 \$/MWh, $(MV)_h$ = 17.0 \$/MWh, $(MV)_l$ = 5.0 \$/MWh, N_T = 0.6,

N_M = 1.0, H = 8000 h/y, C_I = 0, p = 2y and T_H = 473 K

$$C_{HPE} = C_{HPC} = 60 \times 10^3 (Q_H/N_T)[(T_C/T_H) - (1 - N_T)], \quad (3)$$

$$C_{CO} = 30 \times 10^3 (Q_H/N_T)[(T_C/T_H) - (1 - N_T)], \text{ and} \quad (4)$$

$$C_C = 400 \times 10^3 (Q_H/N_T)[1 - (T_C/T_H)] \quad (5)$$

Substitution in Eq. (2) gives $T_C = T_{min}$ = 431.4 K

Case (b): In this scenario, $(MV)_l$ = -4.0 \$/MWh and

$$C_{CO} = 12 \times 10^3 (Q_H/N_T)[(T_C/T_H) - (1 - N_T)] \quad (6)$$

Subsitution in Eq. (2) gives T_{min} = 406.7 K

The maximum economic lifts for Cases (a) and (b) are 41.6 K and 66.3 K, respectively, indicating more scope for heat pumping for Case (b).

Special Cases of the General Equation

Case 1 Payback period $p \rightarrow \infty$

For the case with very large payback period, one obtains the limiting value of the lift or the minimum theoretical coefficient of performance. For this case

$$(T_H - T_C)_{max} = \{[(MV)_h - (MV)_l](N_T N_M T_H)\}/[(MV)_d - N_M(MV)_l] \quad (7)$$

or

$$(COP)_{min} = [(MV)_d - N_M(MV)_l]/N_T N_M[(MV)_h - (MV)_l] \quad (8)$$

For example, if $(MV)_d$ = 50\$/MWh, $(MV)_l$ = -2.745 \$/MWh and $(MV)_h$ = \$11.717 \$/MWh, N_T = 1.0 and N_M = 0.95, $(COP)_{min}$ = 3.815

i.e. for this system only heat pumps with a COP > 3.815 should be considered in searching for the optimum.

An important application of Eq. (7) is in the procedure for optimal integration of heat pumps. Eq. (7) sets the bounds on the feasible temperature range within which the optimum lift (e.g. the one that results in lowest payback period) must lie. It defines the region of search for the optimization problem. Figure 4 is a schematic representation of the application of Eq. (7) to the options for appropriate placement of heat pumps presented in Figure 2.

Case 2: Open System Heat Pumps

The utility system heat pumps are usually open systems and C_{HPE}, C_{HPC} and C_I as well as $\sum C_{HE}$ and $\sum C_{CO}$ are all zero and Eq. (2) simplifies to a great extent:

$$T_C/T_H = \{C_C \times (N_T N_M/HQ_H p) + (MV)_d - N_M(MV)_l - N_T N_M \times [(MV)_h - (MV)_l]\}/[(MV)_d - N_M(MV)_l] \quad (9)$$

CHARACTERISTICS OF INDUSTRIAL HEAT PUMPS

In the previous sections, different options for integration of heat pumps in total sites were identified and a mathematical equation relating the site economics to heat pump lifts was derived. These two are now combined to further characterize the different roles of heat pumps in total sites. A summary of these characteristics is presented in Table 2.

Correct thermodynamic placement of process to process heat pumps is in the pinch region where they pump heat from below the pinch to above the pinch. By pumping heat around the pinch these pumps save both the hot and the cold utility. In principle, it is always possible to find a closed cycle heat pump for process to process heat transfer applications.

Utility to utility heat pumps are open systems. These pumps are used to create new levels of steam from existing lower levels of steam. They are most likely to be appropriate in situations where in an existing facility purchasing a certain level of steam, due to some process changes, there is a sudden demand for a slightly higher level of steam. In all other situations it may be more appropriate to cogenerate this new level from existing higher levels of steam.

An appropriately integrated process to utility heat pump should take the surplus heat from below the pinch and use it to generate steam. This scheme saves only the cold utility. A utility to process heat pump is applicable only above the pinch and competes with the cogeneration option. The most likely configuration for process to utility and utility to process heat pumps is a semi-open system.

It can be seen from Eq. (2) that the maximum economic lift, in general, depends on the cost of additional heat transfer area resulting from appropriate integration of heat pumps. In fact, in semi-open and closed heat pump systems that involve heat integration with process streams, there is a cost trade-off between the compressor and heat transfer area costs. This is illustrated in Figure 5. As the lift increases, the compressor capital and operating costs increase, however, the heat pump heat exchanger costs may decrease because of increase in temperature driving forces. Results presented in Ref. [9] show some cases with lower payback periods for lower COP heat pumps. Therefore, in general, for process to process, process to utility and utility to process heat pumps, the heat pump COP alone cannot be the sole criteria for evaluation.

CONCLUSIONS

In this paper, it has been shown that correct assessment of heat pumping opportunities in industrial sites requires a clear understanding of the site economics and thermodynamics, especially the utility system - process interactions.

The opportunities for appropriate placement of heat pumps in total sites have been identified and characterized. A general equation relating site economics and heat pump thermodynamics has also been derived.

Specific conclusions are as follows:

(1) It is possible to set, apriori, bounds on the temperature range within which the optimum heat pump lift must lie.

(2) For process to process, utility to process, and process to utility heat pumps, COP of a heat pump alone cannot be the sole criteria for evaluation.

(3) For a site with a single process, heat pumping competes with cogeneration as a scheme for generating intermediate levels of steam primarily above the process pinch.

ACKOWLEDGEMENT

This work was accomplished with financial support from the Electric Power Research Institute (EPRI), California (RP 2220-3). However, any opinions, findings, conclusions or recommendations expressed herein are those of the authors and do not necessarily reflect the views of EPRI.

REFERENCES

[1] Loken, P.A., "Process Integration of Heat Pumps," 2nd International Symposium on the Large Scale Application of Heat Pumps, Paper F2, U.K., September 1984.

[2] Linnhoff, B., et.al., "A User Guide on Process Integration for the Efficient Use of Energy," IChemE, England, 1982.

[3] Ranade, S.M., et.al., "Industrial Heat Pumps: Appropriate Sizing and Placement Using the Grand Composite," Proceedings of the Eighth IETC, Vol. 1, pp. 194-203, Public Utility Commission of Texas, Houston, June 1986.

[4] Chappell, R.N. and S.J. Priebe, "Process Integration of Industrial Heat Pumps," ibid., Vol. 2, pp. 463-476.

[5] Hindmarsh, E., et.al., "Maximizing Energy Savings for Heat Engines in Process Plants," Chem. Eng., Feb. 1985, p. 38.

[6] Boland, D., et.al., "Efficiency and Flexibility Improvement in Crude Units," A paper presented at the 1986 NPRA Annual Meeting, Los Angeles, California, March 1986.

[7] Spriggs, H.D. and G. Ashton, "Diverse Applications of Pinch Technology Within Process Industries," Proceedings of the Eighth IETC, Vol. II, pp. 448-454, Public Utility Commission of Texas, Houston, June 1986.

[8] Hindmarsh, E., et.al., "Heat Integrate Heat Engines in Process Plants," ibid., pp. 477-489.

[9] Ranade, S.M., et.al., "Industrial Heat Pumps: A Novel Approach to Their Placement, Sizing and Selection," Proceedings of the 21st IECEC, Vol. 1, pp. 389-397, San Diego, California, August 1986.

[10] ICI, United Kingdom, Heat and Power Course

[11] Townsend, D.W. and B. Linnhoff, "Surface Area Targets for Heat Exchanger Network," 11th Annual Research Meeting, The Institution of Chemical Engineers, April 1984.

APPENDIX A

Heat Transfer Area Penalty Due to Heat Pumping

Figure A-1(a) shows the hot and cold composites for a process. The heat pump evaporator (cold stream) and condenser (hot stream) are shown in Figure A-1(b). The effect of reduction in hot and cold utilities due to heat pumping is clearly seen in Figure A-1(b).

The total heat transfer area of a process consists of heat interchanger area which corresponds to process-process heat exchange, utility heater and utility cooler areas. For simple systems with streams with identical heat transfer coefficients and no constrained matches, it has been shown [11, 6] that the minimum (or target) total heat transfer area for a given minimum approach temperature ΔT_{min} between the hot and cold composites is given by:

$$A_T = \sum_{j=1}^{n} A_j \tag{a-1}$$

where, n is the total number of enthalpy intervals.
For the problem illustrated in Figure A-1, the following equations are applicable:

Before Heat Pump Placement (Figure A-1(a))

$$\text{Target heat interchanger area} = A_1 + A_2 \tag{a-2}$$
$$\text{Target heater area} = A_H \tag{a-3}$$
$$\text{Target cooler area} = A_{CO} \tag{a-4}$$

After Heat Pump Placement (Figure A-1(b))

Target heat interchanger area =
$$A_{11} + A_{12} + A_{13} + A_{21} + A_{22} + A_{23} \tag{a-5}$$
$$\text{Target heater area} = A'_H \tag{a-6}$$
$$\text{Target cooler area} = A'_{CO} \tag{a-7}$$

It can be seen from Figure A-1 and the above equations that

$$A'_H < A_H \tag{a-8}$$
$$A'_{CO} < A_{CO} \tag{a-9}$$

Also, in general, due to reduction in temperature driving forces:

$$A_{11} + A_{12} + A_{13} + A_{21} + A_{22} + A_{23} \geq A_1 + A_2 \tag{a-10}$$

Denoting the total increase in heat interchanger area as ΔA one can write:

$$\Delta A = (A_{11} + A_{12} + A_{13} + A_{21} + A_{22} + A_{23}) - (A_1 + A_2) \qquad \text{(a-11)}$$

ΔA can also be expressed as follows:

$$\Delta A = A_{HPE} + A_{HPC} + A_I \qquad \text{(a-12)}$$

where A_{HPE}, A_{HPC} and A_I are respectively the heat pump evaporator area and condenser area and the change in the interchange area of the remainder of the process.

$$\text{Decrease in heater area} = A_H - A'_H \qquad \text{(a-13)}$$

and

$$\text{Decrease in cooler area} = A_{CO} - A'_{CO} \qquad \text{(a-14)}$$

Assuming α_H and α_{CO} to be the fractions of heater and cooler areas that can be reused, and more than one heater and cooler, the net increase in area can be expressed as:

$$\text{Net increase in heat transfer area} = A_{HPE} + A_{HPC} + A_I - \alpha_H \sum(A_H - A'_H) - \alpha_{CO} \sum(A_{CO} - A'_{CO}) \qquad \text{(a-15)}$$

APPENDIX B

Deriviation of a General Equation for Maximum Economic Lift

Consider a heat pump operating in a temperature region bounded by the temperatures of a low and a high level utility (i.e. $T_{Ul} \leq T_C < T_H \leq T_{Uh}$) with marginal values $(MV)_l$ and $(MV)_h$, respectively. This is illustrated in Figure B-1.

Assuming the mechanical and thermodynamic efficiencies to be N_M and N_T, respectively, one obtains:

$$W_{pi} = Q_H/N_T N_M[1 - (T_C/T_H)] \qquad \text{(b-1)}$$

$$W_p = (Q_H/N_T)[1 - (T_C/T_H)] \qquad \text{(b-2)}$$

and

$$Q_{CP} = (Q_H/N_T)[(T_C/T_H) - (1 - N_T)] \qquad \text{(b-3)}$$

Annual savings at H hours of operation per year for Q_H units of energy delivered per hour at T_H (deg. absolute) can be expressed as:

$$\text{Annual Savings} = \text{Cost of } U_h \text{ Saved} - \text{Cost of Driver Power} - \text{Cost of } U_l \text{ Used} \qquad \text{(b-4)}$$

Substitution and simplification gives:

$$\text{Annual Savings} = \{(T_C/T_H)[(MV)_d - N_M(MV)_l] + N_TN_M[(MV)_h - (MV)_l] - [(MV)_d - N_M(MV)_l]\}(HQ_H/N_TN_M) \quad \text{(b-5)}$$

Let p be the simple payback period in years. Then based on the equation derived for net increase in area (Eq. a-15), one can express the capital cost equation as follows:

$$\text{Annual Capital Expenditure} = [C_{HPE} + C_{HPC} + C_I + C_C - \alpha_H(\sum C_H) - \alpha_{CO}(\sum C_{CO})]/p \quad \text{(b-6)}$$

where C_C is the installed cost of the heat pump hardware excluding the heat exchangers.

For retrofit applications, a conservative value of both α_H and α_{CO} is zero.

Equating (b-5) and (b-6) one can write the general equation to obtain the maximum economic lift (for a certain payback period).

$$T_C/T_H = \{[C_{HPE} + C_{HPC} + C_I + C_C - \alpha_H(\sum C_H) - \alpha_{CO}(\sum C_{CO})] \times (N_TN_M/HQ_HP) + (MV)_d - N_M(MV)_l - N_TN_M \times [(MV)_h - (MV)_l]\}/[(MV)_d - N_M(MV)_l] \quad \text{(b-7)}$$

Electricity price = $51.0/MWh.

IP steam marginal value = $17.0/MWh heat absorbed.

LP steam marginal value = $5.0/MWh heat absorbed.

Cooling water cost $4.0/MWh heat rejected.

Capital cost of compressor = $400/kW installed power.

Capital cost of heat pump heat exchanger = $60/kW absorbed heat (installed).

Capital cost of LP raising heat exchanger = $30/kW absorbed heat (installed).

Capital cost of cooling water cooler = $12/kW rejected heat (installed).

Hours of operation = 8000 h/year.

Simple payback period = 2 years.

COP (practical) = 0.6 COP (based on working medium temperatures).

Table 1
Data for Maximum Economic Lift Calculations

Role	Type	Comments	COP	Significance Heat Transfer Surface Area
Process to Process	Closed	- Appropriate Only Around the Pinch - Saves Both Hot and Cold Utility	Yes	Yes
Utility to Utility	Open	- Demand for Intermediate Steam Level Necessary - Competes with Cogen	Yes	No
Process to Utility	Semi-Open	- Applicable Only Below the Pinch - Demand for Intermediate Steam Level Necessary - Saves Cold Utility	Yes	Yes
Utility to Process	Semi-Open	- Applicable Only Above the Pinch - Competes with Cogen	Yes	Yes

Table 2
Characteristics of Heat Pumps in Total Sites

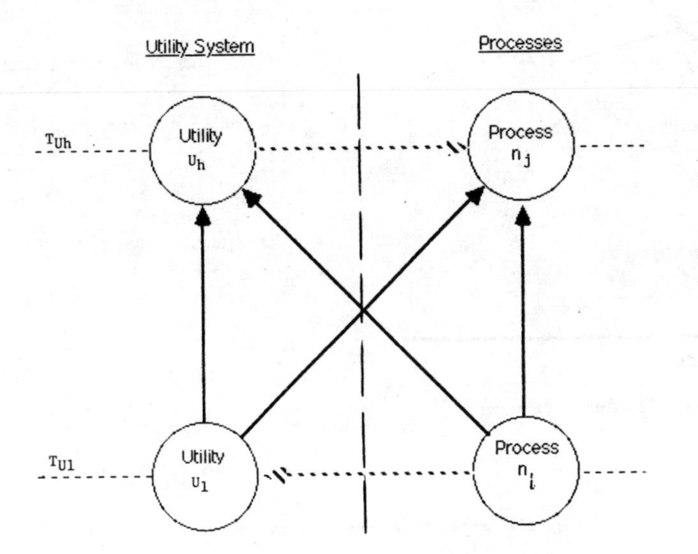

Figure 1. Role of Heat Pumps in a Total Site Context

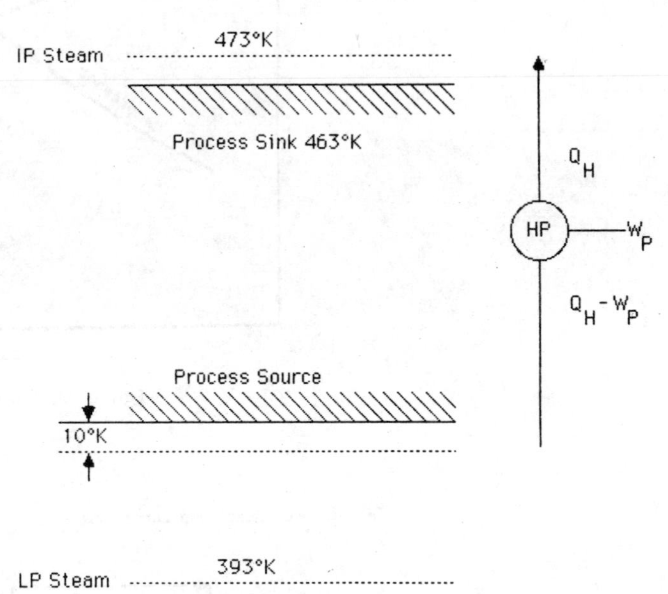

Figure 3. Maximum Economic Lift: Example Problem

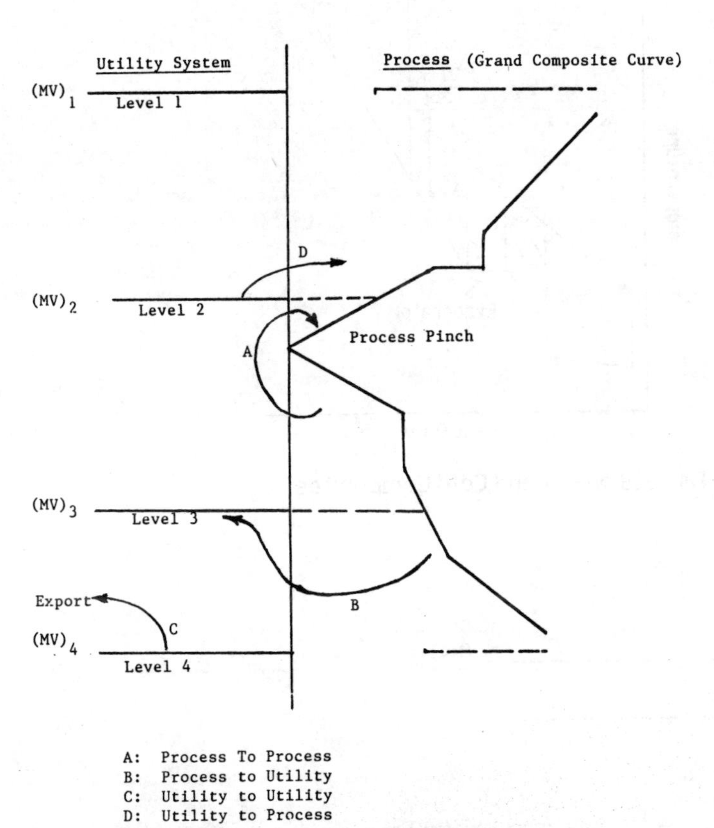

Figure 2. Opportunities for Appropriate Heat Pump Placement in a Total Site Context

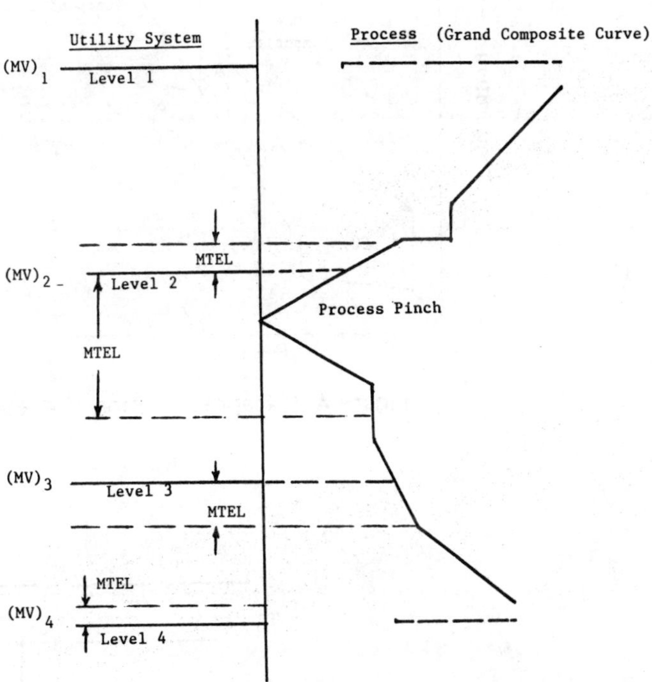

Figure 4. Application of Maximum Theoretical Economic Lift to Define Temperature Range in Which the Optimum Lift Must Lie

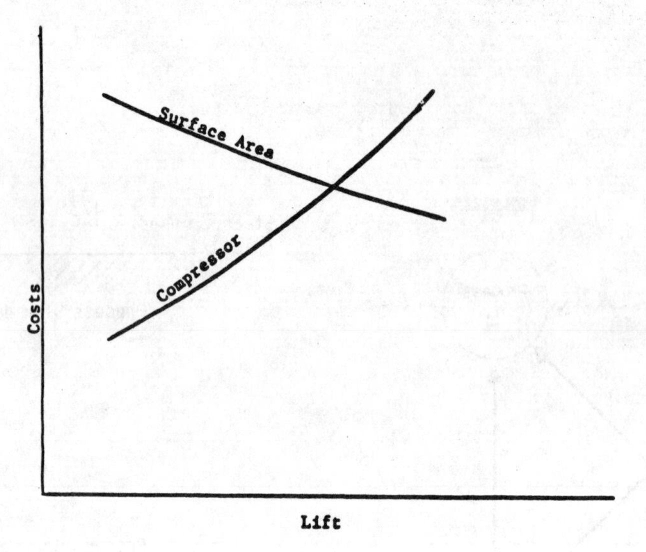

Figure 5. Cost Trade-Offs

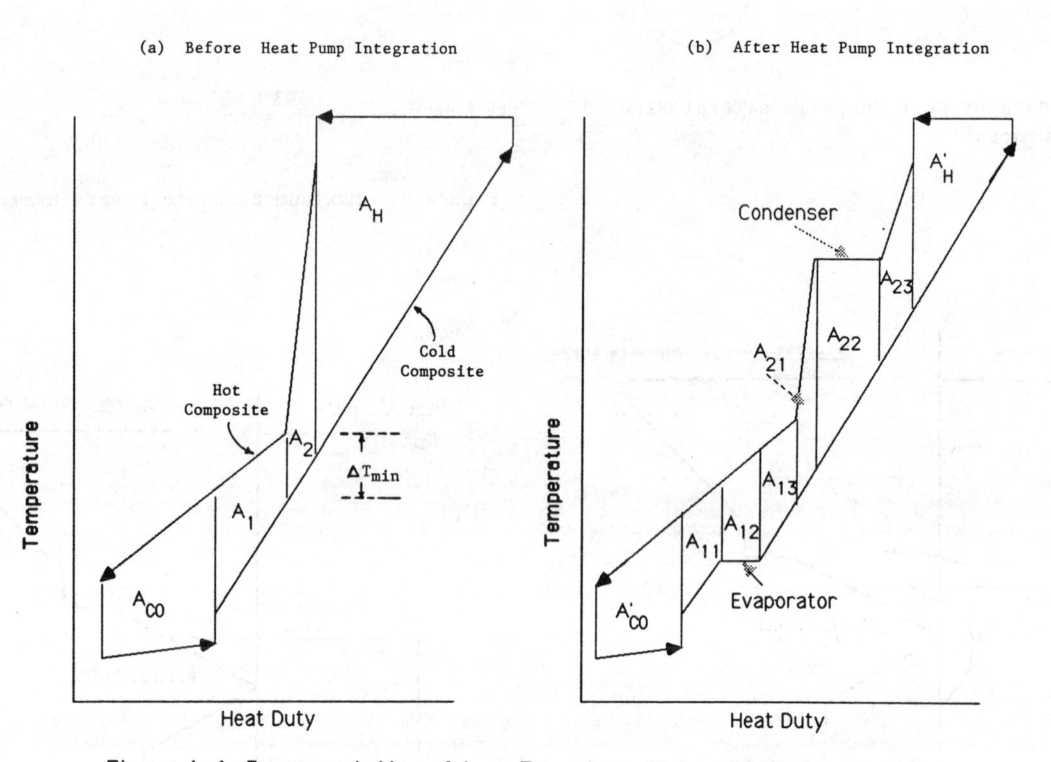

Figure A-1. Representation of Area Targets on Hot and Cold Composites

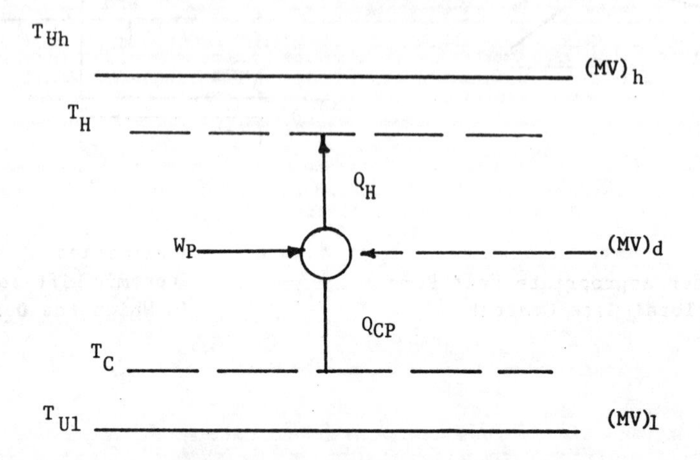

Figure B-1. Heat Pump in a Region Bounded by Two Utility Levels

3rd International Symposium on the

Large Scale Applications of Heat Pumps

Oxford, England : 25-27 March 1987

PAPER D2

HEAT PUMP APPLICATIONS AT BP REFINERY IN GOTHENBURG

M. Brun
T. Berntsson

The authors are M. Eng. Michele Brun and Prof. Thore Berntsson. They are in the Department of Heat and Power Technology at Chalmers University of Technology, Gothenburg, Sweden.

SYNOPSIS

Energy system studies according to the Pinch Technology have shown that one way to reduce the energy consumption at the BP Refinery would be the use of a heat pump for low-pressure (4.5 bar) steam production. As heat source heat from hot products or from a special water system, providing the Volvo factory close to BP with heat, could be used.

The technical and economical calculations have shown, that for raising 7 tons/h steam with an evaporating temperature of 60-70°C, a two-stage heat pump (R114/H_2O) is needed. The total investment cost when hot products are used as heat source is 70% higher than the cost of the heat pump itself and the pay-back period is about 3.5 years.

With the Volvo water as heat source the pay-back period ranges between 2.0-2.6 years (for direct steam compression) with small annual energy recovery and nearly 3 years (two-stage heat pump) with high annual energy recovery.

1. INTRODUCTION

In this paper, an energy system study /1/, performed at BP Refinery in Gothenburg, Sweden, is presented. In this study it was shown that the introduction of a heat pump could improve the energy situation in the refinery.

Due to the low electricity prices in Sweden, 1/3 to 1/2 of those in most other European countries, the interest for heat pumps in this country is high. Today approximately 130000 heat pump units are in operation in Sweden. Of these a very small number, less than 10, are placed in industry. There are several reasons for this, i.e. lack of good hardware for high-temperature applications and lack of experience from experimental and demonstration plants. However, industrial applications of heat pumps must be considered to have a big potential in the future Swedish energy system. The present very low oil prices will of course retard the development of this technique in the next years.

Too many case studies of industrial heat pumps have started with a heat pump study and ended with a suggestion for an improved heat exchanger network instead of a heat pump. In order to avoid such a situation, a study of the whole energy system must be performed, so that the heat pump can be integrated in the process in a thermodynamically and economically correct way. In the BP study the Process Integration method according to Linnhoff /2/, /3/, /4/ (the Pinch Technology) has been used.

2. THE ENERGY SITUATION AT BP, GOTHENBURG

BP Refinery in Gothenburg has today a very good internal energy situation. One of the reasons for this is that waste heat is sold by BP to the Volvo factory, situated close to the plant. This heat is distributed as hot water and is taken from some of the hot products, which previously were cooled by ambient air or a cooling water system.
At the time of the study, 1984-85, the future low-pressure (4.5 bar) steam demand, not passing a back-pressure turbine, was estimated to approx. 7 tons/h, which was practically constant the whole year. Due to the low Swedish electricity prices, the installation of a new back-pressure turbine for this steam amount was not economic. The task of the study presented below was to find a way to reduce the primary energy demand for the production of this steam.

A Pinch Technology analysis of the crude oil unit showed that the theoretical primary energy demand, with reasonable minimum temperature differences in the heat exchangers, was very close to the actual one and that there therefore was no possibility to save low-pressure steam consumption at that part of the plant. Instead the utility system and the Volvo water system were investigated in more detail in the study.

3. THE UTILITY SYSTEM

The utility system is shown schematically in Fig 1. The energy streams which have been taken into account are:

- Ambient air for the furnaces: The air is sucked into the system in many different burners, which makes an air preheater practically impossible. These streams have therefore been deleted.

- The Volvo water: This is a special system which will be treated separately.

- Flue gases: the flue gases are already cooled to the SO_3 dew point.

- Hot products: These streams (excluding the ones for the Volvo system) are now cooled in air or water coolers and can hence be defined as hot streams in the study.

Held at St. Catherine's College Oxford, England. Organised and sponsored by BHRA, The Fluid Engineering Centre, Cranfield, Bedfordshire, MK43 0AJ England.

- Feed water for the low-pressure steam production. Preheating of this water is a cold stream in the study.

According to the discussion above, the remaining streams for a Pinch Technology analysis are the hot products and the feed water.

The result of a Pinch Technology analysis is shown in Fig. 2. The figure shows the net available heat in the plant versus the temperature. For the production of 7 ton/h steam, 4.3 MW, an evaporator load of approximately 2.7 MW is needed. According to the figure an evaporator temperature of 70-80°C (including necessary temperature differences in the heat exchangers) is needed.

4. A HEAT PUMP IN THE UTILITY SYSTEM

To the knowledge of the authors, there is today no commercial heat pump in operation with an evaporating temperature of 70-80°C and a condensing one of approximately 150°C. One of the most important and well-known problems with industrial heat pumps is to find a suitable working fluid at condensing temperatures above approx. 125°C. All the suggested fluids have one or the other disadvantage, they are e.g. flammable, toxic, not stable, or have a low volumetric capacity. As an alternative, some of the larger manufacturers of heat pumps have suggested a two-stage solution, in which low-pressure steam, 100-120°C, is produced by a closed-cycle heat pump and is compressed by an open-cycle stage, i.e. a steam compressor to the desired pressure, see Fig. 3. This solution has been chosen for the study at BP.

There are several possible working fluids for the closed-cycle stage of the heat pump. The most commonly discussed one at condensation temperatures between 100°C and 125°C is R114, which has therefore been chosen in the study.

For the size of the heat pump discussed at BP both the screw and the turbo types of compressors are possible to use in both the closed-cycle and the open-cycle stages. It is impossible to give a general statement of which of these types that is the most suitable one technically and economically in each of the stages.

Calculations and discussions with manufacturers have shown, that the Carnot efficiency for a two-stage heat pump with today's equipment should be between 0.50 and 0.55. With optimized systems in the near future, 0.60 should be possible to reach. This efficiency is of course depending on the possible degree of subcooling in the closed-cycle stage. In the BP plant, the feed water is preheated to saturation before the heat pump and hence no subcooling is possible.

Possible COP:s at BP as a function of the evaporating temperature is shown in Fig. 4 for the two Carnot efficiencies 0.50 and 0.60. In the following the somewhat conservative value 0.50 will be used.

The investment cost for a two-stage heat pump in the size range discussed at BP is approximately 160£/kW heat output. This is nearly double as much as the corresponding figure for "normal" heat pumps with R12, R22, etc. in this size range.

In the BP study, energy cost levels in early 1985 have been used. The electricity cost has been set to 0.0127£/kWh in winter (Nov.-Apr.) and 0.015£/kWh in summer (May-Oct.). This cost includes all fixed costs and tax. The price for oil, the alternative fuel for steam production, is 148£/TOE.

The pay-back period (PBP) versus the evaporating temperature (condensing temperature corresponds to 4.5 bar) is shown in Fig. 5 with the Carnot efficiency and the total investment cost, including piping, heat exchangers for the heat source, control, etc., as parameters. As can be seen in the figure, the PBP is only marginally influenced by the Carnot efficiency in the range discussed here. Also the evaporating temperature has a very small influence. The most important parameter is, of course, the total investment cost.

Due to the geographic and practical reasons the heat from the hot products to the evaporator must be transferred via an intermediate water circuit. The size-dependent cost for the introduction of a hot products-to-water heat exchanger was found to be high, which made it essential to find a solution with as small a number of such heat exchangers as possible without decreasing the evaporating temperature too much. The optimal solution was found to be heat exchangers for two of the hot products, which gave an evaporating temperature of 70°C. The total cost for piping, heat exchangers, an electrical transformer, and improvement of the local electricity distribution grid was calculated to 0.5 million £. With the heat pump cost given above, this means a total investment cost of 1.23 million £. From Fig. 5 it can be concluded that the PBP is 3.5 years, if the Carnot efficiency is 0.5.

5. THE VOLVO WATER SYSTEM

The distribution water to Volvo has a delivering temperature of 130°C in the winter and 140°C in the summer (April to October). The return temperature is 75°C in the winter and 108°C in the middle of the summer. The heat load at Volvo varies between 25-30 MW in the winter to 15-20 MW in the summer. Due to this heat load difference between winter and summer, it would be theoretically possible to use the high-temperature hot products included in the Volvo system for steam production via direct heat exchanging, without any heat pump, in the summer period. Unfortunately this is not possible in practice due to the high temperature of the return water during this period. Instead the heat in the low-temperature hot products are available in the summer, which will be discussed below.

Various possible heat pump systems

Three alternative heat pump solutions connected to the Volvo water system have been identified and investigated. These are:

1. Open-cycle heat pump

Direct steam compression, see Fig. 6, is possible in the summer (April to October). The total heat exchanger network in the Volvo water system is used to produce water of 140°C, both to Volvo and for the steam compression. Saturated steam of approx. 115°C can then be produced and compressed to 4.5 bars. In the winter this heat pump must be shut down.

2. Combined system with heater

In order to make the steam compressor above operate also in the winter, an R114 heat pump can be installed, which can produce the 115°C steam during this half of the year. As heat source the return Volvo water can be used. Due to the decrease of the incoming water temperature caused by this, calculations have shown that 2/3 of the heat to the evaporator can be recovered in the heat exchangers between the Volvo water and the hot products. The evaporating temperature, possible to achieve, has been calculated to 60°C. The heat not recovered can be transferred to the Volvo water via a steam heater with the steam taken from the heat pump, see Fig. 7. This means a net steam production in the heat pump of 5.5 tons/h in the winter. In the summer 7 tons/h are produced with the steam compressor in the same way as in alternative 1.

3. Combined system with penalty cost

This system is the same as in alternative 2 except the steam heater, which is not included. Instead all the produced steam is used in the BP plant. This means that the heat delivery to Volvo is reduced by 1/3 of approx. 2.7 MW, i.e. 0.9 MW, during winter time. This reduction means a reduced income for BP or depending on the actual agreement with Volvo, some kind of penalty cost.

In table 1 the results concerning the calculations of investment costs, COP:s, and PBP for the three alternatives are shown. The corresponding results for the hot-product heat pump, discussed in Section 4, are also included. In the calculations the actual price for the heat delivery to Volvo has been used, the level of which unfortunately is restricted information. The investment cost for the open-cycle heat pump, alternative 1, is uncertain and therefore two levels have been used.

In the table it can clearly be seen that the use of a direct steam compressor in the summer only (alternative 1) has the lowest PBP, well below 3 years. Also the other alternatives for the Volvo water system have PBP:s below or around 3 years. The reasons for the lower investment cost for the heat source compared with the hot-product heat pump are that the Volvo water is situated very close to the suggested site for the heat pump and that only one stream, the Volvo water itself, is used as heat source.

6. DISCUSSION AND CONCLUSIONS

All the heat pump alternatives in the Volvo water system have lower PBP:s than the hot-product one. For heat pump alternatives 2 and 3 the COP is somewhat lower than the corresponding for the hot-product one, due to the somewhat lower evaporating temperature. This is more than compensated by the lower investment cost for the heat source. The open-cycle system has a lower PBP than all the other alternatives but the annual saving of primary energy is only approx. 60% of the others. An economic evaluation using e.g. a dynamic annuity method would therefore probably show a worse economy for this system than for the others and would hence do more justice for the systems working the whole year. Unfortunately the PBP method is still the one used most widely in industry.

As already been pointed out, of the various possibilities to reduce the PBP the investment cost is the most important one. Therefore, further R&D work on technical solutions for this temperature range with lower investment costs must be carried out. One promising such solution is the compression/resorption cycle.

One must also have in mind that the specific heat pump investment cost is size-dependent. A larger size could reduce this cost considerably. Furthermore, in an application with a smaller temperature lift than 80°C and with higher Carnot efficiencies in the future, it would be possible to decrease the PBP.

7. ACKNOWLEDGEMENTS

The authors wish to express their gratitude to Sten-Åke Bergstrand, Nils Kilander, and Mats Lindgren at the BP Refinery for valuable help and interesting discussions.

8. REFERENCES

/1/ Michele Brun & Thore Berntsson: "Heat pump applications at BP Refinery - Studies of different possibilities". Report from the Department of Heat and Power Technology, Chalmers University of Technology, Gothenburg, Sweden, October 1985.

/2/ Bodo Linnhoff & John A. Turner: "Heat recovery networks: new insights yield big savings". Chem. Engineering, November 2 1981, pp. 56-70.

/3/ Tjaan N. Tjoe & Bodo Linnhoff:"Using Pinch technology for process retrofit". Chem. Engineering. April 28 1986, pp. 47-60.

/4/ B. Linnhoff et al.:"A user guide on process integration for the efficient use of energy". The Institution of Chemical Engineers, 1985.

Table 1.

Case	Investment cost (M £)	COP	PBP years
Open-cycle heat pump	0.55-0.77	10.0	2.0-2.6
Combined system with heater	1.14	Winter: 2.5 Summer: 10.0	3.0
Combined system with penalty cost	1.06	Winter: 2.5 Summer: 10.0	2.8
Heat pump with hot products as heat source	1.23	2.6	3.5

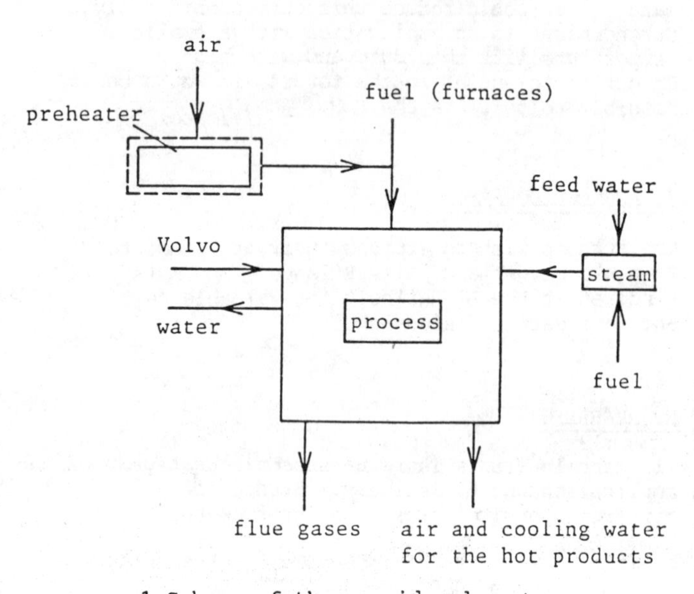

1 Scheme of the considered system

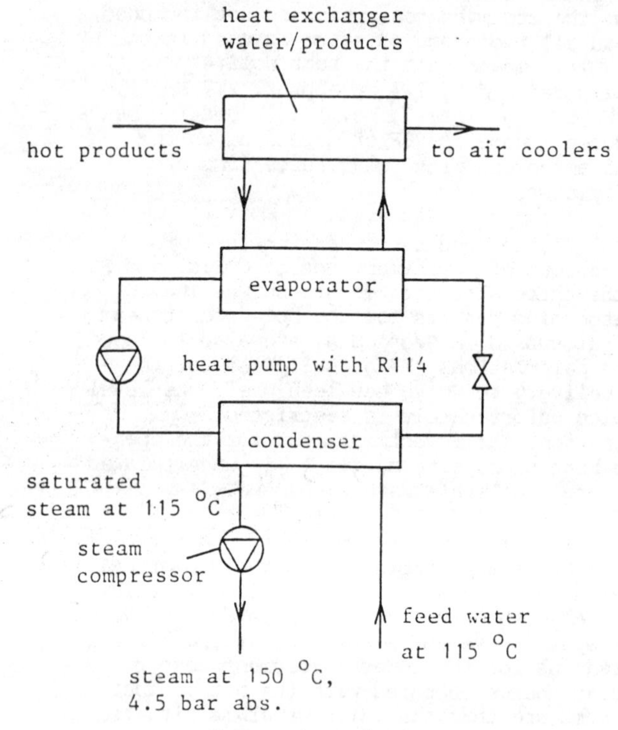

3 Two-stage heat pump

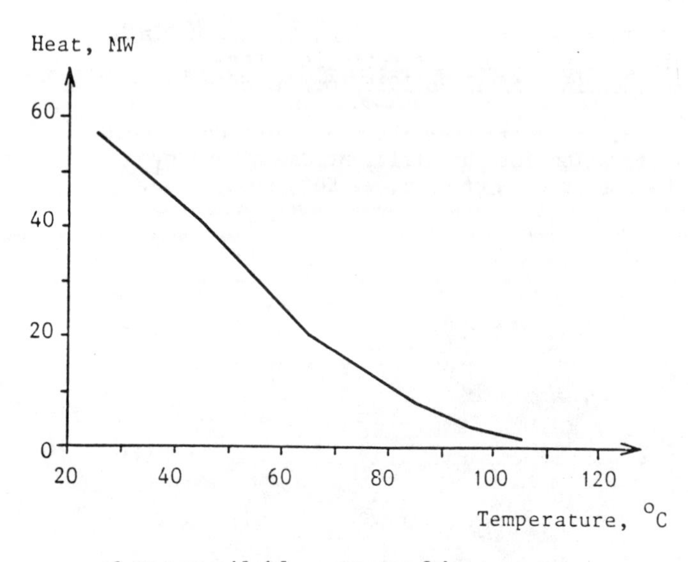

2 Net available amount of heat versus temperature.

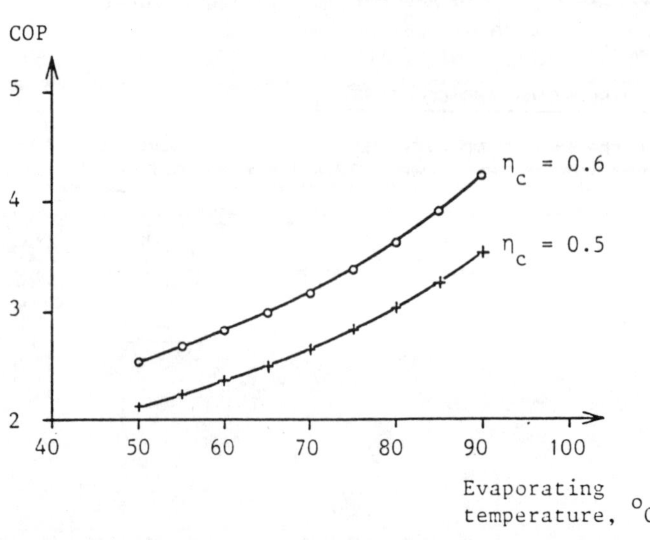

4 COP versus evaporating temperature with the Carnot efficiency as a parameter.

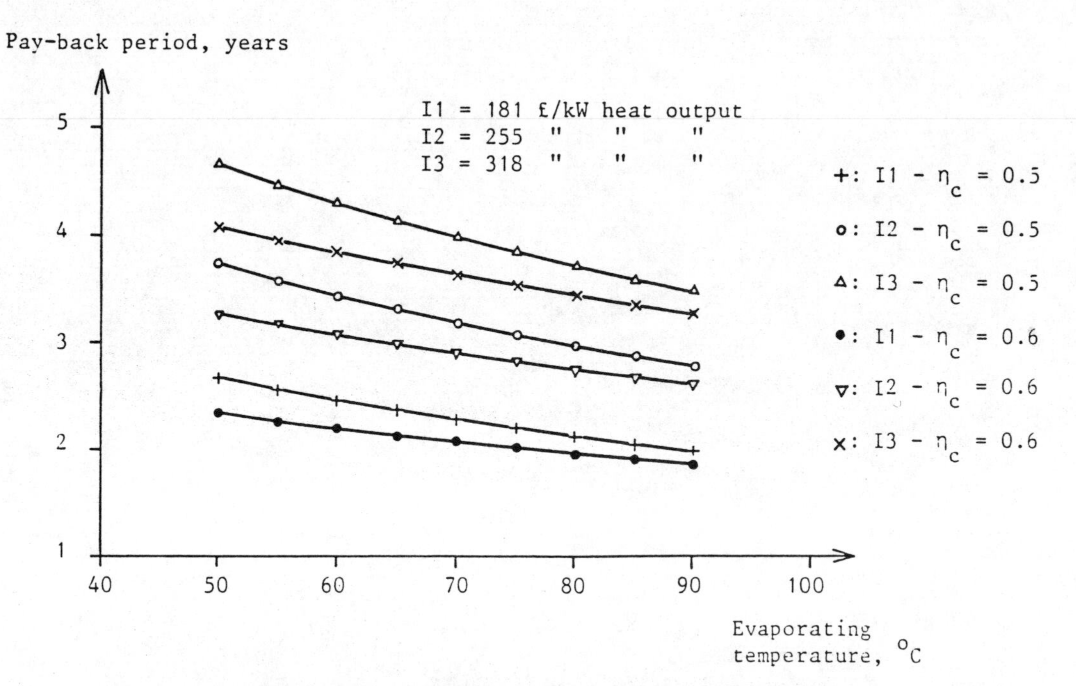

5 The pay-back period versus the evaporating temperature with the total investment cost and the Carnot efficiency as parameters.

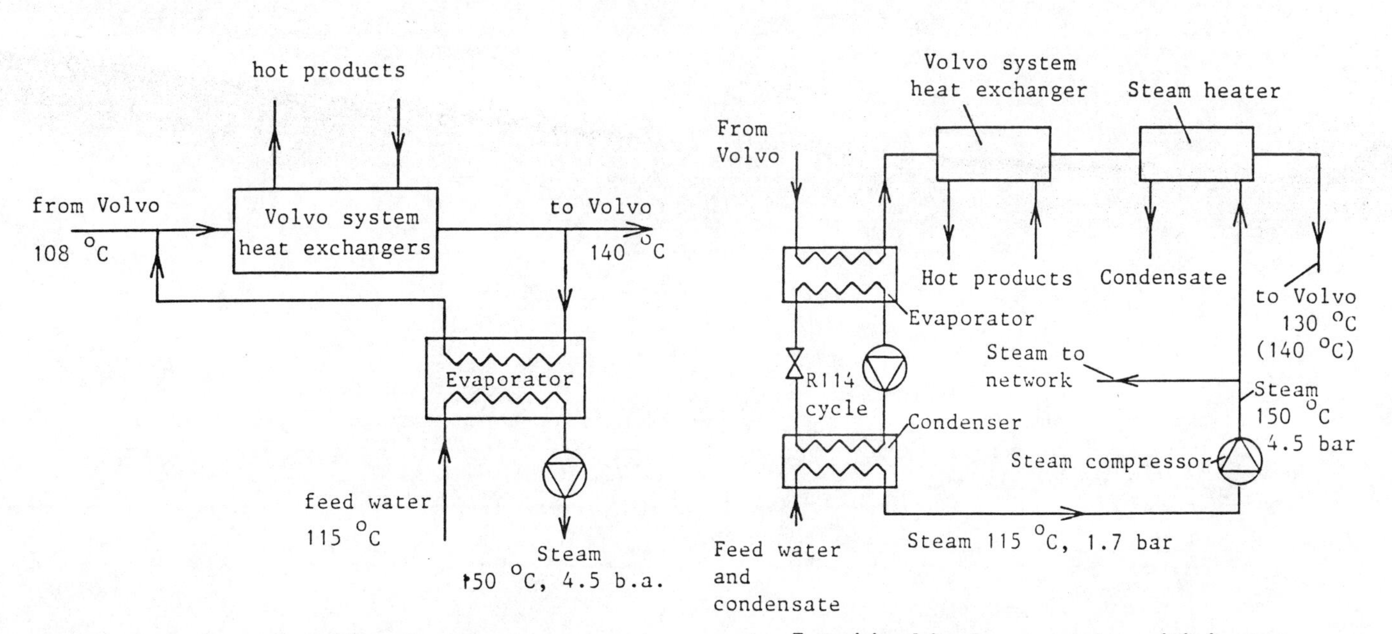

6 Open-cycle heat pump (direct steam compressor)

7 Combined heat pump system with heater.

3rd International Symposium on the
Large Scale Applications of Heat Pumps
Oxford, England: 25-27 March 1987

PAPER E1

HIGH TEMPERATURE HEAT PUMP WITH LIQUID-RING COMPRESSOR USING WATER AS A REFRIGERANT

B. Gromoll

Siemens AG research laboratories, Erlangen, W. Germany

SYNOPSIS

To use water as a refrigerant for high temperature heat pumps would seem obvious. Steam compression presents a problem, however. The liquid-ring compressor offers a promising solution, so a heat pump was designed and built with a liquid-ring compressor. The preliminary results show that steam compression poses no problem for liquid-ring compressors. A thermal oil was used as the working liquid. The high pressure-ratio of the compressor allows a large area of application. Condensation temperatures between 120 - 160°C were reached with evaporation temperatures from 70 - 80°C. The COP lay between 1.6 and 4.4.

1. INTRODUCTION

The technology for industrial heat recovery using heat pumps in the lower and middle temperature ranges is well developed and time-tested. For high temperatures, however (i.e., in producing steam, a source of energy that has many uses for industry), only very little data has been gathered from a few test installations with refrigerants and circuit components that, to a degree, have hardly been tested.

The typical areas of application of high temperature heat pumps are industrial processes in which heat accumulates at temperatures ranging from 60 - 90°C and in which heat is required at temperatures of 120 - 160°C. Examples of this are drying, evaporation, and distillation processes in paper, brewing, and food industries.

In such processes, it would make sense to incorporate a high temperature heat pump for steam generation. Practically speaking, standard refrigerants can only be used at condensation temperatures of up to $< 115^{\circ}C$ because of their limited thermal stability. Various other refrigerants which have been investigated for use at high temperatures mostly reveal thermodynamic disadvantages and are too expensive, explosive, poisonous, or have other disadvantages /1, 2/.

Therefore, the use of water as a refrigerant for high temperatures is logical. Steam compression presents a substantial problem, however. It was therefore the goal of this investigation to develop a high temperature heat pump that uses water as a refrigerant and has a compressor that is suitable for steam.

2. HIGH TEMPERATURE HEAT PUMP USING WATER AS A REFRIGERANT

Water fulfills many requirements for the ideal heat pump refrigerant.

It is neither poisonous nor flammable; it is thermically stable, chemically inert, available everywhere, cheap, and environmentally safe in every sense. And in regard to thermodynamic characteristics, water has many advantages over other refrigerants for high temperature heat pumps which have been examined up to now.

The theoretical COP is high, the vapour pressure curve is favourable in regard to the critical temperature and the condensation pressure, and the material characteristics are good regarding heat transfer.

Steam is also suitable for both open and closed processes, is easy to manipulate, easy to replenish and is a familiar source of energy for industry.

On the other hand there are the various known disadvantages which to a great degree have till now prevented water from being used as a refrigerant in heat pumps. For example, given the usual approximately isentropic compression, the compressed gas overheats at the compressor outlet up to 3 times the condensation temperature, Fig. 1a. This requires that the compressor be directly or indirectly cooled during compression to avoid problems with the materials or lubricants.

At lower evaporation temperatures, a large intake volume of the compressor is necessary on account of the low

Held at St. Catherine's College Oxford, England. Organised and sponsored by BHRA, The Fluid Engineering Centre, Cranfield, Bedfordshire, MK43 0AJ England.

density of the steam, Fig. 1b.

There are also problems with compressor lubrication since water does not dissolve in lubricating oil.

Hence the main problem when using steam as a refrigerant is to find a compressor which can most easily overcome the stated problems.

Oil-free reciprocating compressors with a correspondingly large intake volume and a corresponding cooling facility are not available. Centrifugal or Roots-type compressors are only suitable for conditions of relatively low pressure ratios and would require multi-stage systems with intermediate cooling.

Promising experiments have been carried out with oil-free screw compressors /3/. An apparent problem, however, which limits the field of application, is cooling the compressor under conditions of higher pressure ratios.

The subsequently described liquid-ring compressor offers an encouraging solution since its operating principle appears specially suitable for the problems of steam compression.

3. DESCRIPTION OF THE LIQUID-RING COMPRESSOR

Liquid-ring compressors are usually used in industry to draw off gases and vapors from the sub-atmospheric pressure range. In this they have proved themselves for decades, and their outstanding characteristics are as follows:

- a high suction capacity
- no problems with liquid injection for cooling in suction vapor (no danger from hammer effect)
- large load regulation range by means of a speed regulator
- suitable for open and closed processes
- simple and robust compressor with few wearing parts

The compressor envisaged here works in two stages in order to deal with the pressure ratio necessary for the desired area of application of heat pumps using steam. One impeller each is fitted in the two cylindrical casings which are arranged off centre to the shaft and partly filled with working liquid (Fig. 2). The working liquid forms a rotating ring which pulsates in the pockets between the blades as the impellers revolve. With outward movement the steam is drawn into the compressor through sickle-shaped discharge ports in the port plates (Fig. 2a). With inward movement the steam is compressed and forced out of the likewise sickle-shaped discharge ports in the port plates.

The attainable pressure ratio is determined by the density of the working liquid ρ_L, the speed u of the rotating liquid-ring, a geometry factor $\varkappa$ for the impellers, and the suction pressure P_e.

$$\left(\frac{P_c}{P_e}\right)_{max} \approx \frac{2}{3}\left(\frac{\rho_L\ \varkappa^2\ u_{max}^2}{2\ P_e} + 1\right)$$

This equation shows, why the liquid-ring compressor is specially suitable for steam compression. With decreasing suction pressure, the maximum attainable pressure ratio increases. Therefore high pressure ratios can be achieved especially at sub-atmospheric suction pressures (Refrigerant water: $\vartheta_e < 100\,°C$, $P_e < 1$ bar).

Fig. 2b shows the cross-section of the type of compressor used here with the layout of the impellers and port plates.

The gas forced out of the low-pressure side (LPS) is fed to the intake port of the high-pressure side (HPS). The working liquid is partially forced out with the compressed steam and must be fed to both the low-pressure side and high-pressure side via separate connections. The shaft is sealed at the endshields with bearing ring seals. The compressors have cylindrical roller bearings and a fixed deep-groove ball bearing which eliminates play.

Water is normally used as the working liquid. However, water is unsuitable as a working liquid when used as a compressor for steam since some of the water evaporates out of the working liquid on intake which reduces intake capacity and efficiency. In a test installation, a thermal oil was found to be a liquid suitable for steam compression /4/. In this test installation, condensation temperatures up to 120°C with evaporation temperatures of 90°C were attained using a single-stage compressor and oil as the working liquid. The fundamental suitability of the compressor for steam was thereby substantiated. However, the area of application was limited due to the single-stage compression.

Using oil as the working liquid, the compressor works with a more or less constant intake volume over a wide range because of the large vapor pressure difference between oil and water.

For the test installation, a compressor with a smaller intake volume was selected at first to keep the installation small and to allow the necessary changes to be carried out quickly and easily. Compressors of this type are available with up to ten times the intake volume. Single-stage compressors are built with a substantially higher output /5, 6/.

4. DESIGN OF THE TEST INSTALLATION

A test installation was designed and built to experimentally substantiate the suitability of the liquid-ring compressor for conditions of higher pressure and correspondingly higher temperatures ($>120^{\circ}C$). The goal of the investigation was to determine the area of application and efficiency of the individual components, and to find a thermodynamically suitable heat-pump process for a liquid-ring compressor which uses water as a refrigerant.

The test installation consists of a high temperature heat pump with a closed refrigerant circulation system and a supply unit with which the temperatures of the heatsource and the heatsink for the heat pump can be set. Fig. 3 shows the schematic design of this test installation.

The heat pump circuit consists of the following main components:

Compressor: 2-stage liquid-ring compressor

ELMO-F 2BG, Siemens AG

Pressure ratio (max.): $Pc/Pe = 16$

Intake volume rate: $\dot{V} = 120\ m^3/h$ (n = 2920 rpm, $Pc/Pe = 6$)

The compressor corresponds to the standard design of the product line except that bearing ring seals were installed for the closed refrigerant circulation system instead of stuffing boxes. No special modifications were necessary for steam compression. A standard thermal oil was used for the working liquid. Normal water was used as a refrigerant.

Drive: the compressor is driven by a directly coupled d.c. electric motor whose rotational speed is adjustable. The drive motor was specially selected for the test design to allow the easy setting of different working points. In principle, every motor which can be used for normal heat pump compressors is suitable as a drive unit.

Oil-vapor separator: the oil-vapour mixture discharged from the compressor is guided from the top centrally into a vertical container and reversed at the bottom. The oil is then separated at several strainers, and the vapor above these strainers is again radially expelled from the container. The separated oil is then collected on the bottom and fed via an oil cooler to the low and high pressure sides of the compressor again.

Condenser: the condenser consists of a standard horizontal tubular heat exchanger which is followed by a sub cooler.

Evaporator: the water is fed to the evaporator, i.e., a horizontal tubular heat exchanger, from below and evaporated in the chamber around the tube bundle. Residual oil is deposited on the surface, drawn off by an oil pump via a connecting sleeve and fed to the compressor. The refrigerant vapor is drawn off from above via a manifold.

Expansion valve: an electrically driven valve is used as expansion valve. It is controlled via a level switch. The level is set so that the tube bundle in the evaporator is always covered by refrigerant.

Hence the design of this high temperature heat pump with a liquid-ring compressor and water as the refrigerant corresponds in principle to the usual design of a heat pump circuit.

Fig. 3 also displays the arrangement of the measuring points. The values for pressure, temperature, flow and power are measured via pick-ups and correspondingly processed by a computer.

In order to avoid heat loss, the installation was insulated with 5 cm thick rock wool. Fig. 4 gives a view of the installation (which has not yet been insulated) at the erection site.

5. TESTING AND EVALUATION

The tests were carried out under conditions of steady-state operation. The evaporator inlet-temperature and the condenser output-temperature were set and held constant via the external water circulation systems of the supply unit. The heat pump operating data were determined via the measuring and processing system.

When calculating the operating data, the following definitions of efficiency were used:

- Volumetric efficiency (λ_{vol}): The volumetric efficiency is derived from the relationship between the actually drawn-in volumetric flow and the theoretical volumetric flow delivered. The theoretical volumetric flow delivered is calculated for the liquid-ring compressor from the geometry of the impeller and the rotational speed.
- Isothermic compressor efficiency (η_{iso}):

The isothermic compressor efficiency is the relationship between the theoretically required isothermic compression work for the actually drawn-in volumetric flow and the work actually measured at the shaft of the compressor.

COP:

The COP is the relationship between the sum of the heat flows rejected at heat sink temperature (condenser and oilcooler) and the shaft power of the compressor.

Exergetic efficiency (ζ_{ci}):

The exergetic efficiency is the relationship between the measured COP and the Carnot COP (COP_c). The Carnot COP is determined from the relationship between the saturation temperature corresponding to the discharge pressure of the compressor and the difference of the saturation temperatures corresponding to the suction and discharge pressures at the compressor inlet and outlet.

6. TEST RESULTS AND DISCUSSION

The liquid-ring compressor of the high temperature heat pump deals with steam compression without any problem. The thermal oil used as the working liquid is easily separated in the oil separator and fed back into the compressor. At the operating temperature, there is more or less no mixing of the oil with water so that no problems arise for the pump from water evaporating when the oil is being fed back. The compressor operates very smoothly and quietly. The facility is easily switched started up and shut down.

The initial measurement results show that the compressor is well suited for the intended temperature range of over 120°C because of its high pressure ratio. Fig. 5 shows the COP for various condensation temperatures plotted against the evaporation temperature. The highest temperature lift with 80 K was attained with a evaporation temperature of 80°C and a condensation temperature of 160°C with a COP of 1.6. This corresponds to a pressure ratio of Pc/Pe = 12. Given this compressor, larger pressure ratios are possible; however, the COP would then be lower. At evaporation temperatures of 70 - 80°C and condensation temperatures of 120°C, COP up to 4.4 were attained. All in all, the COP are below the expected values as the comparison with the theoretically calculated values in Fig. 5 shows. One cause of this is that, the compressor itself was not insulated at first although the installation was for technical reasons; hence large losses arose due to heat radiation. Given the small compressor used here, this is especially conducive to a relatively lower performance in relation to volume. An additional reason is also that the compressor's operating mode was not optimized in regard to oil cooling, and the volumetric flow of the oil fed back.

The expected high discharge temperatures during steam compression pose no problem for the compressor since a sufficient amount of heat can be dissipated to the liquid-ring during compression.

In the first tests, the refrigerant temperature at the compressor outlet was adjusted via the oil cooler so as to be only slightly higher than the condensation temperature to prevent steam from condensing in the oil (Fig. 6). Due to a good heat exchange during the compression, the outlet temperature of the working fluid is the same like the outlet temperature of the refrigerant. Therefore, the temperature level of the heat rejected in the oil cooler is comparable to the level of the steam condensing temperature.

The volumetric efficiency of the compressor was also calculated from the experimental values. Since, in the case of liquid-ring compressors, the volumetric efficiency refers to the overall cylindrical volume formed by the impellers, the values are comparatively small on the whole. Fig. 7 shows a strong reduction of the volumetric efficiency and hence a corresponding reduction of heat output under conditions of greater pressure ratios. The exergetic efficiency of the heat pump was plotted in Fig. 8 together with the isothermic compressor efficiency. With the compressor used here, optimum efficiency is attained at a pressure ratio in the vicinity of Pc/Pe = 5. At this pressure ratio, an exergetic efficiency of 45 % is evident.

7. CONCLUDING REMARKS

The initial results with the high temperature heat pump show that the projected condensation temperatures of over 120°C can be attained with no problem when using a liquid-ring compressor with water as the refrigerant. The high pressure ratio of the compressor allows the installation to attain a very large area of application.

The thermal oil used proved itself to be successful as the working liquid. Excessive steam temperatures can be avoided during compression by cooling the oil.

Up to now, an operating time of over 150 hours has been reached without any problems for the compressor and the working fluid. Long term tests are planned concerning the lifetime of the compressor and the constancy of the oil.

On the basis of these results which are encouraging for a start, the compression process will be optimized in further tests to improve the efficiency of the liquid-ring compressor and hence the COP of the high temperature heat pump.

8. REFERENCES

/1/ Taylor, B.J., Bertinat, M.P.: "The design and evaluation of a 150°C heat pump". The large scale application of heat pumps, BHRA York, England: 25. - 27. Sept. 1984, S. 155 - 166

/2/ Yamazahi, T., Kubo, Y.: "Development of a high temperature heat pump". Newsletter of the IEA Heat Pump Center, Vol. 3 No. 4, Dec. 85, S. 18 - 21

/3/ Degueurce, B., Banquet, F., Denisart, J.-P., Favrat, D.: "Use of twin screw compressor for steam compression". The large scale application of heat pumps, BHRA York, England: 25. - 27. Sept 1984, S. 189 - 196

/4/ Strop, H., Hoedt, B.: "Elmo-F Vakuumpumpe als Hochtemperaturwärmepumpe", Schlußbericht vom 1.11.1984, Siemens AG, ZBA Berlin

/5/ Betriebsanleitung ELMO-Kompressoren ZBG, NMA 3301 DE, Siemens AG, NMA Nürnberg

/6/ Elmo-F Verdichter Katalog (EPK3E/KG-P25-K), Siemens AG, NMA Nürnberg, 1979

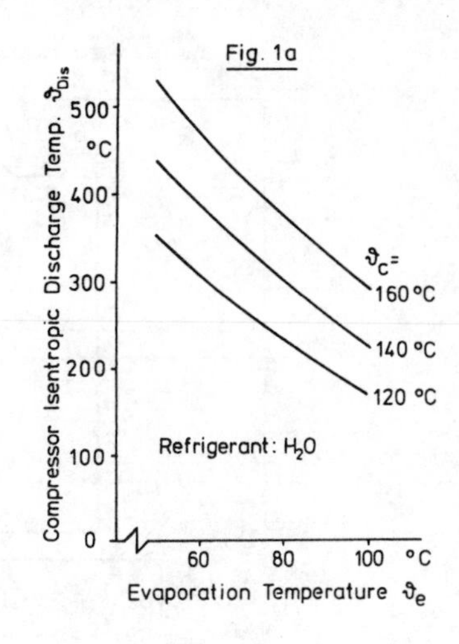

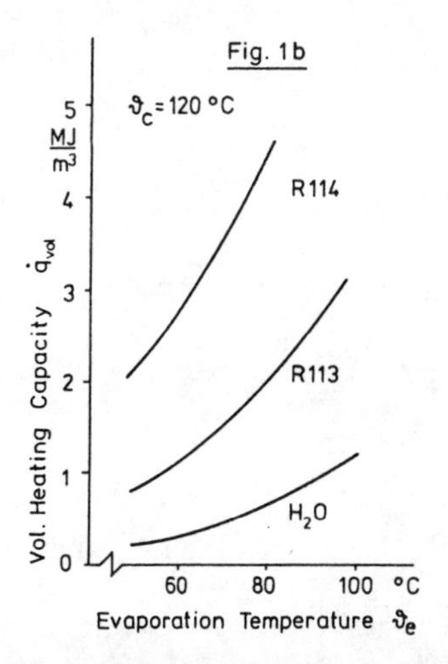

1 Compressor isentropic discharge temperature and volumetric heating capacity for water as the refrigerant

a) Mode of operation:

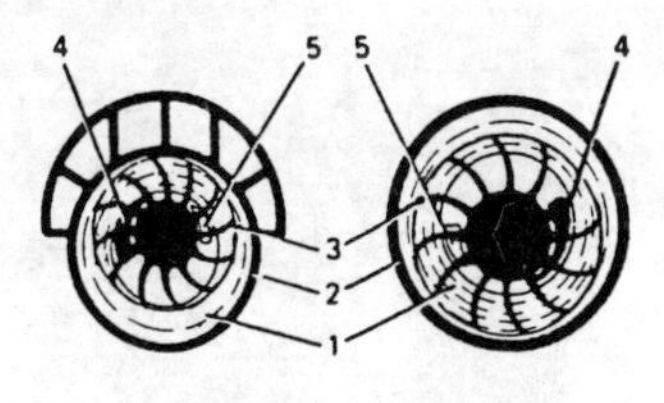

b) Basic construction:

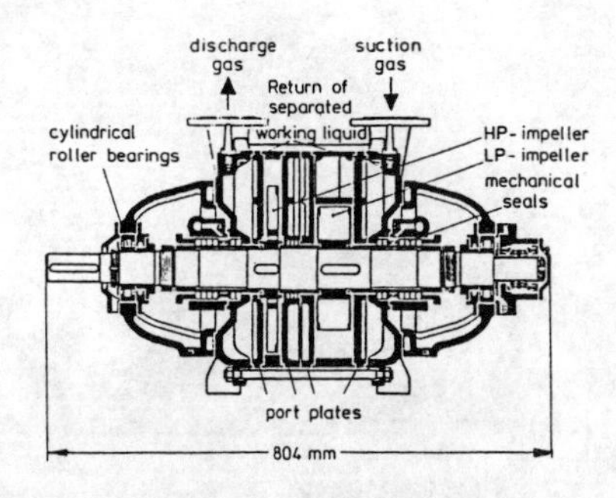

2 Mode of operation and basic construction of the double-stage liquid-ring compressor

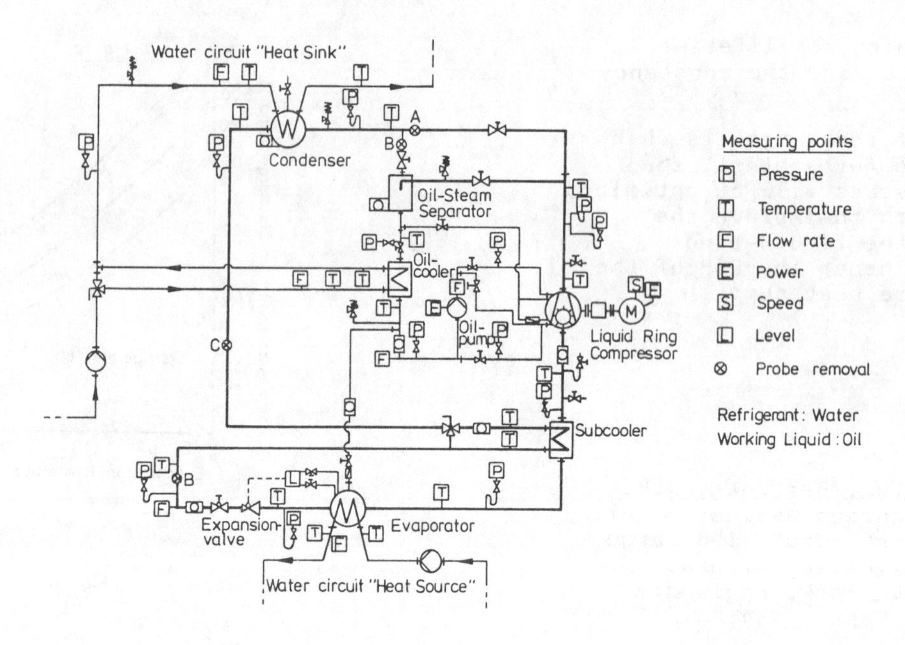

3 Schematic of the high temperature heat pump test facility with liquid-ring compressor

4 View of the high temperature heat pump test installation at the erection site (not insulated)

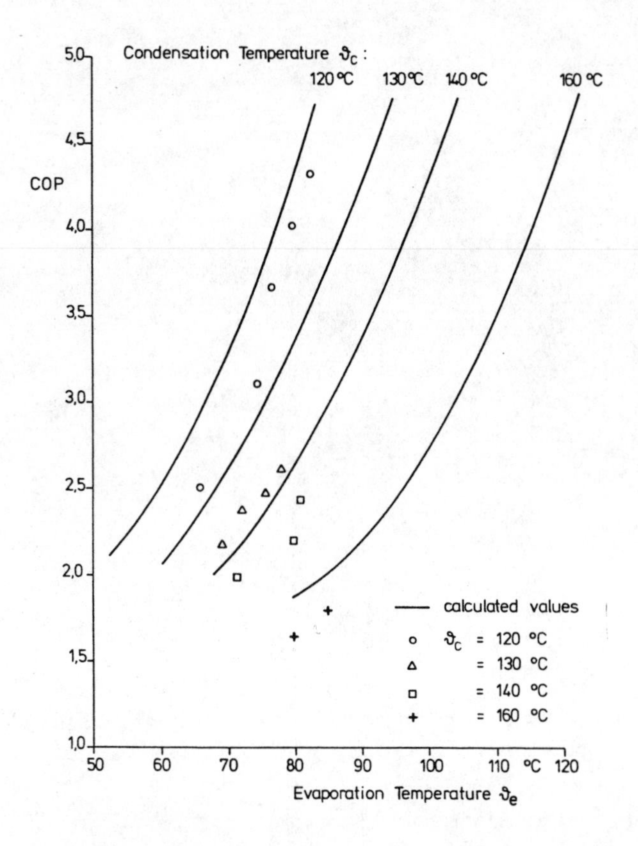

5 COP against evaporation temperature for various condensing temperatures

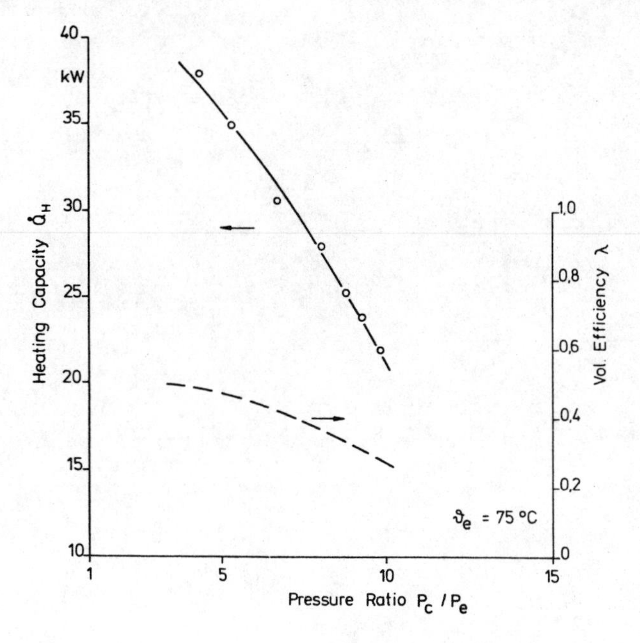

7 Heating capacity and volumetric efficiency against pressure ratio at const. evaporation temperature

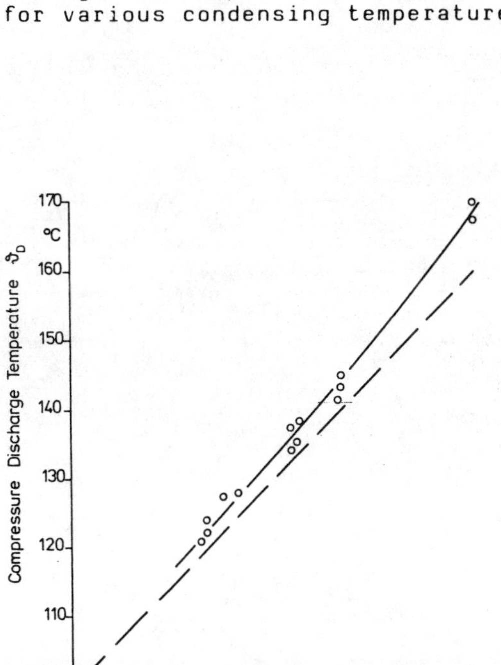

6 Compressor discharge temperature against condensation temperature

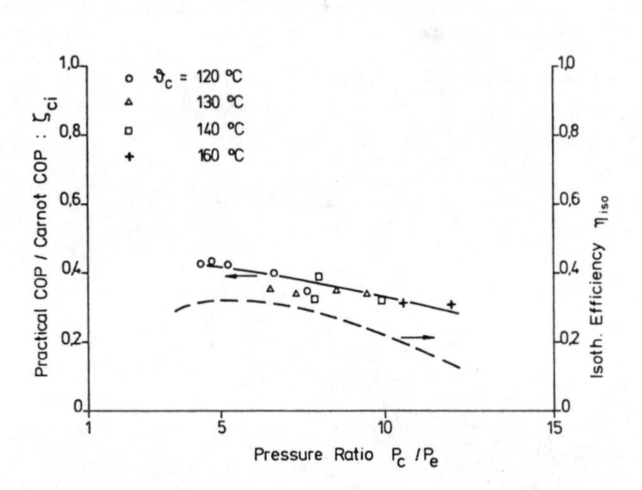

8 Exergetic and isothermic efficiency against pressure ratio for various condensing temperatures

3rd International Symposium on the

Large Scale Applications of Heat Pumps

Oxford, England : 25-27 March 1987

PAPER E2

A LINEAR ELECTRODYNAMIC COMPRESSOR FOR OIL FREE HEAT PUMPS.

P.B.Bailey,
Cryogenics Group,
Department of Engineering Science,
Oxford University.

SYNOPSIS.

A linear compressor gives a direct mechanical link from an electric drive to a reciprocating piston without the need for any intermediate rotating components. Transverse forces between the piston and the cylinder are minimal, allowing oil-free operation, and the piston stroke can be easily altered, giving continuous capacity modulation. This paper looks at previous linear compressors, and compares the various types of electrical drive. Several design problems are discussed, notably the control of the piston stroke, heat dissipation from the drive coils and the piston mounting arrangement. The prospects for linear compressors are discussed with respect to domestic applications, oil-free refrigerators and high temperature heat pumps. This work is a development of the successful Stirling cycle miniature refrigerator that has been developed by the Cryogenics Group at Oxford for aerospace applications.

NOTATION.

B = gap flux
c = equivalent damping rate
F = drive force
i = current
k = equivalent spring rate
L = coil inductance
l_e = effective wire length
m = moving mass
R = coil resistance
R_m = effective motional resistance
t = time
V = coil voltage
X_m = effective motional reactance
z = displacement
η_{em} = electromagnetic efficiency
ω = frequency (rad/s)

INTRODUCTION.

The conventional reciprocating compressor dominates the market for refrigerating machinery. Millions are produced, in sizes ranging from 50W to 500kW refrigeration capacity, and can be regarded as being close to the limit of development given the constraints of materials and cost. Despite this, the conventional reciprocating compressor is far too complex. A reciprocating piston is driven, via a rotating shaft, by a device whose action is invariably linear, such as an electric motor or an internal combustion engine. Any change of direction of motion in such a device will require some sort of rotary bearing - a precision assembly that usually requires lubrication to reduce friction to an acceptable level.

As well as the losses involved in any bearing, the need for lubricating oil has several detrimental effects on the performance of refrigeration plant. These are fully documented elsewhere [1], but the main effects can be summarised as :-

a). Limitation of maximum discharge temperature due to the catalytic decomposition of refrigerants in the presence of oil.
b). Increase in system complexity and pipework frictional losses to ensure adequate oil return to the compressor.
c). Reduction of heat transfer coefficient in the heat exchangers.

A purely linear (or "free piston") compressor offers solutions to these problems. The only bearing serves to locate the piston within the cylinder, and carries virtually no load, hence has little requirement for lubrication. An oil free compressor would reduce frictional losses, simplify heat exchanger design and improve heat transfer coefficients. Without oil, the halocarbon refrigerants can be used at much higher temperatures than at present (R114 has been successfully used at temperatures up to 250°C [2]).

Such a temperature range would enable heat pumps to be used for high temperature heating and steam raising, though there still remains a temperature limitation on the insulation of the windings. In a domestic heat pump, higher temperature operation could lead to the development of a unit that is directly compatible with existing hot water central heating systems (available heat pumps are limited to a flow temperature of about 55°C).

An oil free compressor has effectively only one moving part (excluding the valves), and should therefore be inherently more reliable and cheaper to manufacture than a conventional design. A further advantage is the potential for capacity modulation by changing the amplitude of the piston stroke, thereby reducing the losses caused by on/off cycling.

Held at St. Catherine's College Oxford, England. Organised and sponsored by BHRA, The Fluid Engineering Centre, Cranfield, Bedfordshire, MK43 0AJ England.

THE DEVELOPMENT OF FREE PISTON MACHINES.

The first spark ignition (petrol) engine was built by Count Otto in Germany in 1863. A piston was propelled upwards by an explosion in a combustion chamber. On falling back down the piston engaged a rack and pinion mechanism to produce rotary motion. Though not strictly a free piston machine, the variable stroke and absence of a connecting rod are significant differences from the conventional reciprocating machine.

The next significant development was the free piston gasifier, developed in the 1940's. This consisted of two opposed pistons within a common cylinder. Operating on the diesel cycle, the pistons were driven apart in a central cylinder by the combustion of fuel, compressing the exhaust gas in the outer cylinders. This compressed gas was then used to power a gas turbine.

From the 1950's onwards, various groups have experimented with electrically driven linear compressors in the 20W to 200W range [3,4]. These machines consisted of an axially moving shaft, mounted to the casing of the compressor on a pair of helical compression springs. One end of the shaft formed the piston, often with an integral suction valve, and the other end supporting the moving coil. Excitation was by permanent magnet, and the power supply was low voltage single phase a.c., typically 30 volts, 50 Hz. The furthest development of these was a machine designed by Doelz in Germany, which was commercially produced under licence in Japan. This was a single cylinder compressor with a power input of about 50W [4].

One of the major centres of interest in linear compressors has been Purdue University in the U.S.A., where a considerable amount of work has been done since the late 1960's. Most of the work has been on smaller machines, but they have also developed one of the largest and most advanced linear compressors ever made [6]. This was a double acting induction machine of about 10.5kW cooling capacity, unusual in that it oscillated along a horizontal, rather than a vertical axis. It featured piston stroke control by varying the pressure in gas springs, and the suspension of the compressor body by mounting it on the spiral suction and discharge tubes.

Linear Stirling engines have received much attention with a variety of generators and water pumps. In the U.S.A. Benson [7] has proposed a whole range of linear direct fired heat pumps using a variety of cycles. Also in the U.S.A. the Gas Research Institute has developed a Stirling-Rankine direct fired heat pump to an advanced state [8].

MINIATURE REFRIGERATORS.

The Cryogenics group at Oxford has developed a miniature refrigerator for cooling infra-red sensors in aerospace applications [9]. These machines give approximately 1W of cooling at 80K with a power input of about 30W, and work on the Stirling cycle with helium at 10 bar as the working fluid. The displacer and the power piston are independently driven by linear electrodynamic motors (similar to a moving coil loudspeaker). The magnetic field in the motors is produced by a permanent magnet made from samarium cobalt.

Each motor is single acting, and the spindle is supported at each end by a set of spiral disc springs, which flex easily in the axial direction, but are very stiff radially. Using this means of suspension it is possible to maintain a small clearance between the piston and the cylinder.

These refrigerators are designed to have a maintenance-free life of 10 years and to withstand accelerations of 100 m/s^2 (10g) in any direction.

The oil free heat pump compressor is a development of this miniature refrigerator, using the same type of drive motor and suspension, and operating in a conventional reversed Rankine cycle heat pump.

HEAT PUMP COMPRESSOR DESIGN.

For design considerations, the compressor was envisaged as part of a simple air to water heat pump with R12 as the working fluid, evaporating at $-5^{o}C$ (2.6 bar absolute) and condensing at $60^{o}C$ (15.2 bar absolute). The compressor was sized to give approximately 1kW of cooling with a refrigerant mass flow rate of about 0.01 kg/s. At the U.K. supply frequency of 50 Hz., the desired throughput can be achieved with a double acting machine of 30mm. bore and a stroke of ±10 mm. At 60% isentropic efficiency of compression, this requires a mechanical power input of 600W, and a corresponding electrical input of about 1kW.

The first choice faced by a designer is the type of drive; whether electrodynamic, electromagnetic or induction.

An electrodynamic, or moving coil, drive consists of a moving coil suspended in a static magnetic field, which can be produced by either d.c. or permanent magnet excitation. An a.c. current is fed through the coil, which moves due to the force acting on the current carrying conductor in the magnetic field. This type of drive is commonly used in loudspeakers. The efficiency of such a drive depends on the maximisation of the magnetic flux in the air gap, the tight packing of the conductors in the gap, and the minimisation of the copper resistive loss. The size of such a drive is limited by the necessity to cool the coil, which is usually wound on a light former. The design is complicated by the need to feed the current to and from the moving coil, and by the small clearances between the coil and the pole pieces which are necessary to maximise the gap flux.

The second type is the electromagnetic, or moving iron drive. A coil is wound on a soft iron stator, the magnetic circuit being completed by a moving iron armature which is spring mounted. When a current is passed through the coil the armature will move so as to reduce the length of the air gap. An a.c. current in the coil will thus cause an oscillation of the armature at double the excitation frequency. An electromagnetic drive is very simple both in operation and construction, but is relatively inefficient in that it has a large and variable air gap.

The third type, the linear induction motor, operates on a similar principle to an electrodynamic drive, but it differs in that the current in the moving coil is induced by a surrounding a.c. coil in the stator. It has the advantages of the efficiency of the electrodynamic drive without the mechanical complexity of supplying current to the moving component. The nature of the losses in the induction machine lend it to larger sizes, and like an electrodynamic drive it can have either permanent magnet or d.c. excitation.

Given the experience within the group of electrodynamic drives, and their efficiency, this was chosen for the first prototype heat pump compressor. Although the power requirement is close to the limit for the type of drive, the problems of cooling the coil can be overcome by carefully designed suction gas cooling.

The drive coil and magnetic circuit were sized using the procedure described by Cadman and Cohen [3]. The load diagram for the compressor is calculated from the design evaporating and condensing pressures and the stroke length (fig. 1). This curve is then subjected to a first order Fourier analysis to give an equivalent sinusoidal load of the same stroke length, and of equal area (the smooth curve in fig. 1). The compressor can now be modelled as a mass, spring and damper system (fig. 2), where the spring stiffness and damping coefficients are the constants obtained from the Fourier analysis.

The mechanical and electrical equations for the compressor can now be written as :-

$$V = Bl_e \frac{dz}{dt} + Ri + L \frac{di}{dt} \quad (1)$$

$$m \frac{d^2z}{dt^2} + c \frac{dz}{dt} + kz = F = Bl_e i \quad (2)$$

where B = gap flux
- c = equivalent damping rate
- F = drive force
- i = current
- k = equivalent spring rate
- L = coil inductance
- l_e = effective wire length (i.e. length of wire in the air gap)
- m = moving mass
- R = coil resistance
- t = time
- V = coil voltage
- z = displacement

These two equations can be combined to give an effective steady state impedance for the compressor, given by : -

$$Z_e = R + j\omega L + (Bl_e)^2 \left\{ \frac{c\omega^2 + j\omega(k - m\omega^2)}{(k - m\omega^2)^2 + (c\omega)^2} \right\} \quad (3)$$

This can then be re-written as :-

$$Z = R + j\omega L + R_m + j\omega X_m \quad (4)$$

where R_m = effective motional resistance
X_m = effective motional reactance

At the mechanical resonance of the system X_m becomes zero, and R_m simplifies to

$$R_m = (Bl_e)^2/c \quad (5)$$

The electromagnetic efficiency, η_{em}, can be defined as the power used to drive the load divided by the total power input, i.e.

$$\eta_{em} = R_m/(R+R_m) \quad (6)$$

The value of R, the coil resistance, will depend on the total length of wire in the coil, and its cross-section. Thus maximum efficiency will be obtained when the ratio R/R_m is minimised. In simple terms this can be interpreted as getting as much copper and magnetic flux in the air gap as space allows.

The value of R/R_m depends on factors such as the wire packing factor, the gap fringing factor, the stroke length, the gap dimensions, and many other factors. A computer program was written to evaluate these factors, and thus calculate the efficiency of any given design.

PROTOTYPE DESCRIPTION.

The prototype is a double ended machine, with a single coil and magnetic circuit, and pistons mounted at either end of a common shaft (fig. 3). The machine is mounted with its axis vertical.

The magnetic circuit is excited by a barium ferrite ring magnet of dimensions 213 mm o.d. × 111 mm i.d. × 25.4 mm thickness. The circuit is completed by a soft iron core and pole piece both made of high permeability, low carbon iron. The resulting air gap is an annulus of dimensions 111 mm o.d. × 99 mm i.d. × 25 mm axial length. This gives a gap length of 6mm over an area of 8250 mm^2, with a design gap flux of 0.42 Tesla. This magnetic circuit is not optimal for the driver, but is based on the most suitable commercially available magnet.

The coil is held in a cup-shaped former made of non-magnetic stainless steel to AISI 303. The coil itself consists of 8 layers of polyurethane insulated copper wire of 0.5mm diameter potted in Sty-cast resin, giving a design total of 488 turns, a cold resistance of 13.1 ohms, and current of 6.0 amps at 211 volts r.m.s.

The coil holder is supported on a stainless steel shaft mounted on disc springs. Each spring has three spiral arms, shaped to optimise the bending stress, and is photo-etched from flat beryllium copper sheet 0.3 mm thick. The springs are arranged in two sets of six, the springs separated from each other by 0.3 mm thick brass spacers. The springs have an individual resonant frequency of 37 Hz.

The magnetic circuit and springs are mounted inside an aluminium casing, onto which are bolted the cylinder assemblies. The cylinders and cylinder heads are made of aluminium, and the valve plate is brass, with proprietary valve reeds. The cylinder is fitted with a liner made from a sintered mica filled PTFE.

The live electrical supply to the coil is taken through the upper set of mounting springs, which is insulated at both points of support. The neutral return from the coil is taken through the lower set of springs, which is earthed to the compressor housing.

In order to study the off-design performance of the compressor, especially when looking at the resonant frequency of the mounting springs, it is necessary to have a variable frequency power supply, which is obtained from a motor-alternator set. Rectified a.c. mains is supplied to a shunt-wound d.c. motor, the speed of which is controlled by changing the field and/or armature voltages. The d.c. motor is close-coupled to a three phase alternator, one phase of which is used to drive the compressor. The alternator voltage is controlled by varying the alternator field, which is d.c. excited. One of the supply terminals on the generator is connected to the building earth, and this is used as the a.c. neutral.

EXPERIMENTAL PROGRAMME.

At the time of writing manufacture of the compressor components is almost completed. The final assembly is awaiting the results of tests on heat transfer from the drive coil. Once assembled, the first phase of testing will be to confirm the design efficiency of the drive, and to check that the drive coil can be cooled sufficiently by the suction gas.

The second phase of testing will be to evaluate the performance of the compressor under off design conditions, and at varying frequencies and voltages. Two aspects of the programme are to see how the resonant frequency of the whole system varies with changing evaporating and condensing pressures, and to look at the efficiency under partial stroke operation.

FUTURE DEVELOPMENT.

The aim of the present work is to scale up the original miniature refrigerator to the size of a heat pump compressor. One of the earliest conclusions reached was that this design would be far too expensive in comparison with conventional reciprocating machines. Three particular aspects of the design are prohibitive : - the permanent magnet, the spring suspension and the fine tolerances and alignment required for the moving coil drive. In addition, the moving coil drive is close to its limits for dissipating resistive heat at an acceptable temperature.

For the next stage of the process it will be necessary to review the types of drive available, and to ascertain the most suitable for the particular application. At present the induction drive appears promising, and seems to lend itself to easy cooling, and simple and robust construction.

A further problem that must be tackled is the control of the piston stroke. Whatever control strategy is used must ensure that the piston does not strike the valve plate at the end of each stroke. Testing of the first prototype will indicate the degree of self regulation of the stroke. If external control is required, it can be provided either by a gas spring, or by an electronic method that will vary the supply current. The latter lends itself readily to capacity control, and need not be prohibitively expensive, given the trend towards microprocessor controlled heating systems.

CONCLUSIONS.

The linear compressor has potential advantages over the conventional compressor in temperature capability, system complexity, first cost and efficiency. Successful development could lead to a cheap high temperature heat pump suitable for steam raising and high temperature hot water heating systems.

The linear induction type of drive appears most suitable for a heat pump compressor, and further development is needed to develop this type of drive for compressor applications.

REFERENCES.

1). McMullen, J.T., Hughes, D.W., Morgan, R., "The Influence of Lubricating Oil on Heat Pump Performance", Proc. E.E.C. Contractors Meeting on Heat Pumps, Brussels, 28th-29th April and 12th-13th May 1982, pp 24-39.

2). Strong, D.T.G., "Development of a Directly Fired Domestic Heat Pump", D. Phil. thesis, Oxford University, 1980.

3). Cadman, R.V., Cohen, R., "Electrodynamic Oscillating Compressors", Trans. A.S.M.E., Jnl. of Basic Engineering, Dec. 1969, pp. 656-670.

4). Nagaoka, J., "Regarding the Performance of Electro-dynamical Oscillating Compressors", Proc. 10th Int. Cong. of Refrigeration, Copenhagen, 1959, Vol 2, pp. 90-95.

5). Anon., "An Experimental Oscillating Compressor", Jnl. of Refrigeration, Vol. 7, No. 2, March/April 1964, p. 33.

6). Curwen, P.W., Liles, A.W., "Development of a Free Piston Variable Stroke Compressor for Load Following Electric Heat Pumps", Proc. German/American Conf. on Technology and Applications : The Electric Heat Pump, Dusseldorf, June 18th-20th 1980.

7). Benson, G.M., "Free Piston Heat Pumps", Proc 12th I.E.C.E.C., Washington D.C., vol. 1, paper no. 779068, pp. 416-425, 1977.

8). Meier, C., "Development and Demonstration of a Stirling/Rankine Heat Activated Heat Pump", Proc. 9th Energy Technology Conf., Washington D.C., pp. 413-423.

9). Orlowska, A.H., "An Investigation of some Heat Transfer and Gas Flow Problems Relevant to Miniature Refrigerators", D. Phil. thesis, Oxford University, 1985.

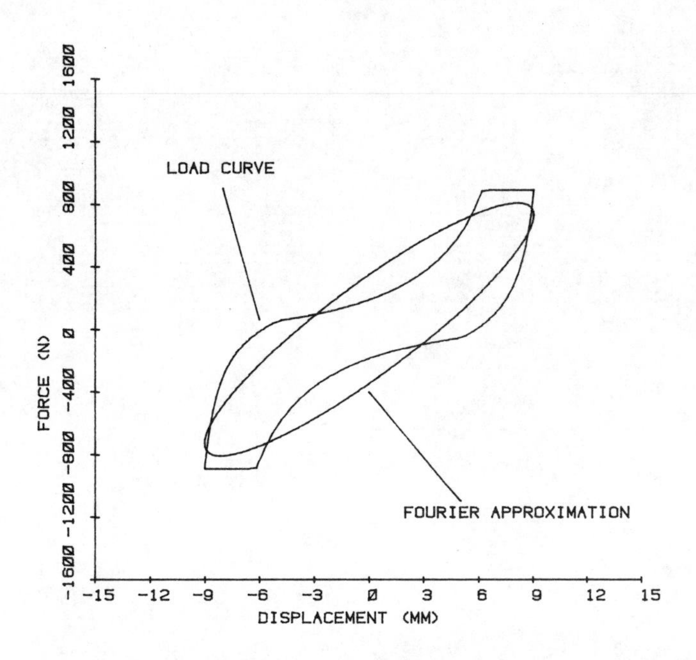

Fig. 1. COMPRESSOR LOAD DIAGRAM.

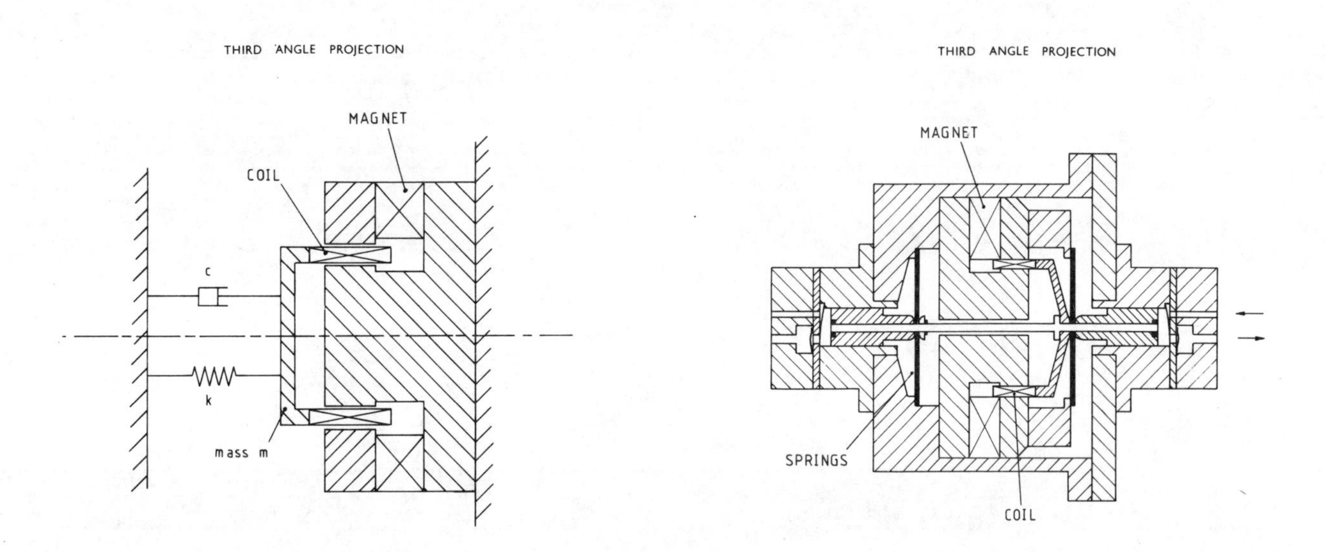

Fig. 2. MASS-SPRING-DAMPER MODEL OF COMPRESSOR.

Fig. 3. CROSS-SECTION OF PROTOTYPE COMPRESSOR.

3rd International Symposium on the

Large Scale Applications of Heat Pumps

Oxford, England : 25-27 March 1987

PAPER E3

DEVELOPMENT AND TEST OF A HIGH SPEED CENTRIFUGAL COMPRESSOR FOR MECHANICAL VAPOUR COMPRESSION

F. BANQUET
ELECTRICITE DE FRANCE - Direction des Etudes et Recherches - FRANCE

J.M. MERIGOUX
ALSTHOM - RATEAU
FRANCE

Synopsis

The paper describes the development of a prototype manufactured by ASTHOM-RATEAU and tested by ELECTRICITE DE FRANCE on a steam compressor rig.

The aim of the program is to develop a two-stage centrifugal vapour compressor with high pressure ratio and low capacity for extending the uses of centrifugal type compressors.

The design of the high speed (50 000 RPM) compressor will be described : impeller, bearings, gear box, inter stage desuperheating ... Tests made and performances measured will be explained.

Potential uses in industrial processes of mechanical vapour compression will be discussed.

1 - INTRODUCTION

Because of its non toxicity and its availability, steam is frequently used as an energy carrier in industrial processes. Steam is also the by-product of energy-hungry processes such as drying, concentration, evaporation and distillation. Although the upgrading concept of open cycle heat pump, commonly called mechanical vapour compression (MVC), has been known and applied from the beginning of the century, low pressure steam from many industrial processes has traditionally been vented away with its valuable latent heat content.

As part of its efforts to promote a rational use of electricity and the substitution of imported fossil fuels, ELECTRICITE DE FRANCE has been involved for many years in a significant research program to stimulate the development of mechanical vapour compression. Part of this research has been focussed on the key component, the steam compressor, and several prototypes of different technologies have been tested on EDF steam facilities described in previous papers (Ref. 1, 2, 3).

Tests have been concentrated on the low capacity range of volume flow up to 3500 m^3/h, where very few compressor alternatives, if any, were available at the beginning of our research program. In this capacity range and according to the potential need for compression ratios of 3 and above, depending on the application several promising alternatives were recognized at an early stage :

- rotary positive displacement compressors,
- high speed centrifugal compressor.

The main part of this paper describes the results obtained with a high speed centrifugal compressor manufactured by ALSTHOM-RATEAU and tested by ELECTRICITE DE FRANCE.

2 - VAPOUR COMPRESSOR

There are two main compressor families :

- volumetric compressors,
- dynamic rotaring compressors.

Volumetric compressors are mainly used for small flows, from a few m^3/h up to about 20 000 m^3/h. Dynamic rotating compressors can be decided in two main categories :

- axial compressors,
- centrifugal compressors.

Should we have to define in two sentences, the characteristics of these two kinds of machine, we could say that :

- with the same overall diameter, the axial compressor flow can be up to 6 times the centrifugal compressor flow,

- the pressure ratio of a single stage centrifugal compressor may equal the pressure ratio of a six-stage axial compressor.

The applicational field of axial compressors is hence high flows, usually above 200,000 m^3/hr. The applicational field of centrifugal compressors covers flows from 2500 to 200,000 m^3/hr, corresponding to most mechanical vapour compression needs ; the pressure ratio capability being 2 to 2.5 per stage (corresponding to a differential temperature of saturated steam of about 20°C), with a polytropic efficiency close to 0.8.

A single stage centrifugal compressor is able to reach, with steam a pressure ratio of 3 but with very high tip speed (600 m/s ; see ref. 6). Usually, centrifugal compressor pressure ratio is limited to around 2 per stage with standard design and materials.

In the range of centrifugal steam compressors developped by ALSTHOM-RATEAU, we selected for testing the lowest flow model (3000 m^3/h), which is

Held at St. Catherine's College Oxford, England. Organised and sponsored by BHRA, The Fluid Engineering Centre, Cranfield, Bedfordshire, MK43 0AJ England.

close to the lower limit of this type of turbomachine, with a high pressure ratio (4.5) obtained with two stages in series.

This machine is quite representative of the technical choices made by this manufacturer for his overall range, i.e :

- Compression stages fitted in an overhung position, at the ends of the gear box high speed shaft. In our case, the impellers are made of titanium with a tip speed of 500 m/s, due to the high pressure ratio per stage.
- Gear box with a single set of parallel shafts capable of high gear ratios (17 in this application). The gear box housing also forms the support for the compressor casings as well as the lubrication oil tank. The gear box may be fitted with a second high speed shaft that may receive either more compressor stages or a recovery turbine.
- High speed shaft with tilting pad bearings, to obtain good stability at high rotation speed (51 200 RPM for the machine tested).
- Reduction of the mechanical losses which, as we shall show during our tests, are of utmost importance (particularly in the small machines) compared to the aerodynamic power. To achieve this the high speed shaft axial thrust is transferred by means of collars to the low speed shaft.
- Flow control system adapted to circuit conditions.
  . variable frequency electric motor,
  . adjustable inlet guide vanes and non bladed diffuser,
  . adjustable diffuser blades.

  In our case, owing to the experimental character of the equipment, a diffuser with fixed blades was fitted.
- Inter stage desurperheating system by means of nozzles spraying water or condensate at the inlet of the second compressor stage. During our first tests the desuperheating devices were located in the inter stage duct near the first stage discharge flange.

3 - TEST FACILITIES

The facilities which are described in Ref. 2, allow conditions of testing very similar to the operating conditions of an industrial heat pump. They include a desuperheater, condenser, expansion valves, evaporator and superheater. The main difference being that the heat is supplied to the evaporator by direct contact of part of the superheated steam throttled at the outlet of the compressor. When ever possible, critical measurement data, such as mass flow, shaft torque, pressures and temperatures were duplicated.

Mass flow, pressure and temperature of the inter stage injection water were also measured.

Data acquisition and processing were computerized in agreement with Ref. 4.

4 - TESTS

The objectives of the tests were to verify the good mechanical strength of the shaft and the bearings, the efficiency of the seals and the inter stage desuperheater and to measure the performance within a range of parameters compatible with Mechanical Vapour Compression process.

The cumulative testing time for this compressor was 32 hours during which compressor speed was built up progressively to 48150 RPM.

The pressure ratio varied from high values corresponding to the proximity of the surge pressure ratio to low values corresponding to the limits of our electric motor.

The following definition have been used :

- Polytropic efficiency :
  The polytropic efficiency of a compression stage is the ratio of the theoretical polytropic compression power, to the power transmitted to the fluid.
- Global isentropic efficiency :
  The global isentropic efficiency is the ratio of the theoretical isentropic compression power calculated for the discharge flow of the compressor, to the shaft power.
- Coefficient of performance (COP) of the heat pump associated with the compressor : the practical coefficient of performance is defined as the ratio of the power that would be available at the condenser of the heat pump associated with the compressor to the compressor shaft power. The CARNOT theoretical coefficient of performance is the ratio of the absolute saturation temperature at discharge to the difference between saturation temperatures at discharge and intake.

Table 1 and Figures 3, 4, 5, 6 summarize the main results obtained by varying the pressure ratio at nearly constant speed of rotation (48 150 rpm). Pressure ratio and efficiencies are represented as a function of reduced flow which is equal to

$$\frac{Qma \quad Ta}{Pa}$$

where Qma is the suction mass flow (kg/s),
Ta the suction temperature (K)
and Pa the suction pressure (10^5 N m^{-2}).

These results are compared to those obtained with air by ALSTHOM-RATEAU and transposed to steam. The characteristic curve was not entirely described because shaft power was higher than that available from our electric motor.

The polytropic efficiencies of the first and second stages are satisfactory : 0.75 and 0.74. The global isentropic efficiency is 0.58. It could be improved if mechanical losses (bearings, gear box) were reduced. The practical COP is 4.60 and the ratio of this COP to the CARNOT theoretical COP is 0.49.

Steam (3140 m^3/h) arrives at compressor intake at a pressure of 104 KNm^{-2} and a temperature of 104°C, and is compressed adiabatically in the first stage up to 221 kNm^{-2} and 220°C. Then water is injected (116 kg/h) in the inter stage duct for desuperheating the steam. Only 25 % of the liquid injected has time to vaporize and cool the steam to 193°C in the inter stages duct.

In the second stage the rest of the liquid injected is vaporized and the steam is compressed to 419 kNm^{-2} and 238°C.

The global pressure ratio is 4.03 which corresponds to a saturated temperature difference of 45°C.

Observations during tests :

- the maximal tip speed reached was 470 m/s,
- the vibration level was acceptable,
- the leading edge of the second impeller could be eroded. as 75 % of the liquid injected as a mist of droplets passes across the second stage. But nothing was observed on the titanium impeller.

5 - INDUSTRIAL APPLICATIONS

Concentration by evaporation is the operation consisting in removing part of the water from the liquid to be concentrated, while bringing the liquid to its boiling point in an evaporator basically consisting of a heat exchange surface.

The liquid to be concentrated is brought to its boiling point by the heat released as latent condensation heat by the heating steam. In order for heat exchange to occur between the heating steam and the liquid to be concentrated, boiling temperature To of the liquid at evaporator pressure Po must be less than condensation temperature T_1 of the steam in the heater.

The concentrated liquid is separated from its steam in a separator and this steam is "enriched" by compression before being recycled in the evaporator circuit.

Consequently, the compressor makes it possible to recover the heat contained in the steam at the end of the cycle and to reuse it in the evaporation process.

The type of configuration influences the saturated steam temperature difference and compressor output :

- with a parallel configuration, the entire steam flow passes through the compressor and the saturated steam temperature difference of compressor is the same as this of each evaporator.

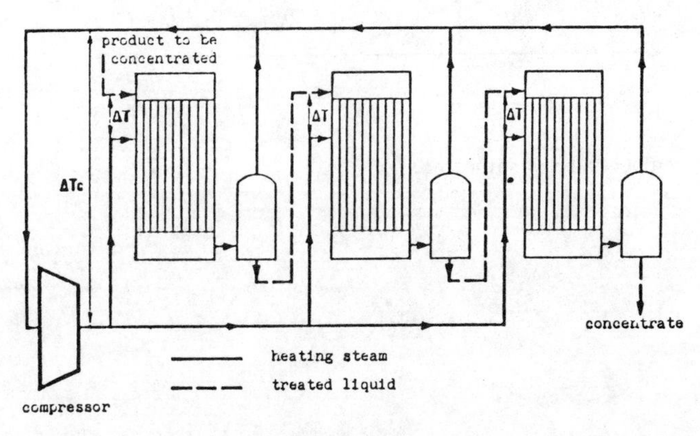

Fig. 1 : Parallel configuration of MVC

- with a three-stage series configuration, one third of the evaporated product flows through the compressor and the saturated steam temperature difference is three times as great as this of each evaporator.

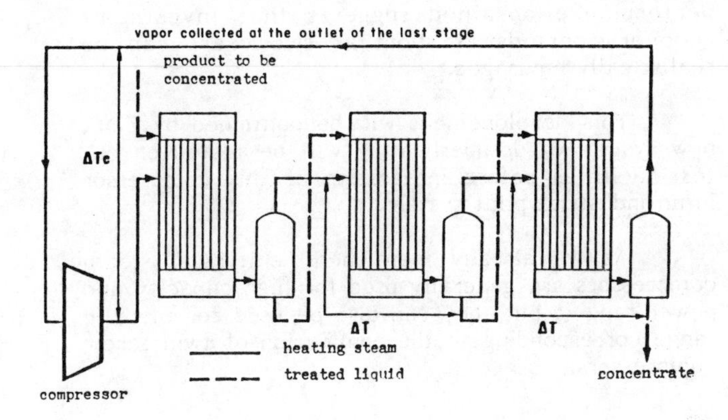

Fig. 2 : Series configuration of MVC

In this case industrial processes need compressors with high pressure ratio and medium capacity.

For an existing installation the interest of a higher saturated steam temperature difference is to increase capacity, and for a new installation, its interest is to reduce the heat exchange surface.

Operation other than concentration by evaporation, such as crystallizing and drying (Ref. 5), also require high pressure ratio : because saturated steam temperature difference is traditionaly high on drying operations and the vaporization interval for highly concentrated solutions may reach 20°C, and more for crystallizing operations.

So, analysing the MVC market, it is possible to associate the industrial sector concerned with the performance level needed :

- chemical and petrochemical industries for supplying intermediate pressure steam networks. In these cases, the required pressure ratios vary from 1.7 to 5.
- paper mills with pressure ratios varying widely from 1.4 to 18, with compressors from single stage to 5 stages (2 machines).
- distilleries using compressors
  - for steam , usually in 2 stages with pressure ratios between 3 and 4.
  - for ethanol vapour with 1 or 2 stages for pressure ratios between 2 and 3.
- desalinisation plants with pressure ratio about 1.7.
- salt works from 1.7 to 3.
- dairies with a saturated steam temperature increase of about 13°C corresponding to a pressure ratio of 1.8.
- sugar mills with pressure ratios between 1.5 and 2.
- effluent treatment, more particularly in distilleries and sugar plants with moderate pressure ratios (1.4 to 1.7) but with very agressive vapours that require the use of appropriate materials.

6 - CONCLUSIONS

These test results are intended to give the process designer some of the previously missing key information required for evaluation of mechanical vapour compression installations. The satisfactory performance obtained suggests that investment recovery periods as short as two years can be realistically envisaged.

This developement will be continued by EDF, new system of desuperheating will be tested on our test facilities before integration of this compressor in an industrial pilot system.

As we already mentioned centrifugal steam compressors are generally used for high capacity and power ranges, but they can also be used for medium range corresponding to the application of twin screw compressors.

REFERENCES

1 - DEGUEURCE B. AND TERSIGUEL C. : "Problems encountered and results obtained in the applications of compressors in industrial processes for energy savings". International Symposium on the Application of Heat Pumps, p. 179, BHRA, Warwick, 24-26 March 1982.

2 - DEGUEURCE, PASCAL, ZIMMERN : "Compression directe de vapeur d'eau par un compresseur Monovis" Int. Congress of Refrigeration, Venise, Sept. 1979, Proc. Vol. II.

3 - DEGUEURCE, BANQUET, DENISART, FAVRAT "Use of a twin screw compressor for steam compression" 2nd Internation Symposium on the large scale applications of heat pumps, York. England, 25-27 september 1984.

4 - Norme ISO/TC 118
Turbo compresseurs - Code d'essais des performances

5 - COSTA MISSIRIAN
La compression mécanique de vapeur transfert sectoriel et évolution technique - Rapport EDF

6 - J.J. TUZSON
High pressure ratio centrifugal compressor development and a vapor compression application in the dairy industry
2nd International symposium on the large scale applications of heat pumps, York 25-27 september 1984.

Speed (rpm)	Suction		Discharge		Pressure ratio	Suction flow (m^3/h)	Shaft power (kW)	Polytropic efficiency		Global isentropic efficiency
	Temp. (ºC)	pressure (KNm^{-2})	Temp. (ºC)	Pressure (KNm^{-2})				1rst St.	2nd St.	
47 618	104	108	229	452	4.20	2 984*	261	0.73*	0.85*	0.60
48 115	116	102	235	364	3.56	3 229	258	0.72	0.65	0.53
48 161	116	103	238	396	3.85	3 190	259	0.73	0.72	0.56
48 196	117	104	238	419	4.03	3 142	259	0.75	0.74	0.58

TABLE 1 : Test results

* Values less accurate because suction temperature is saturated.

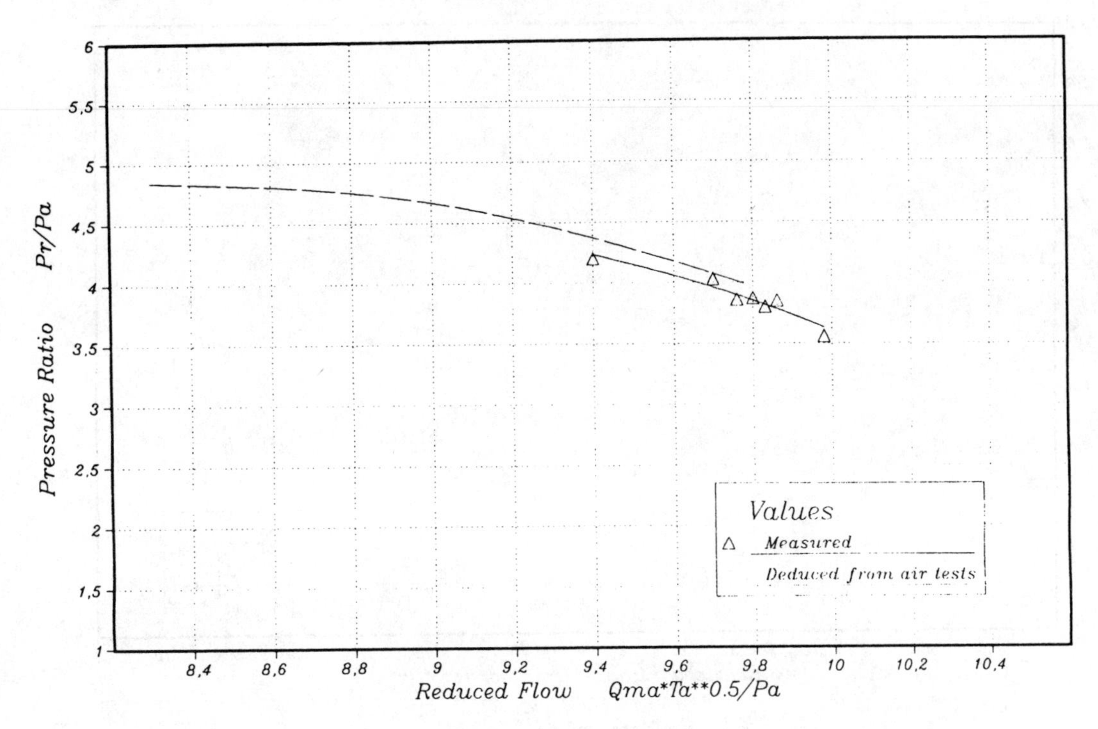

Fig. 3 : Test results

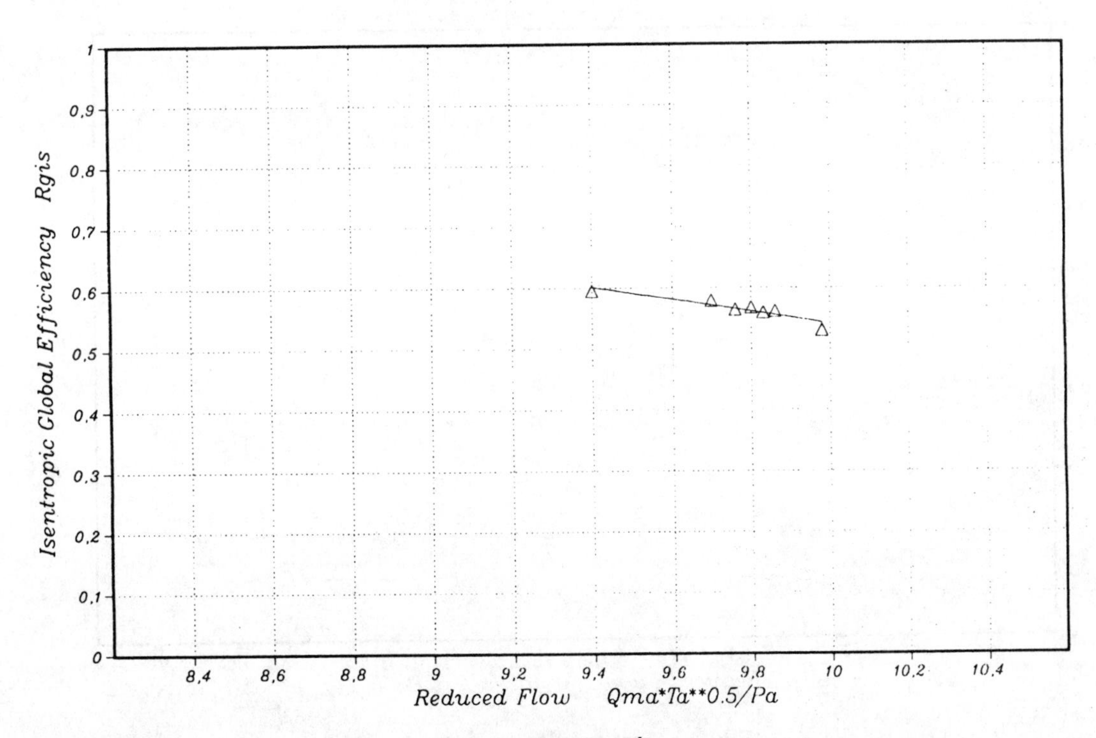

Fig. 4 : Test results

Stages 1 and 2

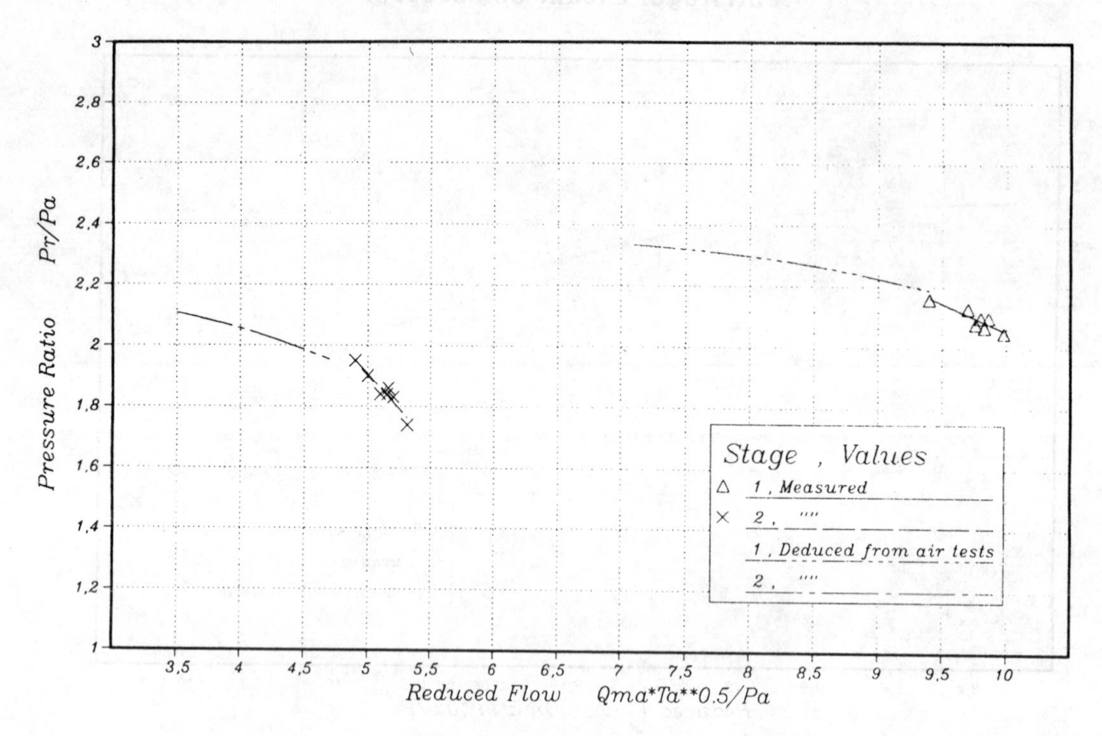

Fig. 5 : Test results

Stages 1 and 2

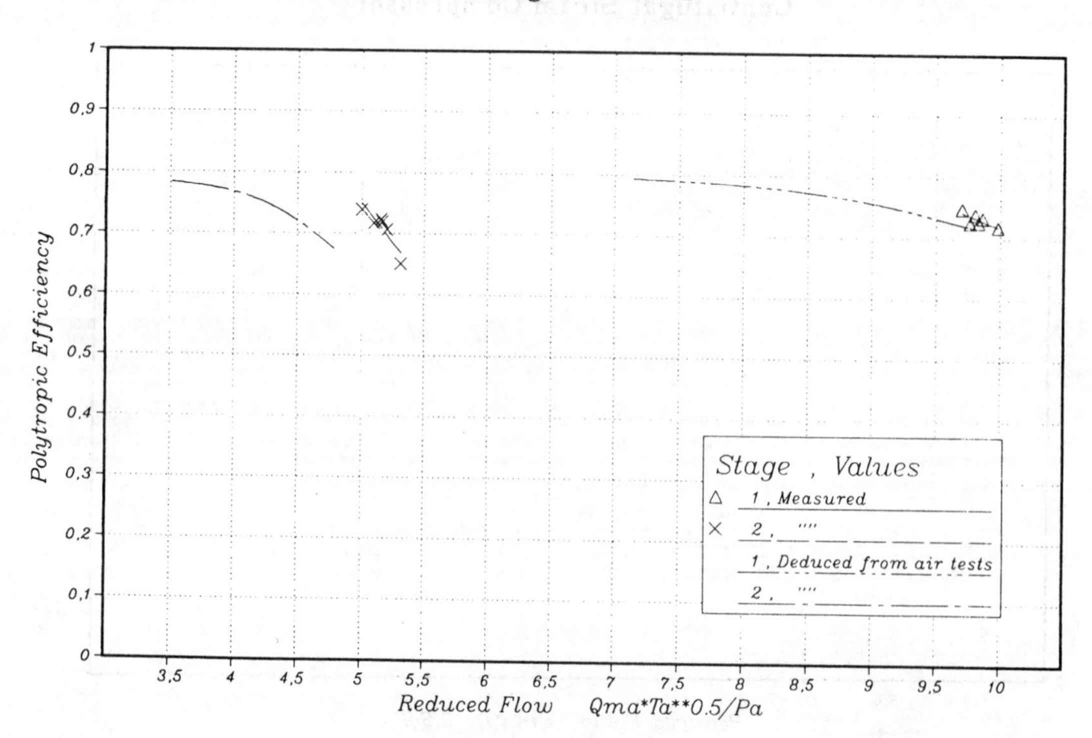

Fig. 6 : Test results

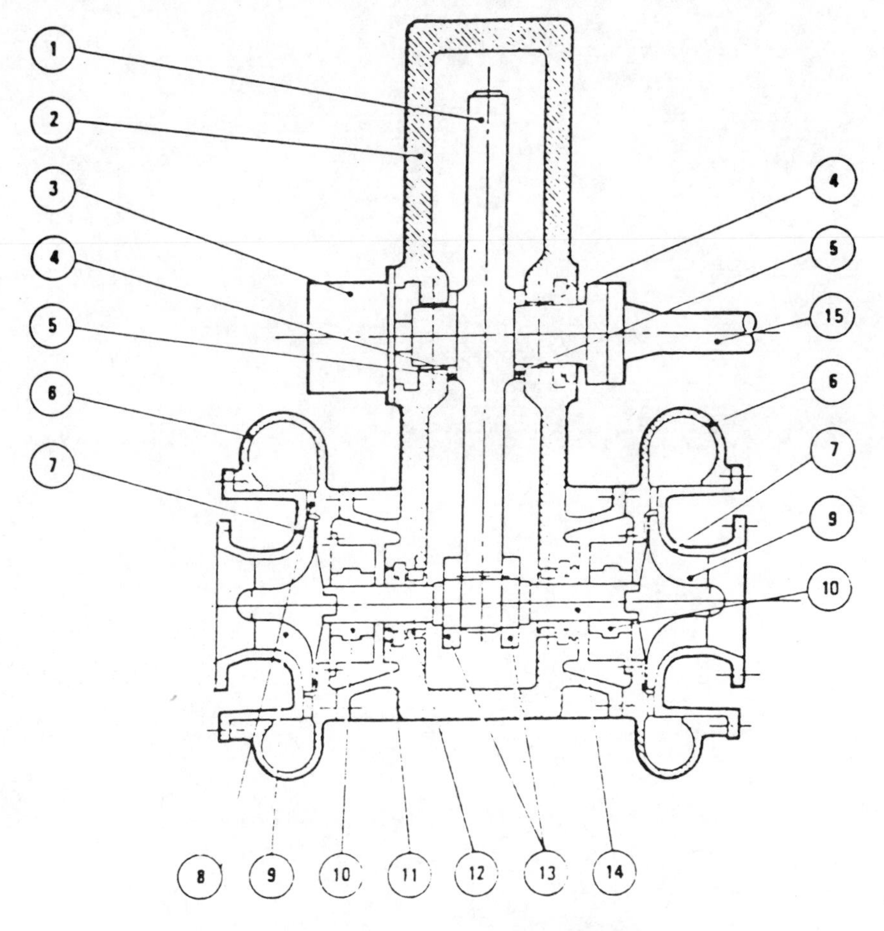

1–Strep - up gear impeller
2 Strep - up gear casing
3 Coupled pump
4 Plain bearing bush
5 Basin thrust bearing
6 Volute
7 Intake casing
8 First stage diffuser
9 Open radial impeller
10 Steam seal
11 Oil seal
12 Tilting pad bearing
13 Collar
14 Pinion shaft
15 Flexible shaft

FIG. 7 SM STEAM COMPRESSOR SECTIONAL VIEW

FIG. 8 HIGH–SPEED ROTOR

3rd International Symposium on the

Large Scale Applications of Heat Pumps

Oxford, England : 25-27 March 1987

PAPER F2

LARGE CAPACITY HEAT PUMP USING VACUUM ICE PRODUCTION AS HEAT SOURCE

Knud Andersen, M.Sc. (Mech.Eng.), HD, Project Manager
Flemming V. Boldvig, M.Sc., Manager, Corporate Communication

A/S THOMAS THS. SABROE & Co.
P.O.Box 1810
DK-8270 Højbjerg
Denmark

SYNOPSIS

The paper describes a pilot plant comprising the newly invented Vacuum Ice Maker utilized as an industrial Heat Pump with cold water as a heat source, in a Danish Central Heating System.

The technical principle is described, and the economical benefits are calculated.

It is concluded that the invention is well suited as large industrial heat pump under extreme ambient conditions, and economically attractive under the conditions existing in Denmark.

INTRODUCTION

Many inhabited areas around the world have significant fluctuations in ambient temperatures preventing use of the same heat source during both summer and winter.

The new vacuum ice maker (VIM) can operate on all water temperatures, and is thus well suited as evaporator in industrial heat pumps for low ambient or even subzero climates using cold water as heat source.

To demonstrate the efficiency a pilot plant has been installed with a Danish district heating company. The plant was commissioned during autumn 1986.

This paper will explain the technical principles and the economical benefits of the Vacuum Ice Heat Pumps (VIH).

VIM BASIC THEORY

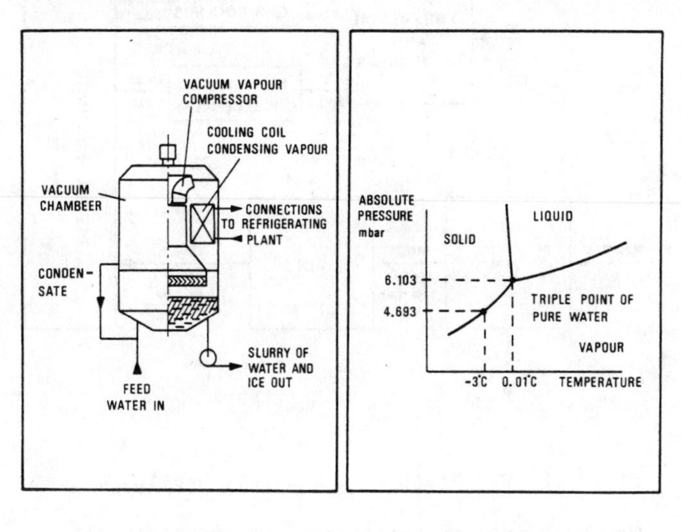

Figure 1: Phase Diagram for Pure Water Sketch of VIM

Figure 1 shows the relative areas for the three phases of water - liquid, vapour and solid (ice).

At the triple point of pure water the pressure is 6,103 mbar abs. (4.58 mm Hg), and the temperarure is 0.01°C. The actual location of the triple point is a function of the salinity (S) of the water as shown below:

$$P_t = 6.103 - 0.246\ S\ \text{(mbar)} \quad (1)$$

The basic idea is to establish a vacuum corresponding to the triple point, in the lower part of the VIM.

When water is injected into the freezer area of the VIM, part of the water will evaporate instantly causing a part to freeze.

The amount frozen is closely related to the amount that evaporates, which in turn is given by the capacity of the Vacuum Vapour Compressor (VVC) mounted as an integral part of the VIM.

In the VVC the vapour is compressed to a saturation pressure of approx. 7,857 mbar abs. (5.97 mm Hg), equal to a saturation temperature of 3.5°C., at which temperature the vapour is condensed, and the heat from the ice generation rejected.

VIH Pilot Plant

Figure 2 shows a schematic of the pilot plant in Denmark.

The vacuum vapour compressor and the reciprocating compressors used in the VIH are, together with all other electric motors powered from a 4-pole generator, driven by a direct coupled LNG motor. The energy transferred from the generation of ice is boosted to condensing temperature by means of 4 reciprocating compressors.

Held at St. Catherine's College Oxford, England. Organised and sponsored by BHRA, The Fluid Engineering Centre, Cranfield, Bedfordshire, MK43 0AJ England.

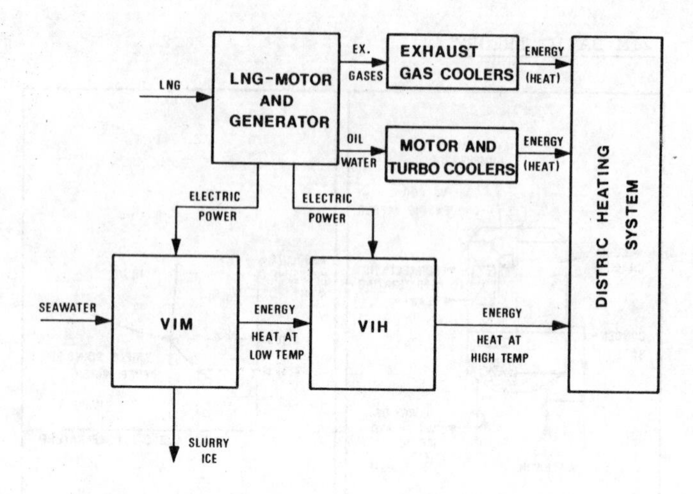

Figure 2: VIH Plant in a District Heating System

The heat rejection from the condensers to the district heating system is divided in two temperature levels - one at approx. 64°C. and one at approx. 75 Deg.C. to optimize the coefficient of performance of the compressors.

The LNG motor is equipped with cooling of oil, -motor, -turbochargers, and -exhaust gases. The exhaust gas is cooled in three stages. First stage is a dry cooling by means of water, second stage is a condensing cooling by means of water, and third stage is a condensing cooling by means of evaporating R12. Thus the exhaust gas leaves the process at a low temperature.

VIH Operating Conditions and Capacity

For the actual installation in Denmark figure 3 showns the supply diagram.

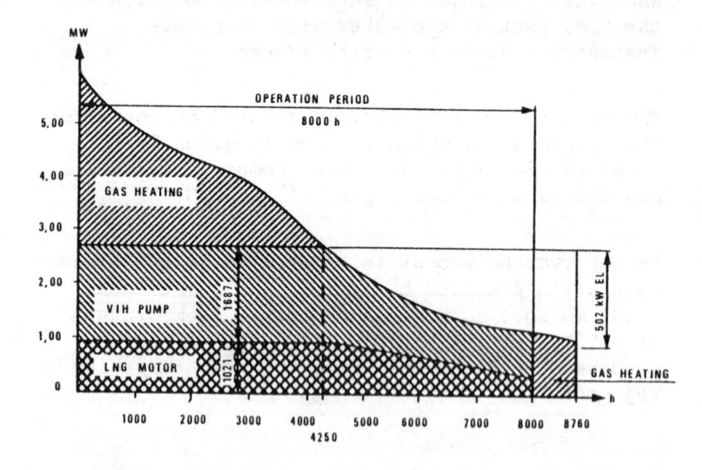

Figure 3: Supply Diagram for the Thermal Energy Sale as Average of 1982-1984 inclusive

The supply diagram shows that the VIH condensers reject 1687 kW, the LNG motor cooling including exhaust gas cooling reject 1021 kW, and that the VIH will operate at full load for 4250 hours/year and at part load for 3750 hours/year.

The LNG motor supplier states the approx. correlation between the load and the consumption as:

$$C=30.058+1.2987\ \frac{P \times 100}{565} + 2.2900\times10^{-6}\left(\frac{P \times 100}{565}\right)^3 \ (Nm^3/h) \qquad (2)$$

As several of the power consuming components installed in the plant have different variations in consumption with load, it is found most convenient to present the VIH data in a table.

From the actual Supply Diagram and Table I the annual production and consumption of energy can be calculated for 8000 hours of operation:

$$\dot{Q}_A = 2708kW \times 4250h + 2360kW \times 1000h + 1670kW \times 1000h + 1360kW \times 1000h + 1190kW \times 750h = 17792MWh \ (= 64.051\ GJ/year) \qquad (3)$$

$$G_A = 1604kW \times 4250h + 1461kW \times 1000h + 1154kW \times 1000h + 1025kW \times 1000h + 940kW \times 750h = 11162MWh \ (=40.183GJ/year=1,034,000\ Nm^3/year) \qquad (4)$$

This corresponds to an annual average coefficient of performance of:

$$COP_A = \frac{\dot{Q}_A}{G_A} = \frac{17,792\ MWh/year}{11,162\ MWh/year} = \underline{1.59} \qquad (5)$$

Compared with a conventional gas burning unit the VIH reduces the amount of consumed gas by:

$$R_g = \frac{\dot{Q}_A}{\eta_g \times LHV} - \frac{G_A}{LHV} \qquad (6)$$

$$= \frac{64,051 \times 10^9\ J/year}{1.00 \times 38.88 \times 10^6\ J/Nm^3} - \frac{40,183 \times 10^9\ J/year}{38.88 \times 10^6\ J/Nm^3}$$

$$= \underline{613,837\ Nm^3/year\ (23866\ GJ/year)}$$

Economical Analysis

To make a proper evaluation the economy of a VIH plant must be included.

The district heating company was faced with the following investments:

	1000 DKK
VIH incl. VIM	6,000
Pipe lines in/out for VIM	500
LNG motor and generator	1,700
Building/foundations	700
Erection	400
Consulting engineer	350
Total	9,650

The project qualified for support from the Danish government energy programme amounting to 25% of the investment cost.

Governmental support, actually 25%	2,413
Net total investment	7,237

The additional installation of gas burning units is due to the VIH installation subsequently being reduced from 6000 kW to app. 3500 kW (figure 3).

The difference in price including buildings, smoke-stacks, exhaust gas coolers a.o. for the two sizes of gas burning units is app. DKK 2,500,000.

In Table II the economical consequences for a period of 10 years have been calculated.

The consequences could also be calculated as a present value of the investment.

Figure 4 illustrates the key figures.

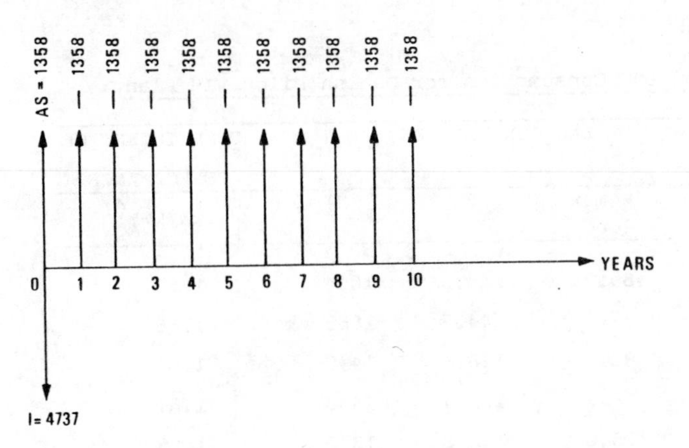

Figure 4: Present Value of an Investment Followed by a Constant Annual Saving (Net)

$$k_o = (1358 - 4737) + \alpha_{\overline{10|}}\, 0.07 \times 1358 \qquad (7)$$

$$= -3379 + \frac{1 - (1 + 0.07)^{-10}}{0.07} \times 1358 = 6159$$

As the present value of the marginal investment is positive, it is an attractive investment to establish the VIH.

Finally the internal rate of interest for the project is approx. 38.7% p.a.

Conclusion

The invention of a VIM operated at a pressure below 6.103 mbar abs and thus being able to produce a slurry of ice and water has made it technically feasible to operate heat pumps under extreme ambient conditions.

The first plant has been made in Denmark, and it consists of 1 VIM, 5 reciprocating compressors, 1 LNG gas motor, and various heat exchangers. The installation covers approx. 68% of the annual total heat production while the remaining 32% are produced by normal gas boilers with exhaust gas coolers.

An economic calculation shows that it will cost the District Heating Company an additionally DKK 4,737,000.00 to install the VIH, and that it will result in annual savings of DKK 1,358,000.00. These figures lead to a present value of approx. DKK 6,159,000.00 and an internal rate of interest of approx. 38.7% p.a.

It is thus clear that the VIH under the said conditions is economically attractive when installed in Denmark.

Symbols

AS: Annual saving by VIH operation (DKK 1000/year)

C : LNG consumption by gas motor (Nm^3/h)

G_A: Annual gas consumption by VIH (MWh)

k_o: Present value of marginal investment (DKK 1000/year)

P : Generator load (kW)

P_t: abs pressure at triplepoint (mbar)

$\dot{Q}_A$: Annual heat production by VIH (MWh)

S : Salinity of water (%)

$\alpha_{\overline{10|}}\, 0.07$: Discounted annual saving multiplier for 10 equal years at 7% annual interest rate.

References

1. Fisher, U.: Compressor and System Matching in Vapour Compression Distillation Unit. in: Israel Journal of Technology, Vol. 15, 1977, p. 102-111.

2. Boldvig, F. V.: Vacuum Ice, A New Economical Production Method for Large Quantities of Ice. in: LAURITZEN NEWS, February 1986, No. 15.

3. Andersen, K.; Boldvig, F. V.: Vacuum Ice - A New Methos for Production of Large Quantities of Ice for Cooling Deep Mines. in: Cooling - Its Vital Role in Agriculture, Mines, Factories and Buildings, Frigair 86, Vol. 1, Mine Cooling.

4. Lynggaard, P.: Investeringskalkuler, 1st Edition, Erhvervsøkonomisk Forlag S/I, Nyt Nordisk Forlag Arnold Busck, København 1981.

5. Capacity Data from Manufactor of Gas Motor.

6. Capacity Data for SABROE Compressors and Heat Exchangers.

7. Various Computer programs for Regression Analysis and Economical Feasibility Studies.

TABLE I Heat Production, Power Consumption, and LNG Consumption for Danish Pilot VIH Plant

HEAT PRODUCTION		POWER CONSUMPTION			LNG CONSUMPTION			COEFFICIENT OF PERFORMANCE
		VIM[1]	VIH[2]	TOTAL				
%	kW	kW	kW	kW	% max. load	Nm^3/h	kW[3]	(kW/kW)[4]
100	2708	59	449	508	89.9	148.5	1604	1.69
95	2573	57	435	492	87.1	144.7	1563	1.66
90	2437	55	412	467	82.7	138.8	1499	1.63
85	2302	53	389	442	78.2	132.7	1433	1.61
80	2166	51	366	417	73.8	126.8	1370	1.58
75	2031	50	347	397	70.3	122.2	1319	1.54
70	1896	49	324	373	66.0	116.4	1257	1.51
65	1760	48	301	349	61.8	110.9	1197	1.47
60	1625	47	277	324	57.3	104.9	1133	1.43
55	1489	47	254	301	53.3	99.6	1076	1.38
50	1354	47	233	280	49.6	94.8	1023	1.32

1) Turbo compressor, pumps, agitator, NLG-removal, fans a.o.

2) 2 x 16 cylinder, 2 x 12 cylinder and 1 x 8 cylinder reciprocating compressor

3) LHV = 38.88 MJ/Nm^3. Informed by the LNG supplier.

4) Related to the use of primary energy.

Table II: Cash Flow for the Actual Investment in a VIH at a Danish Distric Heating Company

	1987	1988	1989	1990	1991	1992	1993	1994	1995	1996	1997
Investment (net)	4737										
Operating cost	110	110	110	110	110	110	110	110	110	110	110
Energy consumpt.[1]	2471	2471	2471	2471	2471	2471	2471	2471	2471	2471	2471
Rate of interest[2]		236	158	74	-16	-112	-215	-325	-443	-569	-704
Total cost	7318	2817	2739	2655	2565	2469	2366	2256	2138	2012	1877
Energy sales[3]	3939	3939	3939	3939	3939	3939	3939	3939	3939	3939	3939
Financing demand	3379	-1122	-1200	-1284	-1374	-1470	-1573	-1683	-1801	-1927	-2062
Accumulated Financing demand	3379	2257	1057	-227	-1601	-3071	-4644	-6327	-8128	-10055	-12117

1) 40,183 GJ per year at 61.50 DKK per GJ.

2) Rate of interest: 7.0% p.a.

3) 64,049 GJ per year at 61.50 DKK per GJ.

3rd International Symposium on the

Large Scale Applications of Heat Pumps

Oxford, England : 25-27 March 1987

PAPER F3

CENTRAL HEATING AND COOLING PLANT FOR YEAR-ROUND AIR CONDITIONING AND HOT WATER USING SOLAR ASSISTED HEAT PUMP

AUTHOR: ENRIQUE ALAIZ

COMPANY: COBRA, S. A.

ABSTRACT

This paper describes a central heating and cooling plant operation, designed and constructed to supply year round air conditioning and hot water for a 20.000 m^2 residential, academic and office complex distributed in five buildings in Hoyo de Manzanares, Madrid. It also includes an analysis of the expected energy output and consumption.

The system consist basically of a 375 kW water-to-water solar assisted heat pump. The low temperature solar system is composed of 960 flat plate, selective solar copper collectors with a total absorption area of 2050 m^2, two storage tanks with capacities of 700 m^3 and 100 m^3 and absorption chiller units.

In a working operation while in the heating mode, the auxiliary energy required for heating is derived from electric heaters which are connected during the night and the heat is stored in the 700 m^3 tank.

The analysis of the energy consumed and that supplied by solar systems and night-time electrical energy storage shows the scarce utilization made of the daytime electrical energy as well as the high percentage of solar energy available.

The results obtained in the first year of starting up and regulation of the system confirm the energy previsions and show the high efficiency of the solar assisted pump.

DESCRIPTION OF THE OPERATION MODE (fig 1)

the operation of the system is carried out through several different program options which can be executed simultaneously. These programs are:

- Solar energy collection and storage
- Heating
- Cooling
- Domestic hot water
- Heat dissipation. Security and automatic maintenance

SOLAR ENERGY COLLECTION AND STORAGE (Fig. 1 Circuit B1-B2)

In winter the solar energy is collected and stored for heating and domestic hot water supply. These processes are simultaneously carried out whenever the collectors produce useful heat at a temperature which exceeds the temperature of either one of the two tanks. In summer this solar radiation is collected and stored in the 100 m^3 tank for the production of domestic hot water and chilled water production in the absorption unit.

In winter the energy collected is stored in both tanks, priority being given to the low temperature tank of 700 m^3, until it reaches the maximum admission temperature of the heat pump evaporator (23ºC). After that, solar energy is stored in the 100 m^3 hot water tank to be used directly for heating, until the tank reaches its maximum admissible inlet temperature in the fan coils and climatizers of the building (65ºC).

When this temperature is reached, the storage proccess continues in the 700 m^3 tank above. As long as this tank reaches temperatures over 23ºC the inlet temperature of the water in the heat pump evaporator is controlled by the three-way valves installed for this purpose.

HEATING (Fig. 1 Circuit B3-B4)

The 700 m^3 tank is the low temperature heat source for the heat pump. The condenser's heat is delivered in the distribution circuit in which a 100 m^3 tank is placed and where the water flow returns after passing through the buildings. The heat pump has a ten-step graduated power control.

Two electronic thermostats, in response to a maximum temperature in the condenser and/or maximum temperature in the evaporators are in charge of sequencially connecting or disconnecting the 10 steps.

The electric heaters consist of 8 groups of electric resistances distributed in 11 vessels which are connected sequencially by a control which checks the temperature in the heated water for buildings at fixed time intervals. Heaters are used during the night when the electric rates are reduced. The quantity of heat delivered in the storage tank is controlled in function of the temperatures of the tanks and the daily average demand of heat.

Held at St. Catherine's College Oxford, England. Organised and sponsored by BHRA, The Fluid Engineering Centre, Cranfield, Bedfordshire, MK43 0AJ England.

The nominal heating capacity of the heat pump is 1.150.000 kcal/h. The maximum combined heating power of the plant is 2.150.000 kcal/h, when the auxiliary electrical heaters are connected as an occasional supplement of the heat pump.

COOLING (Not represented in figure)

The necessary chilled water for the air conditioning systems of the building is delivered first by the absorption chiller, whenever the solar energy is capable of supplying water at a temperature over 70ºC to the unit.

In addition to the absorption chiller, the heat pump is used as an auxiliary system for the production of chilled water.

The cooling capacity of the absorption chiller unit installed is 33 tons. This power capacity will be increased in the future when the solar-assisted absorption system is proved efficient.

DOMESTIC HOT WATER (Fig. 1 Circuit B6)

The production of domestic hot water is carried out by a flat plate parallel heat exchanger. The primary of these heat exchangers is fed with the water of the 100 m³ tank. One electric heater is used as an auxiliary heating system. The capacity of this hot water system is 8500 liter/hour of water at 50ºC.

NIGHT-TIME ELECTRICAL STORAGE (Fig. 1 Circuit B2)

When solar energy is not available day after day due to weather conditions, or heating demand is over the capacity of the solar energy supply during several days, the 700 m³ storage tank reaches its lowest temperature for the heat pump (7ºC). Under these conditions the night-time electrical storage program will start connecting the electric heaters during the night when low electric rates are in effect.

SECURITY AND AUTOMATIC MAINTENANCE. HEAT DISSIPATION (Not represented in figure)

The automat that controls the system actuates several different security programs, for example:

- Start-up of the reserve pump (in case of main pump failure)
- Valve position check
- Alarms (in case of malfunction, etc.)

Another program establishes a heat dissipation circuit through the cooling tower as a security measure for the tanks, flat plate solar collectors and in several hot fluid pipes.

Still another group of programs permits an automatic system check and a periodic maintenance check of the proper functioning of each of the components in the entire system.

ENERGY BALANCE

This diagram shows the breakdown of the energy inputs used to satisfy the heating load. To determine the solar energy inputs the simulation program of the (1) Ø F-Chart metod has been used. A daily temperature model for the tanks was used to calculate the heat pump working conditions.

Due to the difficulty of a rigorous simulation of the total system several simplifications were necessary and hence the end results must be considered approximate. However, these previsions are starting to be confirmed by the data available after the first year of start up and regulation of the system.

CONCLUSIONS

This solar energy assisted heat pump system is able to supply 80% of the total energy demand in combination with the electric night-time heat for the five building military complex in Hoyo de Manzanares, Madrid. It is interesting to point out the low daytime electric heat input estimated to be 5%, which together with another 20% of electric energy absorbed by the compressor of the heat pump makes a total of 25% of daytime energy consumption compared to conventional systems, which accounts for the low energy cost of the installation.

REFERENCES:

(1) * Ø F-Chart represents the fraction of the total load provided by solar energy

* Solar Heating Design by the F-Chart published by John Wiley, 1977, New York. Authors: William A. Beckman Klein and John Duffie.

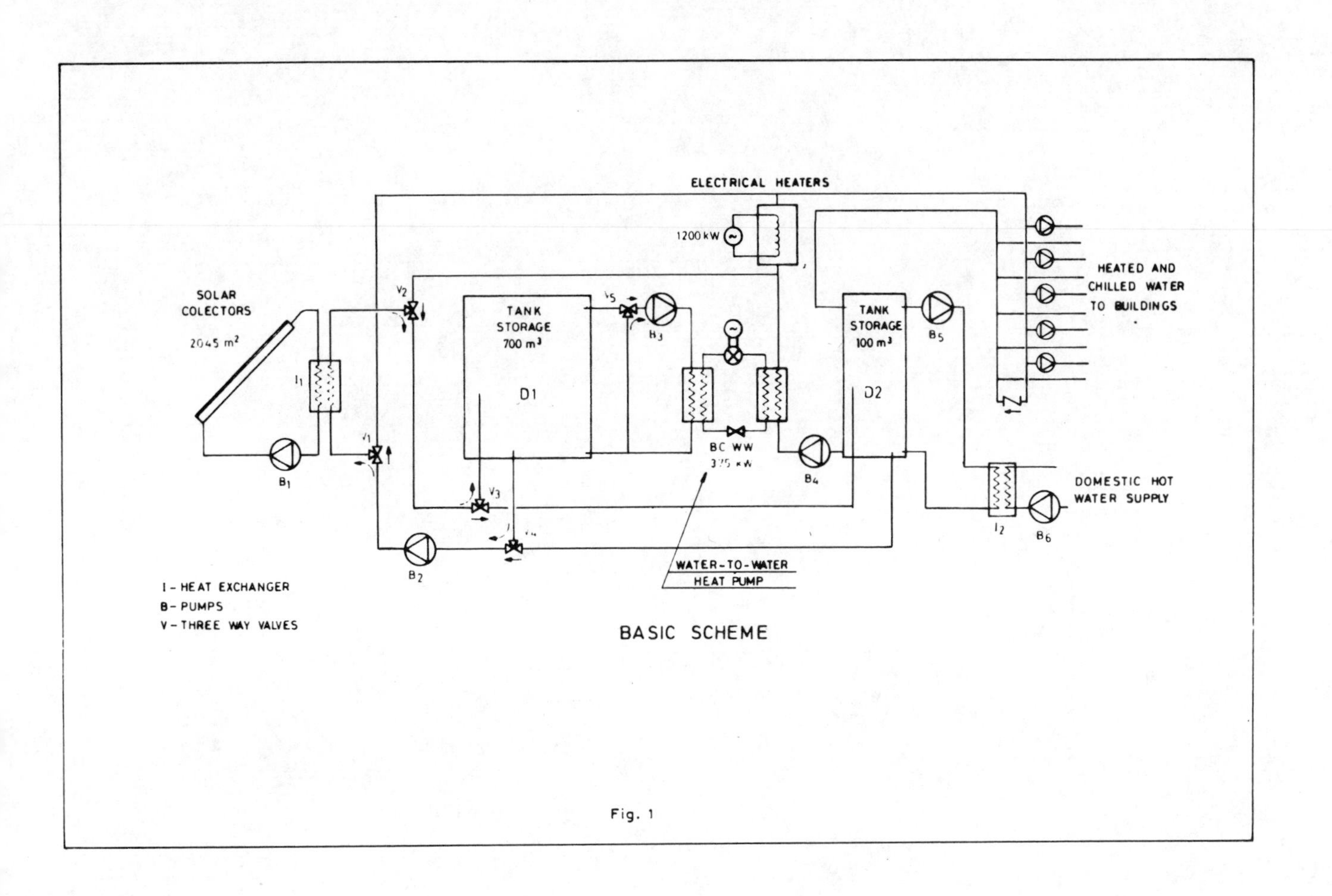
ELECTRICAL HEATERS
1200 kW
SOLAR
COLECTORS
2045 m²
TANK
STORAGE
700 m³
D1
TANK
STORAGE
100 m³
D2
HEATED AND
CHILLED WATER
TO BUILDINGS
BC WW
375 kW
DOMESTIC HOT
WATER SUPPLY
WATER-TO-WATER
HEAT PUMP
I - HEAT EXCHANGER
B - PUMPS
V - THREE WAY VALVES
BASIC SCHEME

Fig. 1

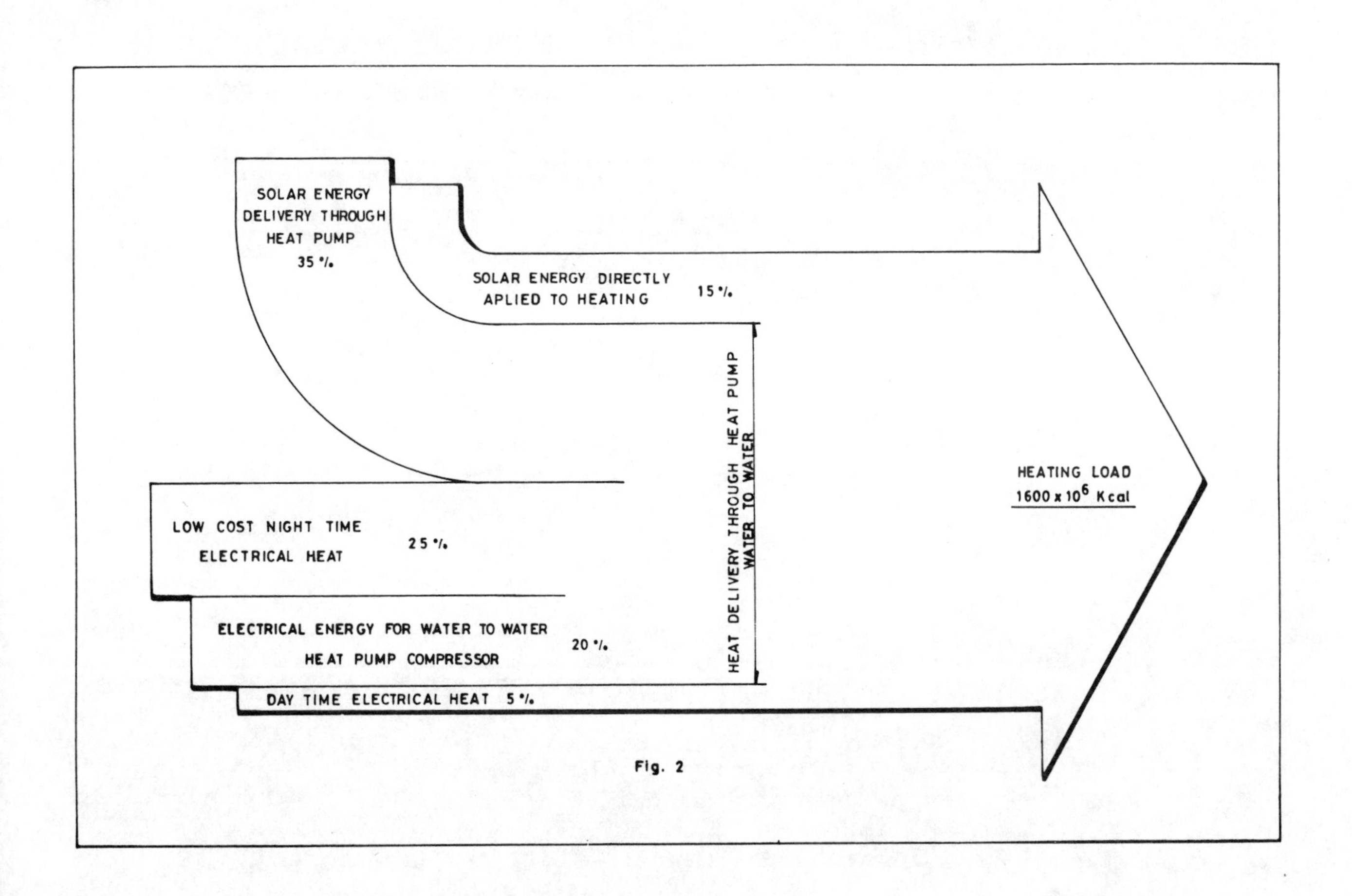
SOLAR ENERGY
DELIVERY THROUGH
HEAT PUMP
35 %
SOLAR ENERGY DIRECTLY
APLIED TO HEATING
15 %
LOW COST NIGHT TIME
ELECTRICAL HEAT
25 %
ELECTRICAL ENERGY FOR WATER TO WATER
HEAT PUMP COMPRESSOR
20 %
DAY TIME ELECTRICAL HEAT 5 %
HEAT DELIVERY THROUGH HEAT PUMP
WATER TO WATER
HEATING LOAD
1600 x 10⁶ Kcal

Fig. 2

3rd International Symposium on the
Large Scale Applications of Heat Pumps
Oxford, England: 25-27 March 1987

PAPER F4

THE GROUND COUPLED HEAT PUMP WITH A MULTIPLE WELL SYSTEM AS HEAT SOURCE FOR HEATING OF APARTMENT BUILDINGS.

L. Spante, M. Larsson and R. Österberg.

L. Spante and M. Larsson, Swedish State Power Board, Alvkarleby Laboratory, S-81071 Älvkarleby, Sweden.

R. Österberg, SwedPower, Box 34, S-10120 Stockholm, Sweden.

SYNOPSIS

The Swedish State Power Board (SSPB) is actively encouraging the introduction of heat pumps in Sweden. To enlarge this work SwedPower, to a major part owned by SSPB, will work for an increased international exchange of heat pump experiences and for a transfer of Swedish know-how to other markets.

The programme carried out by SSPB includes today 450 demonstration plants of which 30 are ground coupled heat pumps for apartment and school buildings, with vertical pipes in a multiple well system as the heat source. In order to decrease the required bore-hole depths some of the systems are equipped with simple air convectors or solar collectors for recharging of the wells.

The Kragstalund plant, near Stockholm, is presented. It consists of three 100 kW heat pump systems installed in identical new apartment buildings. Bore-hole depths and degree of recharging differ between the three systems.

The plant shows good annual performance and is more cost-effective than a corresponding system with electric boilers.

Measurements and computer simulations for one of the three systems show that the positive effect of recharging with air convectors is small. In Sweden it is usually more cost-effective to design a multiple well system with optimal bore-hole depth, than to design it with recharging.

INTRODUCTION

The Swedish State Power Board (SSPB) is actively encouraging the introduction of heat pumps for heating and hot water production in Sweden. The goals are a reduction of the use of oil and a more efficient use of electrical energy.

The production of heat from heat pumps in Sweden is today approximately equal to the production from 2 nuclear units of the size of 1000 MW_e each.

The SSPB heat pump programme started in 1980 and includes now 450 demonstration plants with heating capacities from 1.5 kW to 11 MW.

To enlarge the work in the heat pump field SSPB has decided to work for an increased international exchange of experiences. This will be made through a consulting company, SwedPower, in which SSPB owns the majority.

The programme carried out by SSPB includes about 30 ground coupled heat pump systems for apartment and school buildings, with vertical pipes in a multiple well system as the heat source. In order to decrease the required bore-hole depth and thereby the costs for drilling some of the systems are equipped with simple air convectors or solar collectors for recharging of the wells.

In this paper the Kragstalund plant in Vallentuna, just outside Stockholm, is presented. The plant consists of three 100 kW heat pumps installed in identical new apartment buildings. Bore-hole depths and degree of recharging differ between the three systems. The design of the systems consciously implies a very high heat extraction rate.

The main purpose of the evaluation is to find out how the charging influences the annual mean and the winter minimum temperature of the well and to estimate its economic value.

The heat pump systems are equipped with meters for power supply, operating time, heat delivery and auxiliary heat. In addition the temperatures in the well and on the brine circuit are recorded. The analysis is based on weekly mean values gathered during 1984-86. For comparison and further analysis, especially regarding the long term temperature development, the computer program SBMA /1/ is utilized.

DESCRIPTION OF THE PLANT

The three heat pumps were installed during the autumn of 1983 in a newly built district with apartment buildings in Vallentuna, 16 km north of Stockholm. Each heat pump is placed in a separate heat station which supplies 50 apartments (fig. 1).

The heat pump, TETAB 18L, is of liquid-to-water type with R22 as refrigerant and is manufactured by Thermia AB, Sweden. The compressor is an eight cylinder semi-hermetic reciprocating compressor manufactured by Copeland. The compressor capacity is regulated in three steps; 50, 75 and 100 % capacity. The heating capacity is 93 kW at a brine temperature of -1°C and a condensing temperature of 55°C. The evaporator is of tubular type with the brine circulating outside the tubes. The brine (antifreeze solution) is a mixture of ethyl alcohol (28 %), propylene glycol (2%) and water.

The heat from the heat pump is delivered by a hot gas heat exchanger of tubular type and a

Held at St. Catherine's College Oxford, England. Organised and sponsored by BHRA, The Fluid Engineering Centre, Cranfield, Bedfordshire, MK43 0AJ England.

condenser of shell and tube type to the radiator system and a domestic hot-water accumulator. The heating system is of low temperature type with a supply line temperature of 60 ^{o}C at -20 ^{o}C ambient air temperature. The design condensing temperature implies a maximum supply line temperature of 55 ^{o}C from the heat pump. Auxiliary heating during cold periods is produced with an electric boiler with a heating capacity of 72 kW in each heat station. The annual heat requirements are about 470 MWh for each group of apartments (A,J,N).

The multiple well system consists of 8 bore-holes (114 mm diameter) inclined 20° to the vertical direction (fig. 2) with a closed-loop system of polyethylene pipes. There are three pipes (40 mm diameter) in each bore-hole, two for the upward direction, and the bore-holes are connected in parallel. The distance between the bore-holes at the ground level is 0.7 metre. In this way a large volume of rock is used but the occupied surface area is kept at a minimum. The difference in bore-hole depth and degree of recharging between the three heat pump systems is described in table 1. The bore-holes are drilled with a down-the-hole hammer equipment and the drilling costs, including set-up time, are 140 SEK/metre (1 Pound = 10.3 SEK, August 1986). The investment for the well system with pipes is about 200 SEK/metre bore-hole which means 190000 SEK for the A-house well system (1986 price level).

The energy roof is of low temperature type and consists of aluminium strips with copper tubes placed under the tile roof (fig. 2). This type of energy roof works mainly as a convection ambient-air heat exchanger. The brine is circulated through the copper tubes when the air temperature under the tile roof is higher than the temperature in the well system. The annual heat output related to the aluminium strips area is about 470 kWh/m^2. The total costs for this type of energy roof system is not less than 450 SEK/m^2 strips area.

Energy from the exhaust air system is continuously delivered either to the inlet brine circuit or to the well system. About 25 MWh/year is in this way recharged from the exhaust air to each heat pump system.

The investment for each heat station, including heat pump, well system, recharging devices and electric boiler, is about 600000 SEK. The whole plant with three heat pump systems is more cost-effective than a corresponding system with electric boilers. The price for electricity, based on a power subscription, was in August 1986 0.32 SEK/kWh. The oil price was, at the same time, 1600 SEK/m^3 which means 0.20 SEK/kWh if the oil boiler efficiency is presupposed to be 80 %. The oil price was about 2500 SEK/m^3 at the time for the projecting of the plant.

RESULTS AND DISCUSSION

Depending on varying performance and availability for the heat pumps, the heat extraction from the well system differs a lot for the three systems. It is therefore difficult to analyse the effect of recharging and varying bore-hole depths from measured data only. Because of this the analysis is based on measured data from the A-house system and on computer simulations /1/.

The heat production during the period March-84 to Feb-85 for the A-house heat station is presented in figure 3. The heat load in the system was 473 MWh of which the heat pump delivered 408 MWh (86 %) with a seasonal performance factor (SPF) of 2.6. The SPF is defined as the annual heat production from the heat pump divided by the annual electrical energy consumption of the compressor.

The total heat to the evaporator is 253 MWh/year. The recharging system produced about 72 MWh/year during the first two years of operation. This is less than expected.

In figure 4 recorded mean temperatures in the well water during the first two years of operation are shown, together with results from the computer simulations. The simulated result is a mean temperature at the bore-hole wall. The temperature difference between the wall and the water is estimated to be about 1°C when the water is frozen to ice which will be valid for the period November to April. Taking this into account the computer simulation is seen to be in good agreement with observed temperatures. This has also been the result from other analyses with the SBMA-programme /2/.

The simulated result is based on measured heat extraction from the well system. Results for the A-house system without recharging are also presented in figure 4. As expected the temperature is lower in this case.

After some time the cooling of the ground around each bore-hole will influence the temperature at the other bore-holes. This influence increases with time and depends on the spacing between the bore-holes.

The long term temperature development for the A-house system is simulated and the results during the 15th year of operation are presented in figure 5 (Case 1). The annual heat extraction from the well system is presupposed to be constant and with the same monthly distribution as for year 1.

The monthly minimum temperature in the well will be about 3°C lower in the 15th year than in the second year of operation. This means that the evaporation temperature for the heat pump will be very low during winter periods, which could cause operation disturbances. The evaporation temperature limit is for this heat pump -21 ^{o}C.

Two ways of solving this problem have been analysed with computer simulations.

Case 2. Two extra bore-holes and recharging with the exhaust air system only. Total active bore-hole depth H=1120 metres. Recharged energy 25 MWh/year.

Case 3. A new recharging system with fan-driven ambient-air heat exchangers and the existing exhaust air system. Total active bore-hole depth H=909 metres and recharged energy 125 MWh/year.

The results are presented together with Case 1 in figure 5 as monthly mean temperatures in the 15th year of operation, and in table 2 as minimum temperatures and energy weighted annual mean temperatures in years 5 and 15.

The simulated results show that the difference in the well temperature during the heating season between Case 2 and 3 is small.

Compared to the existing system (Case 1) the minimum temperature will be about 2°C higher after 15 years of operation if the Case 2-solution is selected. This will decrease the risk of too low evaporation temperatures and also increase the heat pump performance. The saving in the total electrical energy consumption will be in the range of 5 - 10 MWh/year compared to the existing system.

The investment for two extra bore-holes with pipes will be 45000 SEK. The solution with fan-driven ambient-air heat exchangers will cost more than 45000 SEK.

This means that the best way of improving this system is to increase the total bore-hole depth (Case 2). As the performance is the same for Case 2 and 3, the lower investment cost and simplified system solution of Case 2 is conclusive in making this solution the better choice.

CONCLUSIONS

The Kragstalund plant shows good annual performance. This installation demonstrates that the ground coupled heat pump, in the range 50-200 kW, with a multiple well system as heat source is a competitive option for a large number of apartment buildings in Sweden, both for new buildings and for retrofit.

Results from measurements and computer analyses presented in this paper indicate that it is more favourable to add bore-length than to add recharging equipment. This will give a more cost-effective system with a simpler and more reliable system design.

The only exception from this is when exhaust air is available as a complementary heat source to the multiple well system.

REFERENCES

/1/ Eskilson, P. (1985). Superposition Bore-hole Model (SBMA). Simulation model for oblique or vertical bore-holes. University of Lund, Sweden.

/2/ Larsson M. ,Spante L. Solar charging of energy wells for small heat pump systems. Swedish State Power Board, Alvkarleby Laboratory. Report UL-FUD-B 85:12.

TABLE 1. Description of the three heat pump systems in Kragstalund, Vallentuna.

House	A	J	N
Bore-hole configu-ration	8*120 m	8*100 m	8*120 m
Effective bore-hole depth	909 m	748 m	925 m
Recharging system	Exaust air and energy roof 100 m^2	Exaust air and energy roof 100 m^2	Exaust air only

TABLE 2. Simulated bore-hole temperatures for three different system solutions of the A-house heat pump system.

Case	1 No change	2 Two extra bore-holes No energy roof system	3 Recharging with fan-driven air heat exchangers
Active bore-hole depth	H= 909 m	H= 1120 m	H= 909 m
Annual rechar-ging	Q= 72 MWh	Q= 25 MWh	Q= 125 MWh
Annual mean heat extrac-tion W/metre bore-hole	23	23	16
Year 5			
Monthly minimum temp.	-4.0 °C	-2.2 °C	-3.3 °C
Energy-weighted annual mean temp.	-1.8 °C	-1.0 °C	-0.1 °C
Year 15			
Monthly minimum temp.	-6.0 °C	-4.2 °C	-4.6 °C
Energy-weighted annual mean temp.	-3.7 °C	-2.9 °C	-1.5 °C

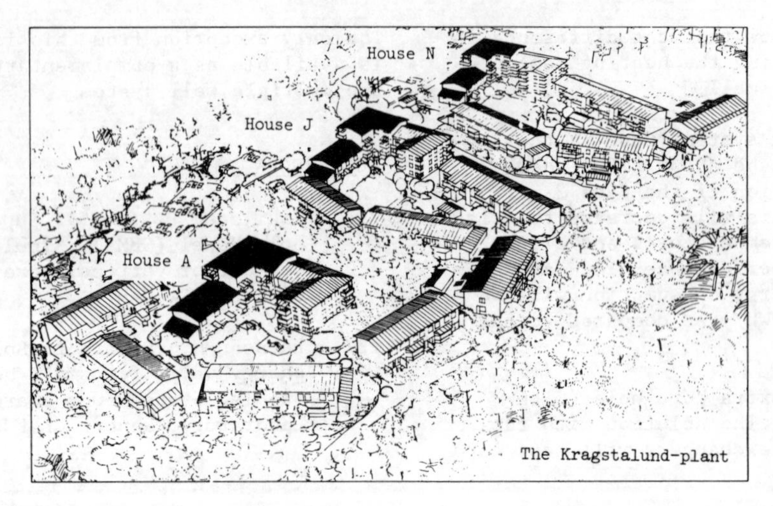

Fig. 1 View over the Kragstalund plant with three 100 kW heat pumps placed in the houses marked A, J, N.

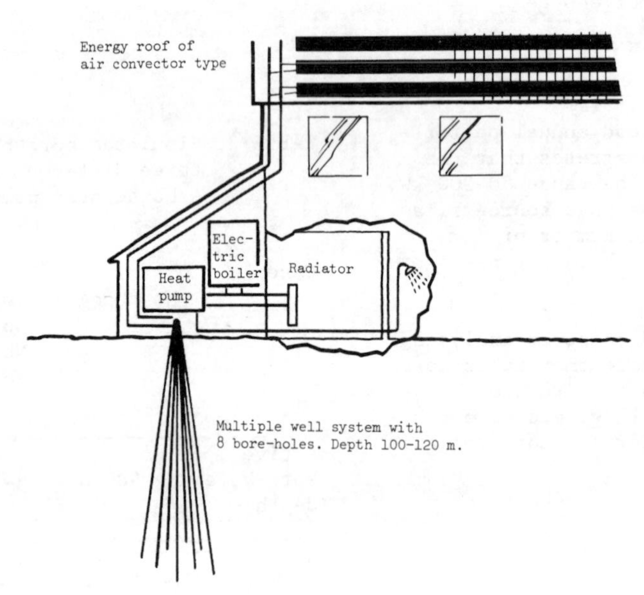

Fig. 2 The heat pump is connected to a multiple well system with 8 bore-holes. The well is recharged with air convectors.

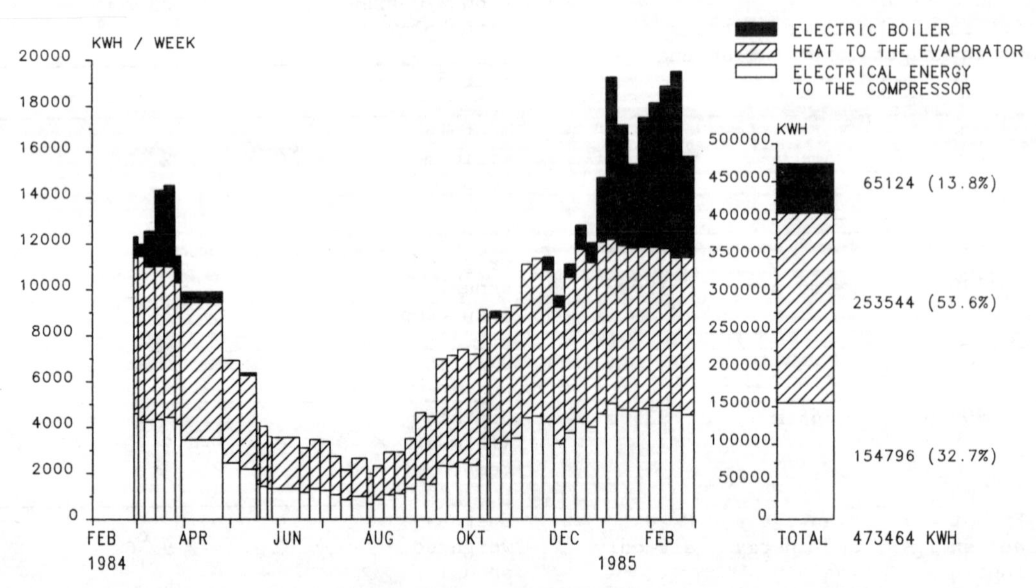

Fig. 3 Heat production during the period 1984/85 for one of the three heat stations (A). Auxiliary heating with an electric boiler

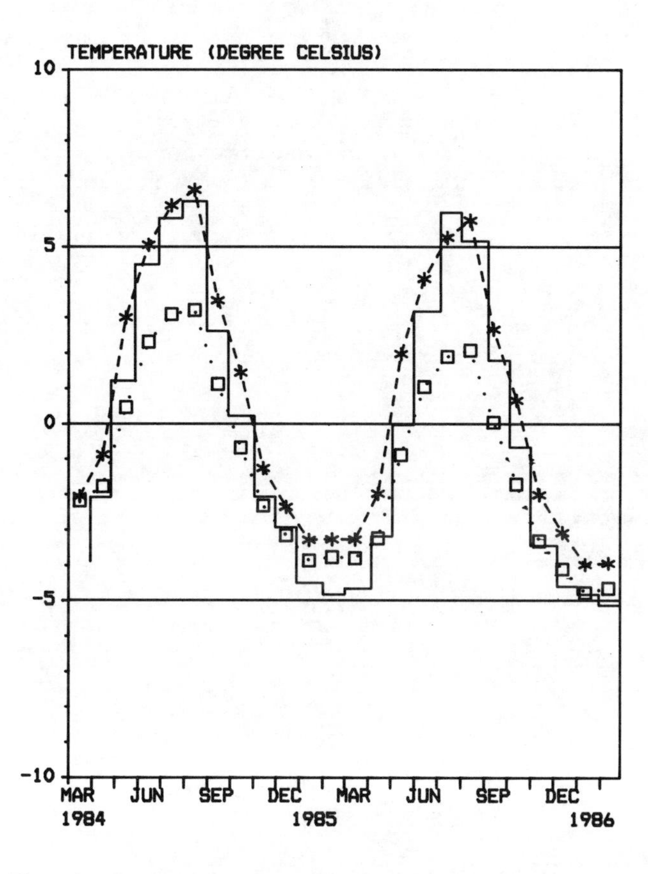

Fig. 4 Measured and simulated monthly mean temperatures in the well system during the first two year of operation.

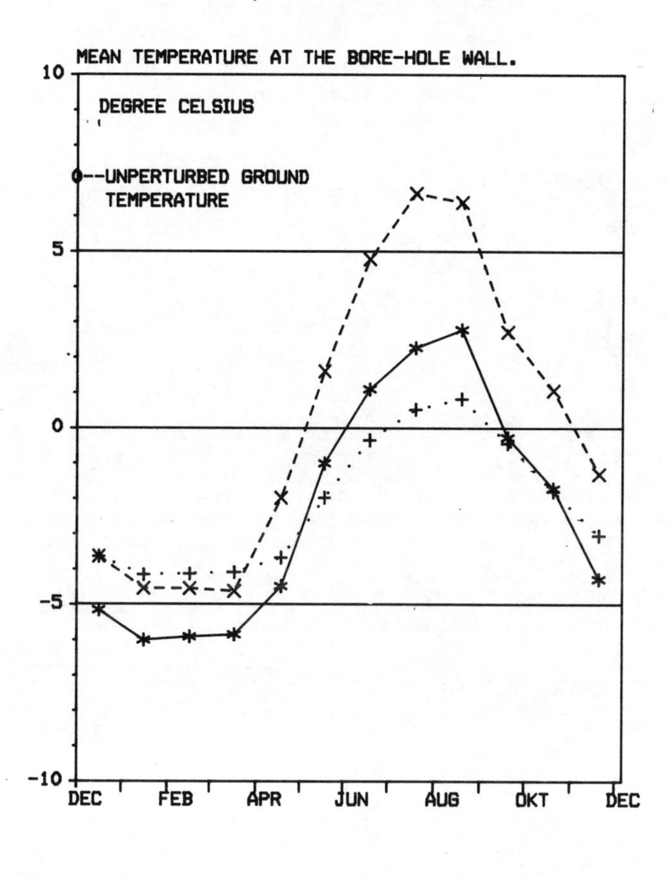

Fig. 5 Simulated temperatures in the well system during the 15th year of operation. Three different systems have been analysed.

3rd International Symposium on the

Large Scale Applications of Heat Pumps

Oxford, England : 25-27 March 1987

PAPER G1

BEHAVIOUR OF ABSORBERS WITH FALLING FILMS OF SALT SOLUTIONS IN HEAT PUMP APPLICATIONS

M. Pflügl and F. Moser

Dr. M. Pflügl is in the Institute of Verfahrenstechnik at the Technical University of Graz, Austria
Prof.Dr.F.Moser is head of this institute

SYNOPSIS

A mathematical model allows the numerical simulation of a falling film absorber for salt solutions being based only on the conservation laws, a turbulence model and correlations for the thermodynamics. Thereby the behaviour of absorbers using falling films of aqueous Lithium Bromid solutions are examined leading to various consequences for the simulation of heat pump and heat transformer cycles. The heat transfer coefficient predicted by this simulation is in good agreement with experimental correlations.

NOMENCLATURE

a	thermal diffusivity of solution, $a=\lambda/\varrho c$ (m^2/s)
c	heat capacity of solution, (J/kgK)
D	diffusion coefficient in solution, (m^2/s)
Fr	Froud number, $Fr=\dot{M}_f{}^2/(g\delta)^3$ (-)
g	gravity, (m/s^2)
h	enthalpy of solution, (J/kg)
Δh_a	heat of absorption, (J/kg)
h_{gg}	enthalpy of vapour at T_g, (J/kg)
h_{gs}	enthalpy of vapour at T_s, (J/kg)
Ka	Kapitza number, $Ka=(g\varrho/\sigma)^3\Lambda^6$ (-)
L	absorber length, (m)
Le	Lewis number, Le=a/D (-)
$\dot{m}_a$	flux of absorbed mass, (kg/sm^2)
$\dot{M}_a$	rate of absorbed mass, $\dot{M}_a=\dot{m}_a L$ (kg/sm)
$\dot{M}_f$	film mass flow rate, (kg/sm)
$\dot{M}_{fo}$	film mass flow rate at absorber inlet, (kg/sm)
p	pressure in absorber, (bar)
Pr	Prandtl number, $Pr=\nu/a$ (-)
$\dot{Q}_s$	increase of sensible heat of solution from inlet to outlet, (kW/m)
$\dot{Q}_w$	rate of heat transfered to coolant, (kW/m)
Re	Reynolds number, $Re=\dot{M}_f/\varrho\nu$ (-)
Sc	Schmidt number, $Sc=\nu/D$ (-)
T	film temperature, $T=T(y,z)$ (K)
$\overline{T}$	average film temperature, $\overline{T}=\overline{T}(z)$ (K)
T_o	film temperature at entrance, (K)
T_c	temperature of coolant, (K)
T_{co}	temperature of coolant at top of absorber, (K)
T_{cL}	temperature of coolant at bottom of absorber, (K)
T_{eqo}	equilibrium temperature for solution entrance conditions p and X_o, (K)
T_g	temperature of vapour to be absorbed, (K)
$\overline{T}_L$	average solution temperature at absorber outlet, (K)
T_s	film surface temperature, (K)
T_w	wall temperature, $T_w=T_w(z)$ (K)
ΔT	solution temperature increase, $\Delta T=T_o-\overline{T}_L$ (K)
ΔT_c	temperature increase of coolant, $\Delta T_c=T_{co}-T_{cL}$
u	vertical film flow velocity, $u=u(y,z)$ (m/s)
$\overline{u}$	average film flow velocity, $\overline{u}=\overline{u}(z)$ (m/s)
v	transverse velocity in the film, $v=v(y,z)$ (m/s)
X	concentration of absorbate in solution, $X=X(y,z)$ (wt %)
$\overline{X}$	average concentration of absorbate insolution, $\overline{X}=\overline{X}(z)$ (wt %)

Held at St. Catherine's College Oxford, England. Organised and sponsored by BHRA, The Fluid Engineering Centre, Cranfield, Bedfordshire, MK43 0AJ England.

X_o initial concentration of absorbate in solution, (wt %)

X_{eqL} equilibrium concentration of solution at temperature $\overline{T}_L$, (wt %)

$\overline{X}_L$ average concentration of absorbate in solution at outlet, (wt %)

X_s concentration of absorbate at film surface, (wt %)

ΔX increase of concentration of absorbate, $\Delta X=\overline{X}_L-X_o$ (wt %)

ΔX_{eq} concentration deviation from equilibrium, $\Delta X_{eq}=X_{eqL}-\overline{X}_L$

y distance from wall in film, (m)

z coordinate in flow direction, (m)

α_c heat transfer coefficient of coolant, (W/m^2K)

α_g heat transfer coefficient of vapour, (W/m^2K)

α_w heat transfer coefficient of film, (kW/m^2K)

β mass transfer coefficient of absorption, (m/s)

δ film thickness, (m)

ε_M turbulent mass diffusivity, (m^2/s)

ε_Q turbulent heat diffusivity, (m^2/s)

ε_τ eddy diffusivity, (m^2/s)

ξ dimensionsless vertical coordinate, $\xi=z/\delta$ (-)

η dimensionless distance from wall, $\eta=y/\delta$ (-)

η_s dimensionless distance from wall for intersection of inner and surface turbulence region, (-)

η_w dimensionless distance from wall for intersection of wall and inner turbulence region, (-)

λ thermal conductivity of solution, (W/mK)

Λ reference length, $\Lambda=(\nu^2/g)^{1/3}$

ν kinematic viscosity of solution, (m^2/s)

ϱ density of solution, (kg/m^3)

σ surface tension of solution, (N/m)

φ dimensionless film flow velocity, $\varphi=u/\overline{u}$ (-)

Ψ dimensionless transverse velocity, $\Psi=v/\overline{u}$ (-)

INTRODUCTION

The absorber is an essential component in absorption heat pump and heat transformer cycles. To calculate the coefficient of performance for these cycles in almost all simulation programs the absorber and desorber are considered as ideal, that is thermodynamic vapour-liquid equilibrium is assumed for their outlets and the apparatus are considered realizable for the requested solution field. To judge these assumptions the examination of the behaviour of absorbers is necessary.

In heat pump/transformer applications the absorbers task is threefold

- all the vapour of heat carrier has to be absorbed into the solution
- the thereby released rate of heat should be transfered for utilization to an appropriate medium (e.g. warm water supply or steam generation) and
- to garantee the function of a heat pump the temperature of the useful heat has to be boosted to a level essentially higher than that of the heat carrier evaporator.

The first task is realizable by a simple adiabatic absorber, too. The released heat of absorption increases the solution temperature and must be removed by a separate heat exchanger. This method is limited to subcooled solutions and is highly inefficient.

Thus for heat pump applications only an absorber is suitable with simultaneous heat removal. The falling film technique has this possibility and also the advantage to use a mathematical model for the simulation of absorber behaviour.

MATHEMATICAL MODEL

To formulate the problem we consider a thin liquid film of salt solution flowing down a plane vertical wall (which is applicable also for vertical tubes). Due to the gravitational force and according to the liquid mass flow a vertical velocity profile will appear (see fig.1). The absorption of the vapour and the thereby released heat lead to concentration and temperature profiles.

To get the governing equations for this two-dimensional problem we have to obey the conservation laws (Ref./17/)

for total mass
for the volatile component
for momentum and
for heat.

For simplification but not crucial restriction we make several assumptions.

Considering the film mean flow in vertical direction we assume a fully developed velocity profil $\varphi(\eta)$ at any vertical position. Due to sufficient low vapour velocities no interfacial momentum transfer will be taken into account. The liquid film is considered as a Newtonian fluid with constant physical properties across the film in y-direction.

Thus the conservation of momentum including the turbulent transport coefficient ε_τ

$$\frac{\partial}{\partial y}\left[(\nu+\varepsilon_\tau)\ \frac{\partial u}{\partial y}\right] + g = 0$$

becomes integrated the equation for the dimensionless distribution of vertical velocity φ across the film (dimensionless coordinate η)

$$\left[1+\ \frac{\varepsilon_\tau(\eta)}{\nu}\right]\frac{d\varphi}{d\eta} - \frac{Re}{Fr}\ (1-\eta)=0\ . \qquad (1)$$

In the falling film we consider also a transverse velocity v influencing only the transport of mass and heat but not that of momentum. The integration of the conservation equation for total mass

$$\frac{\partial u}{\partial z} + \frac{\partial v}{\partial y} = 0$$

with resepct to the result from integrating (1) leads to the profil of dimensionless transverse velocity Ψ

$$\Psi = \frac{d\delta}{dz}\ \eta\ \varphi(\eta)\ + \frac{\dot{m}_a\delta}{\dot{M}_f}\ \int_0^\eta\ \varphi(s)\,ds \qquad (2)$$

where means $\dot{m}_a$ the flux of absorbed mass in the film of thickness δ and mass flow rate $\dot{M}_f$.

To reduce the conservation laws for volatile component mass and for heat we neglect the molecular and turbulent transport of component mass and heat in vertical z-direction. With the restriction of constant properties across the film (not in vertical direction) we get the system of two partial differential equations.

$$\varphi\frac{\partial X}{\partial \xi} + \Psi\frac{\partial X}{\partial \eta} = \frac{1}{Re}\ \frac{1}{Sc}\ \frac{\partial}{\partial \eta}\left[(1+\ \frac{\varepsilon_M}{D})\ \frac{\partial X}{\partial \eta}\right] \qquad (3)$$

$$\varphi\frac{\partial T}{\partial \xi} + \Psi\frac{\partial T}{\partial \eta} = \frac{1}{Re}\ \frac{1}{Pr}\ \frac{\partial}{\partial \eta}\left[(1+\ \frac{\varepsilon_Q}{a})\ \frac{\partial T}{\partial \eta} + \frac{1}{c}\ \frac{\partial h}{\partial X}\right.$$

$$\left.(1 - \frac{1}{Le} + \frac{\varepsilon_Q-\varepsilon_M}{a})\ \frac{\partial X}{\partial \eta}\right] \qquad (4)$$

In addition to these equations we have to consider the boundary conditions at the top of the film, at the wall and the film surface

$\xi=0$ $\quad X=X_o \quad T=T_o$

$\eta=0$: conservation of mass $\frac{\partial X}{\partial \eta}\big|_o = 0 \quad (5)$

conservation of heat

$$\alpha_c(\xi)\ \left[T_w(\xi)-T_c(\xi)\right] = \frac{\lambda}{\delta}\left[1+\ \frac{\varepsilon_Q(0)}{a}\right]\frac{\partial T}{\partial \eta}\Big|_o \qquad (6)$$

$\eta=1$ phase equilibrium

$$X_s\ (\xi) = f\ (T_s(\xi),\ p) \qquad (7)$$

conservation of mass

$$\dot{m}_a = X_s\ \dot{m}_a + \frac{\varrho D}{\delta}\left[1 + \frac{\varepsilon_M(1)}{D}\right]\frac{\partial X}{\partial \eta}\Big|_1 \qquad (8)$$

conservation of heat

$$\dot{m}_a\ (\Delta h_a+h_{gg}-h_{gs})+\alpha_g\ (T_g-T_s) =$$

$$= \frac{\lambda}{\delta}\left[1+\ \frac{\varepsilon_Q(1)}{a}\right]\frac{\partial T}{\partial \eta}\Big|_1 \qquad (9)$$

Because of the last term of equation (4) - due to the differential heat of solution - and the conditions (7), (8) and (9) for the film surface the differential equations (3) and (4) are coupled and can be solved only simultanously. In all the equations the turbulent diffusivities of momentum, mass and heat are represented by ε_τ, ε_M and ε_Q. For these suitable models have to be selected. We choose the simple concept of mixing length and following Grossman (Ref./1/) we devide the film into three different regions. Similar to the turbulent flow in pipes we define a region near the wall and an inner region. For the turbulent viscosity ε_τ we apply in the wall region the relation suggested by Van Driest (Ref./2/) for

$$0 \le \eta < \eta_w = 30\ \sqrt{Fr}/Re$$

$$\frac{\varepsilon_\tau}{\nu} = -\ \frac{1}{2} + \frac{1}{2}\sqrt{1+576\ (\frac{\eta}{\eta_w})^2\ \left[1-\exp(-\ \frac{30}{26}\ \frac{\eta}{\eta_w})\right]^2}\ . \qquad (10)$$

This formula was prefered to the modified form used by other authors because of simplicity and minimal differences between them (see also Ref./3/).

For the inner region or the turbulent core of the film flow we use the equation proposed by Reichardt (Ref./4/) for pipe flow, but modified for our film model

$$\eta_w \le \eta < \eta_s$$

$$\frac{\varepsilon_\tau}{\nu} = 7{,}73015\ \frac{\eta}{\eta_w}\ \frac{2-\eta}{2-\eta_w}\ \frac{3-4\eta\ +\ 2\eta^2}{3-4\eta_w\ +\ 2\eta^2_w}\ . \qquad (11)$$

Adapting the original Reichardt equation the radius in the pipe flow was replaced by the distance δ-y from film surface and the bulk diffusivity was made equal to that of equation (10) for $\eta=\eta_w$.

How it is seen we don't include the film surface in the bulk region. Following the eddy diffusivity concept of Lamourelle and Sandall (Ref./5/) we use for the third region the generalized correlation of Mills and Chung (Ref./3/ and /6/)

$$\eta_s \leq \eta \leq 1$$

$$\frac{\varepsilon_\tau}{\nu} = 0{,}00663\ (\delta/\Lambda)^2\ Ka^{1/3}\ Re^{1{,}678}(1-\eta)^2. \tag{12}$$

This concept takes into account the damping of turbulence near the film surface by surface tension.

The region of turbulent core intersects with the surface region at the non-dimensional wall distance η_s which can be calculated by equating the eddy diffusivities due to (11) and (12).

The turbulent diffusivities for mass and heat transport we choose here equal to that for momentum transport. The study of various literatures showed a certain amount of disagreement relating to the turbulent Prandtl number (see also Ref./1/). Frequently a constant value of 0,9 is used (e.g. see Ref./3/ and /7/) or a correlation is proposed taking into account the dependence on molecular Prandtl number (Ref./8/) and Reynolds number (Ref./1/ and /9/). Such correlations might be of importance for low Prandtl numbers and very low Reynolds numbers. For our problem, this is the simulation of absorbers with falling films of aqueous salt solutions, we must consider Prandtl numbers in the range from 5 to 15 and Reynolds numbers from 100 to 2000. Following the results of Jischa (Ref./9/) we therefore choose the turbulent Prandtl number and turbulent Schmidt number both equal to 1 as has been done by Siu-Ming Yih (Ref./10/), too.

For the system of equations (1) to (12) a computer program has been developed. It calculates the solution of the parabolic system of partial differential equations (3) to (9) by a method of finite differences. In addition to the boundary condition (6) the coolant temperature T_c can be hold constant, simulating an evaporating coolant, or is allowed to be calculated from heat conservation in the coolant flow so simulating the current or countercurrent heating of a fluid. Similar calculations are provided for the vapour to be absorbed. The vapour/liquid equilibrium, the thermophysical and transport properties for the system Water/Lithium Bromid is calculated by a subroutine package using correlations derived from literature (Ref./11/, /12/, /13/, /14/ and /15/).

RESULTS

To get a better insight in the behaviour of absorbers used in heat pump and heat transformer cycles numerous case studies by comupter simulation were made. Representative for almost all results two cases should be discussed in this contribution.

Case 1

The first case study considers an absorber in the first stage of a heat transformer. The amount of heat released in this absorber is used to generate steam in the evaporator of the second stage of the heat transformer. The absorber consists of a vertical tube bundle. The absorption takes place inside the tubes the evaporation is carried out outside the tubes. As prescribed by the heat transformer cycle the entering solution is subcooled (p, T_o, X_o, T_{eqo}) and the temperature T_g of the vapour to be absorbed and coming from the first stage evaporator is essentially lower than the solution temperature T_o. The temperature T_c of generated steam is nearly equal to T_o. Case 1 is characterized by the following values:

p = 0,785 bar L = 5 m

T_o = 430,43 K T_g = 366,7 K

X_o = 31,36 wt % H_2O T_c = 430,65 K

T_{eqo} = 447,4 K

The simulation of this case we have made for an absorber of fixed length L, various film mass flow rates $\dot{M}_{fo}$ and three values of the heat transfer coefficient α_c at the steam generation side. The one, $\alpha_c=0$, represents an adiabatic absorber, possible because of the subcooled solution at absorber inlet, and we consider this special case for comparison.

Fig.2 and 3 can be understood as performance charts of such an absorber. How it is seen from fig.2 the heat rate $\dot{Q}_w$ transfered through the wall to the evaporating water increases substantially with increasing film mass flow rate and, how to expect, is higher for higher evaporation heat transfer coefficients α_c. For low values of $\dot{M}_{fo}$ almost all of the absorption heat is transported to the coolant. However, as the film mass flow rate increases, quite a good part of it heats up the solution. Hence the solution's sensible heat increase $\dot{Q}_s$ and mean temperature rise ΔT become raised.

For the adiabatic case the heat transfer rate $\dot{Q}_w$ is equal to zero, of course, and the total absorption heat is stored in the solution (see dashed line for $\dot{Q}_s$ at $\alpha_c=0$ in fig.2). Comparing the adiabatic and the cooled cases it becomes obvious that the adabatic absorber is inefficient.

Fig.2 seems to tell us that for high heat transfer rates the absorber should be designed for high film mass flow rates that is for rather thick films. That this is not right in most cases is shown by fig.3.

The desginer of an heat pump or heat transformer tends normally to realize wide solution fields for his cycle to get a good coefficient of performance. Therefore the absorber (and desorber) has to work with rather high concentration differences from inlet to outlet. This, however, is realizable only by absorbers with thin films or with rather low film mass flow rates as indicates fig.3 very clearly. If the absorber can not give sufficiently high concentration increases the rich solution must be recirculated partly to just increase the input concentration of solution. This can lead to a decrease of temperature boost.

The dotted curves of fig.3 for ΔX_{eq} show us an other important fact. The rich solution at absorber outlet is subcooled, it is away from equilibrium. The condition of equilibrium at absorber outlet, however, is the base of almost all simulations of heat pump/transformer cycles.

Fig.4 shows the mean temperature profil of solution along the absorber. In the adiabatic cases (dotted lines) the solution temperature reaches an asymptotic value and that the earlier the lower the film mass flow rate. Indeed for $\dot{M}_{fo}$=0,25 kg/sm an absorber length greater than 1 m is useless.

Cooling the absorber allows to regain a part of the sensible heat. So it can be seen that for high film mass flow rates the simulated absorber of length L=5 m is obviously to short to remove an essential part of the solution's sensible heat (see also fig.2).

The strong increase of solution temperature in the first part of absorber gives the idea to boost the temperature of the useful heat even several degrees above the solution entrance temperature. Simulating that for various film mass flow rates and cooling heat transfer coefficients can lead to an optimum behaviour.

The local heat transfer coefficients are seen in fig.5. In the entrance region the heat transfer to the wall is relatively low. This is because quite a high part of the absorption heat rises the solution temperature (see fig.4). After the film temperature has reached its maximum and begins to decrease the heat transfer coefficient reaches its asymptotic and maximal value. This behaviour differs for different heat transfer coefficients α_c so that the overall heat transfer coefficient of the film depends on the conditions at the side of coolant, too.

The local mass transfer coefficient (fig.6) is almost constant along the absorber. Only for the adiabatic case we see a substantial decrease for positions distant more than 1 m from inlet. This indicates once more the inefficiency of absorbers without heat removal.

The asymptotic value of the local heat transfer coefficient α_w is independent of α_c and can therefore be compared with experimental correlations in literature. At this occasion we must remember that α_w is refered to the temperature difference between film surface equilibrium temperature T_s and wall temperature T_w. The comparision of heat transfer coefficient at absorber outlet (fig.7) with calculated values for condensation (Ref./16/) shows good agreement. The Prandtl number in our case is between 9,12 for the lowest Reynolds number and 8,28 for the highest film mass flow rate. So also a comparison with the experimental data of Chun and Seban can be made (see e.g. Ref./7/). Taking into account the higher Prandtl numbers our calculated curve agrees well with these data. In the first part of absorber the temperature profil in the film already is not developed. Therefore the heat transfer coefficient depends substantially on the cooling as it has been mentioned before and as is seen from the dotted curves in fig.7, too.

Case 2

In contrast to the first case an other application is discussed. The absorber is a component of a heat pump for warm water supply. Thus its working temperature is quite lower and the absorber works on the low pressure level (absorbers of heat transformer cycles are working on the upper pressure level). Hence, the lean solution is output from the expansion valve and therefore almost always on the bubble point.

For this second case we chose the absorption outside of a vertical tube bundle. The cool vapor coming from the evaporator enters the shell and for simulation a large vapor volume at rest is taken into account. The cold water is flowing countercurrently in the tubes, that is from bottom to top of absorber.

At the absorber inlet we prescribe the following values

p = 0,0424 bar	L = 5 m
T_o = 361,85 K	T_g = 303,65 K
X_o = 33,43 wt % H_2O	T_{co} = 357 K

saturated

The simulation of this case was made for constant temperature of generated warm water. The water mass flow rate was varied resulting in various water entrance temperatures that means the results for this case are compared for three different water temperature differences ΔT_c.

Fig.8 and 9 are the performance charts for this case. In contrast to case 1 now we have a decrease of solution temperature from inlet to outlet. Thus a certain amount of the transfered heat rate comes from cooling the solution that is up to almost 40 % for high film mass flow rates. Again it is observed that sufficient concentration increases only can be realized by designing the absorber for thin films. But especially for low film mass flow rates the rich solution at absorber outlet is far away from equilibrium.

CONCLUSIONS

It is useful to study the behaviour of an absorber in heat pump or heat transformer applications by a simulation concept which don't need any transfer coefficients. It is much more better to use a suitable turbulence model. So it is possible to consider in the right manner this part of absorber where the temperature and concentration profiles are developing and where the corresponding transfer coefficients differ from the asymptotic values. Therefore the simulation results in more details and is able to describe all the influences of different boundary conditions.

To verify the turbulence model a comparison of the calculated asymptotic transfer coefficients with experimental data is suitable. An adjustment of the distribution of turbulent diffusivities to rough surfaces or surfaces with turbulence promotors should be possible.

The substance of absorber simulations, represented here only by two cases, are the following conclusions:

1) The falling film absorber with simultaneous heat removal is much more efficient than the adiabatic one.
2) The realizable absorber length is not sufficient to reach equilibrium conditions for the solution at outlet. The simulation of heat pump or heat transformer cycles therefore should take into account this fact.
3) According to the cooling conditions the obtainable concentration increase is limited by the realizable absorber length and the minimal film thickness. If greater solution fields are required recirculating of a part of the rich solution is necessary. In general this decreases the possible temperature boost.
4) To realize sufficient solution fields the absorber must work rather with low film mass flow rates, possibly in the laminar or wavy laminar region. Therefore some provisions for the enhancement of heat and mass transfer - e.g. like turbulence promotors - can improve the absorber and should be taken into account by further investigations.

REFERENCES

/1/ Grossman, G. and M.T. Heath: Simultaneous heat and mass transfer in absorption of gases in turbulent liquid films, Int. J. Heat Mass Transfer 27, No.12 (1984)

/2/ Van Driest, E.R.: On Turbulent Flow Near a Wall, J. Aero.Sci., Nov.1956

/3/ Sandall, O.C., O.T. Hanna and C.L. Wilson: Heat transfer across turbulent falling liquid films, Heat Transfer - Niagara Falls 1984, AIChE Symposium Series 80, No.236

/4/ Reichardt, H.: Vollständige Darstellung der turbulenten Geschwindigkeitsverteilung in glatten Leitungen Z. Angew. Math. Mech. 31 (1951)

/5/ Lamourelle, A.P. and O.C. Sandall: Gas absorption into a turbulent liquid, Chem.Eng.Sci. 27 (1972)

/6/ Mills, A.F. and D.K. Chung: Heat transfer across turbulent falling films, Int. J. Heat Mass Transfer 16 (1973)

/7/ Blangetti, F., R. Krebs and E.U. Schlünder: Condensation in Vertical Tubes - Experimental Results and Modelling, Chem.Eng.Fund. 1, No.2 (1982)

/8/ Malhotra, A. and S.S. Kang: Turbulent Prandtl number in circular pipes, Int. J. Heat Mass Transfer 27, No.11 (1984)

/9/ Jischa, M.: Turbulent heat and mass transfer, Int.Chem.Eng. 25, No.2 (1985)

/10/ Siu-Ming Yih and Jung-Liang Liu: Prediction of Heat Transfer in Turbulent Falling Liquid Films with or without Interfacial Shear, AIChE J. 29, No.6 (1983)

/11/ Brunk, M.F.: Thermodynamische und physikalische Eigenschaften der Lösung Lithiumbromid/Wasser als Grundlage für die Prozeß-Simulation von Absorptions-Kälteanlagen, Klima-Kälte-Heizung 10 (1982)

/12/ McNeely, L.A.: Thermodynamic Properties of Aqueous Solutions of Lithium Bromide, ASHRAE Trans. 85, pt.1 (1979)

/13/ Löwer, H.: Thermodynamische und physikalische Eigenschaften der wäßrigen Lithiumbromid-Lösung, Diss. TH Karlsruhe 1960

/14/ Bromley, L.A.: Thermodynamic Properties of Strong Electrolytes in Aqueous Solutions, AIChE J. 19, No.2 (1973)

/15/ Reid, R.C., J.M. Prausnitz and T.K. Sherwood: The Properties of Gases and Liquids, McGraw-Hill Book Comp., 3rd Ed. (1977)

/16/ VDI-Wärmeatlas: Berechnungsblätter für den Wärmeübergang, Filmkondensation Ja 1-14, VDI-Verlag GmbH (1984)

/17/ Pflügl, M. and F. Moser: Simultaneous heat and mass transfer in technical transfer processes, Proc. 4th Italian-Yugoslav-Austrian Chem.Eng. Conf. Grado (1984)

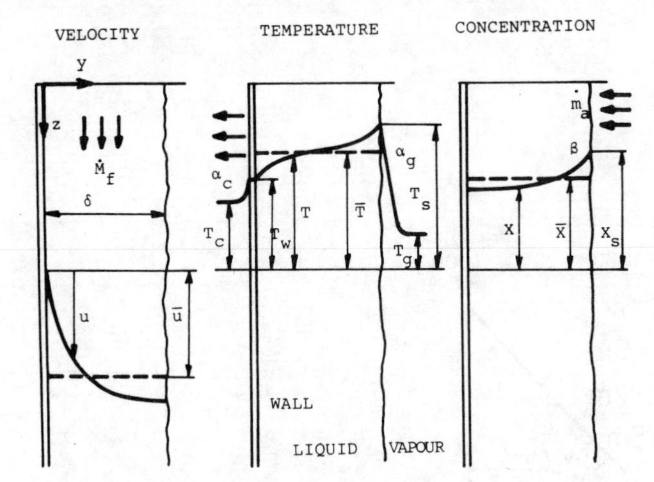

Fig.1: Profiles of velocity, temperature and concentration in a falling film due to absorption and cooling

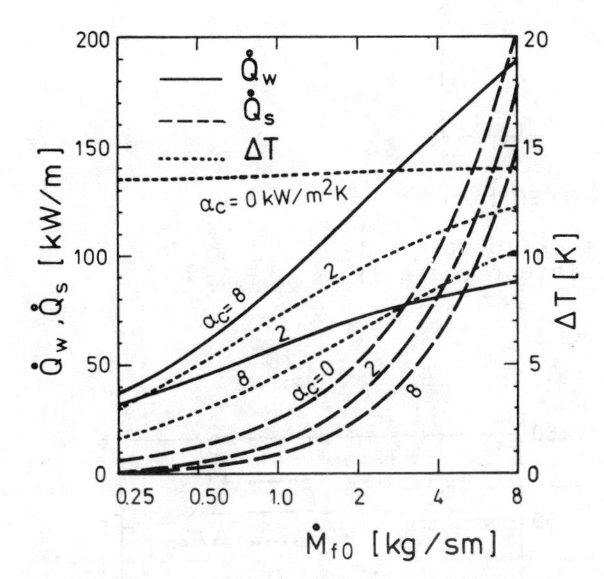

Fig.2: For steam generation transfered heat rate, heating rate and temperature increase of rich solution (case 1)

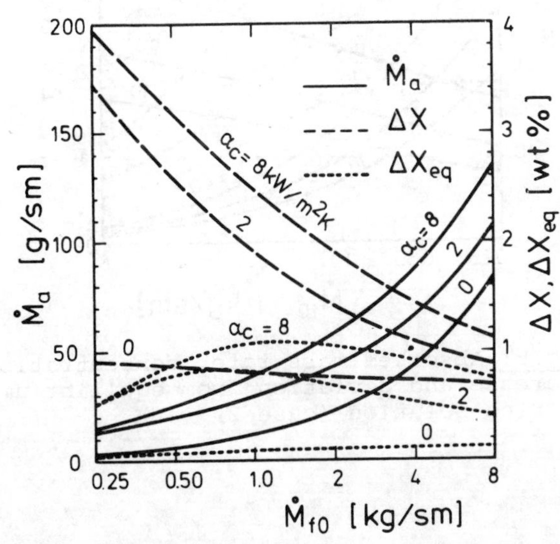

Fig.3: Absorbed mass rate, concentration increase and deviation from equilibrium of rich solution (case 1)

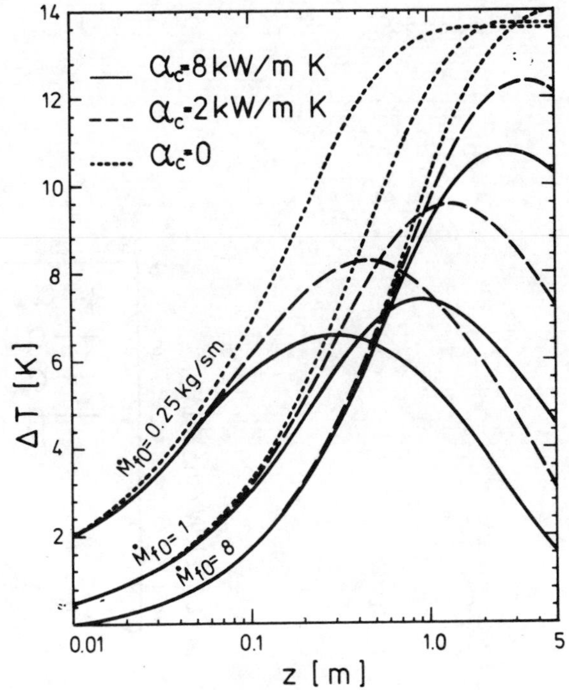

Fig.4: Local temperature rise over solution entrance temperature (case 1)

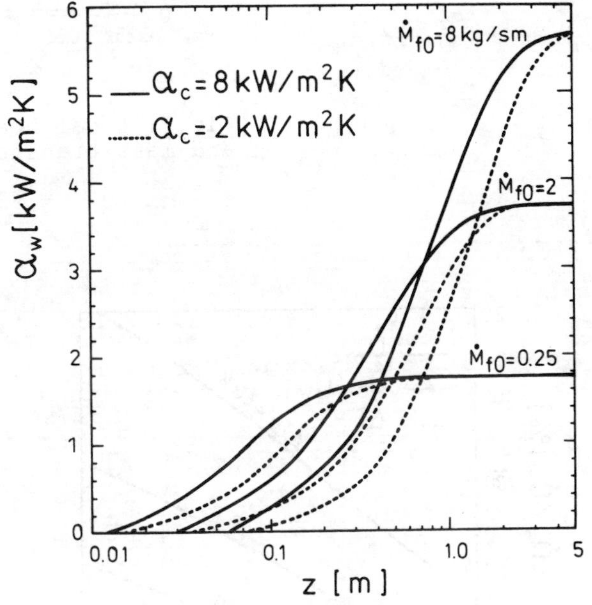

Fig.5: Local heat transfer coefficients (case 1)

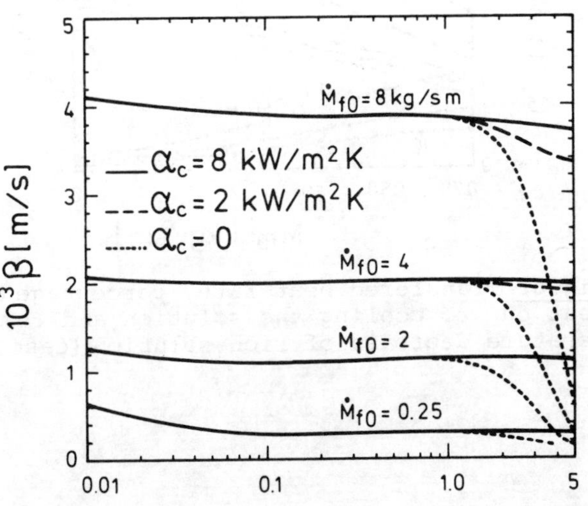

Fig.6: Local mass transfer coefficients (case 1)

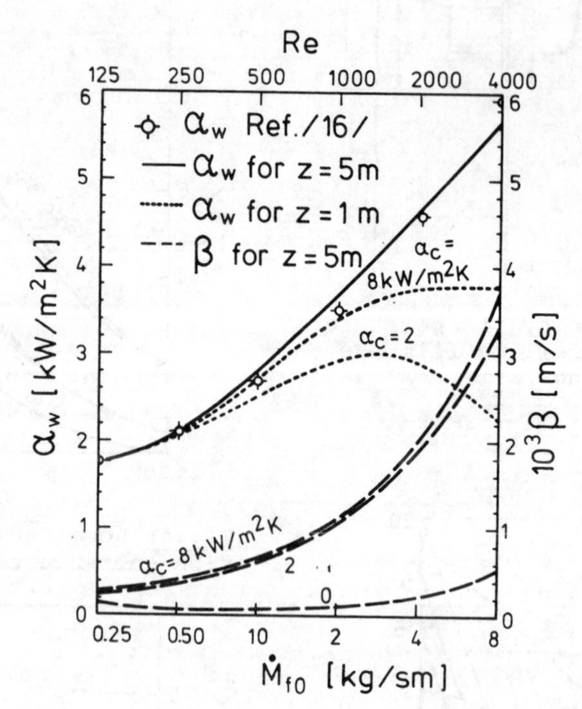

Fig.7: Local heat transfer coefficient and mass transfer coefficient (case 1)

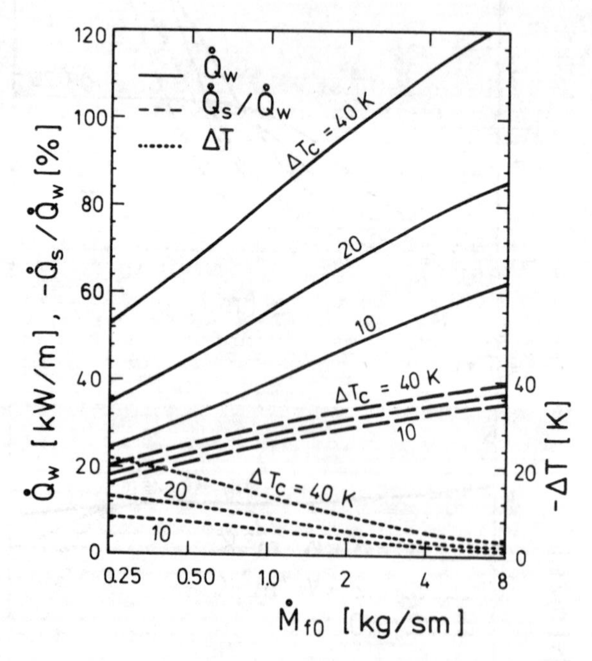

Fig.8: Transfered heat rate, percentage of heat due to cooling the solution and temperature decrease of rich solution (case 2)

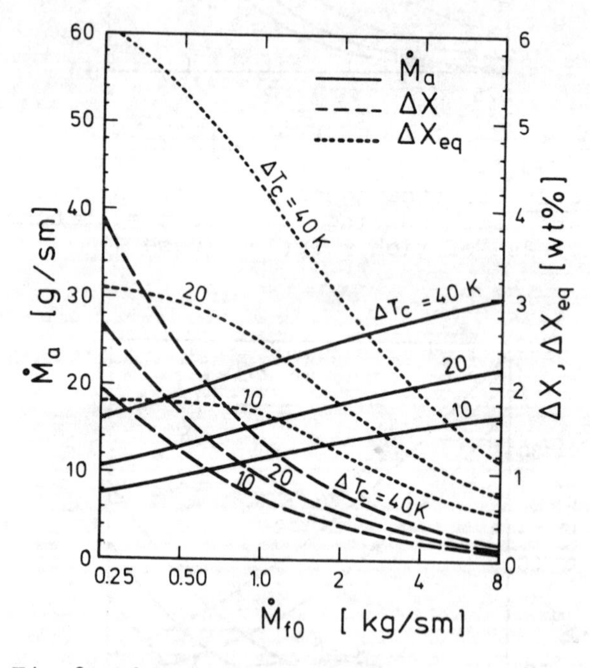

Fig.9: Absorbed mass rate, concentration increase and deviation from equilibrium of rich solution (case 2)

3rd International Symposium on the

Large Scale Applications of Heat Pumps

Oxford, England: 25-27 March 1987

PAPER G2

ECONOMIC CRITERIA FOR APPLICATION OF SINGLE STAGE OR DOUBLE STAGE ABSORPTION HEAT TRANSFORMERS

W.Kern

Physik-Department E19 der
Technischen Universität München
8046 Garching, W-Germany

SYNOPSIS

It will be shown that two stage absorption heat transformers can be more economical than single stage machines. The investment cost per unit heat recovered are estimated for the single stage, the double-effect and double-lift cycle as a function of the temperature lift. For temperature lifts of 20°C, 40°C and 60°C the pay-back time and the savings are compared.

Nomenclature

Symbol	Name	Unit
A	heat exchange area	m^2
COP	coefficient of performance= heat output/heat input	
Δ	mean logarithmic temperature difference	K
k	heat transfer coefficient	$kW/(m^2 K)$
T	internal temperature	°C
t	external temperature	°C
p	price per heat exchange area	DM/m^2
Q	heat output or input	kW
q	heat ratio = Q/Q_a	

Indices

a	absorber
c	condenser
e	evaporator
g	generator
s	solution heat exchanger
x	internal heat exchange

Introduction

The limited capabilities of the absorption plants which have been installed up to the present time, has given rise to serious consideration of the so-called double stage processes. Of these mainly two are of major interest:

- the double-effect cycle which offers an increase of 30 % in COP,
- the double-lift cycle which allows a very high temperature difference between waste heat and recovered heat.

Although these cycles need more components, there is strong evidence that they will prove to be better under certain conditions. This paper is not intended to be an exact economical analysis but a brief and very general consideration of the benefits of the double stage cycles.

Description of the Considered Machines

The calculations are based on the data published by Lowell A. McNeely /1/.

Single Stage Heat Transformer:

A scheme of this well-known cycle in a ln p - 1/T diagram is shown in figure 1.
T_c, T_e, T_g and T_a are the process temperatures (internal temperatures) of the absorption heat transformer in which T_g and T_a are mean temperatures with regard to the solution width. The external temperatures (t_c, t_e, t_g and t_a) are the temperatures of the heat sink and source. For the COP, that is the ratio of heat output in the absorber at the temperature t_a and the sum of heat input at the temperatures t_e to the evaporator and t_g to the generator, values up to 0.48 have been reached.

Double-Effect Heat Transformer:

In figure 2 a scheme of the double effect heat transformer is shown. In addition to the single stage machine the condenser C' and the generator G' for the internal heat exchange and one additional solution heat exchanger S_1 are needed. The cycle yields a temperature lift which is lower than that of a single stage cycle for comparable conditions. However a COP as high as 0.64 can be expected.

Double-Lift Heat Transformer:

Just as the double-effect machine the double-lift heat transformer (figure 3) has a internal heat exchange between the additional components absorber A' and evaporator E'. The temperature lift is larger than that of the single stage machine. Consequently the COP is lower. It is in the range of 0.30 /2/.

Held at St. Catherine's College Oxford, England. Organised and sponsored by BHRA, The Fluid Engineering Centre, Cranfield, Bedfordshire, MK43 0AJ England.

Optimization of the Heat Exchange Areas

As parameter for the comparison of the three cycles we have chosen the ratio of the heat output of the absorber A at the temperature t_a and the investment cost for the machine. In a first approach it is supposed that this investment cost is proportional to the sum of the total installed heat exchange area A_{total}. This area consists of the areas for the heat exchange from each unit to the external heat sinks and sources, the areas for the solution heat exchangers and additional for the two stage cycles the area for the internal heat exchange:

In the case of the single stage cycle:

$$
\begin{aligned}
A_{total} &= A_a + A_e + A_g + A_c + A_s = \\
&= Q_a / (k_a \Delta_a) + \\
&+ Q_e / (k_e \Delta_e) + \\
&+ Q_g / (k_g \Delta_g) + \\
&+ Q_c / (k_c \Delta_c) + \\
&+ Q_s / (k_s \Delta_s)
\end{aligned}
$$

In the case of the double-effect and double-lift cycles:

$$
\begin{aligned}
A_{total} &= A_a + A_e + A_g + A_c + A_{s1} + A_{s2} + A_x = \\
&= Q_a / (k_a \Delta_a) + \\
&+ Q_e / (k_e \Delta_e) + \\
&+ Q_g / (k_g \Delta_g) + \\
&+ Q_c / (k_c \Delta_c) + \\
&+ Q_{s1} / (k_{s1} \Delta_{s1}) + \\
&+ Q_{s2} / (k_{s2} \Delta_{s2}) + \\
&+ Q_x / (k_x \Delta_x)
\end{aligned}
$$

The overall heat transfer coefficients k are assumed to be identical. Though this is a rather restricting assumption the results will only be changed quantitatively but not qualitatively. The temperature differences are the logarithmic mean temperature differences. If we take A_{total} to be given we can find a distribution of the quantities of the individual exchange areas which maximizes the output of useful heat. An optimization procedure has been carried out in order to obtain a maximum in output heat per installed transfer area. In a numerical calculation the heat output per investment cost as a function of the temperature lift was computed for all three cycles. The following values for the heat transfer coefficients, the condenser temperature, the generator temperature, the evaporator temperature and the price of heat exchanger area were assumed for the numerical computation:

$$k_a = k_e = k_g = k_c = k_s = 1 \text{ kW/(m}^2\text{K)}$$

$$t_c = 30^oC \qquad t_e = t_g = 85^oC$$

$$p_a = p_e = p_g = p_c = p_s = \text{DM } 500.\text{-/m}^2$$

In figure 4 you see a graphic presentation of the results obtained by the numerical calculation:
The heat output per investment cost is plotted against the temperature lift $t_a - t_e$.

At first we will discuss the single stage heat transformer:
There is a linear relationship between the heat output per investment cost and the temperature lift. The maximum temperature lift is 67°C, when the internal temperatures T and the external temperatures t become equal. Then, there is no more driving temperature difference for the heat transfer and consequently no heat flux. At a temperature lift of zero the machine has the highest heat output for a given investment. The COP was computed to be about 0.49 and doesn't vary with the temperature lift.
The double-lift cycle yields the highest maximum temperature lift (about 145°C). For a given output at vanishingly small temperature lift the investment cost of the double lift heat transformer is about 1.5 times higher, which is caused by the higher number of exchange units. Consequently there exists a crossover point of the curves of the single stage and the double lift cycle at about 34°C temperature lift. For a temperature lift lower than this value the single stage and in the case of a larger temperature lift the double-lift cycle has the lower investment cost for a given heat output.
Finally we regard the double-effect heat transformer:
At a temperature lift of zero it yields about the same heat output for a given investment cost as the double-lift cycle. This is because the number of exchange units is the same. The small difference is due to characteristics of the solution field. The double-effect heat transformer yields the lowest maximum temperature lift and consequently investment costs of the single stage and of the double-lift cycles are always lower for a given heat output.
But there also exists a difference in the COPs of the three cycles and this can change the judgement about profitability considerably. The COP for the single stage machine was computed to be 0.49, for the double-effect cycle the COP was 0.65 and for the double-lift cycle it reached 0.32, independent of the temperature lift. We will take into account this effect of different COPs when the energy saving as a function of the running time of the plant is considered.
It was possible to minimize the heat exchanger surface (excluding the solution heat exchanger) by analytical methods /3/. The results are shown in figure 5. Comparing with figure 4 it can be seen that the linear temperature dependence is confirmed. The difference in slope and axis intercept must be attributed to the solution heat exchanger.

Pay-back Time and Profit

It is our intention to compare the three cycles by quite simple economic considerations. So we will take into account neither interest nor all the running expenses but only the investment cost and the saving of energy cost. Of course, the value of the heat recovered increases with temperature. Again for our simple comparison we assume a fixed price of 10.- DM/GJ equivalent to about 22.- DM/ton of steam. Moreover we assume that the heat input to the heat transformer as well as the cooling water are for free.
In the following we are considering a process with 1 MW waste heat which may be upgraded by one of the competing devices.
In figures 6, 7 and 8 the saving is presented as a function of the operation time. The three

figures are for temperature lifts of 20°C, 40°C and 60°C.
The axis intercept at operation time zero gives the first cost of the heat transformer. The slope is proportional to heat output times energy price. Since all three curves are taken for the same heat input the slopes are proportional to the COPs, that is 0.49 for the single stage cycle, 0.66 for the double-effect cycle and 0.33 for the double-lift cycle. The pay-back time therefore is the intercept at the abscissa, the time axis. From this time on the heat transformer yields real gain until it has to be dismantled.
In figure 6 the temperature lift amounts to 20°C. According to figure 4 the investment cost per heat output is 1.3 times greater for the double-lift heat transformer than for the single stage machine. Yet, since for equal heat input the heat output is proporional to the COP, the first costs for given heat input are almost equal. An analogous argument accounts for the high first cost of the double-effect heat transformer for given heat input which is apparent in figure 6.
The following conclusion can be drawn: for a temperature lift of 20°C, the single stage machine has the shortest pay-back time. However for running time longer than 6 years the double-effect machine gives higher savings due to its high COP.
In figure 7 the same curves for a temperature lift of 40°C are presented. The double-effect cycle cannot provide such a temperature lift (figure 4 and 5). At this temperature lift and for the given input power of 1 MW the investment cost for the single stage cycle is about twice the cost of the double-lift cycle. As can be seen from figure 7 the double lift heat transformer has a slightly smaller pay-back time compared to the single stage heat transformer. Yet, after about twice the pay-back time the single stage heat transformer has higher savings.
If the temperature lift is still larger e.g. 60°C as shown in figure 8, the investment cost of the single stage cycle is so high compared to the double-lift cycle, that only after a long operating time the single stage cycle becomes more profitable. So the double-lift cycle is the most economic if a high temperature lift is required in spite of its having three additional exchangers.
It should be noted that the pay-back time for this cycle only increases by a factor of 1.5 if the temperature rise is increased from 40°C to 60°C. Therefore the double-lift cycle appears to be the most profitable one for the high temperature rises commonly desired in industry.

Conclusions

In this paper we wanted to show that multi-stage absorption cycles, in general, can be economically superior to the single stage cycle. This fact has already been known for the double-effect chiller. In contrast to chillers for heat transformers the double-lift cycle is favourable for higher temperature lifts, as compared to the single stage cycles. The physical reason for this result is as follows:
For high temperature upgrading the double-lift cycle permits larger driving temperature differences for the heat transfer into and out of the machine. Although there are more components, the total heat transfer area is smaller. In contrast to chillers, the double-effect heat transformer is, in general, economically less attractive in spite of the high COP.
For real applications the range of applicability of the individual devices may be altered for the following reasons: The heat transfer coefficients and prices per unit area have been assumed to be equal for all exchange units. Yet, these may differ drastically for the individual exchange units, depending on e.g. corrosiveness of media, fouling etc.. Also the running cost for cooling water, which has not been taken into account, may render the double-effect heat transformer more competitive.

Acknowledgements

I would like to thank Prof. G. Alefeld for helpful discussions and also J. Scharfe and F. Ziegler for help in the numerical computation and in writing this paper.

References

/1/ McNeely, A. L., Thermodynamic Properties of Aqueous Solutions of Lithium Bromide, ASHRAE Transactions 85, (1979), part 1

/2/ Ikeuchi, M. et al., Design and Performance of a High-Temperature-Boost Absorption Heat Pump, ASHRAE Transactions ,(1985), V.91, part 1

/3/ Alefeld, G., Optimization of Heat Exchange Areas at Absorption Cycles, to be published

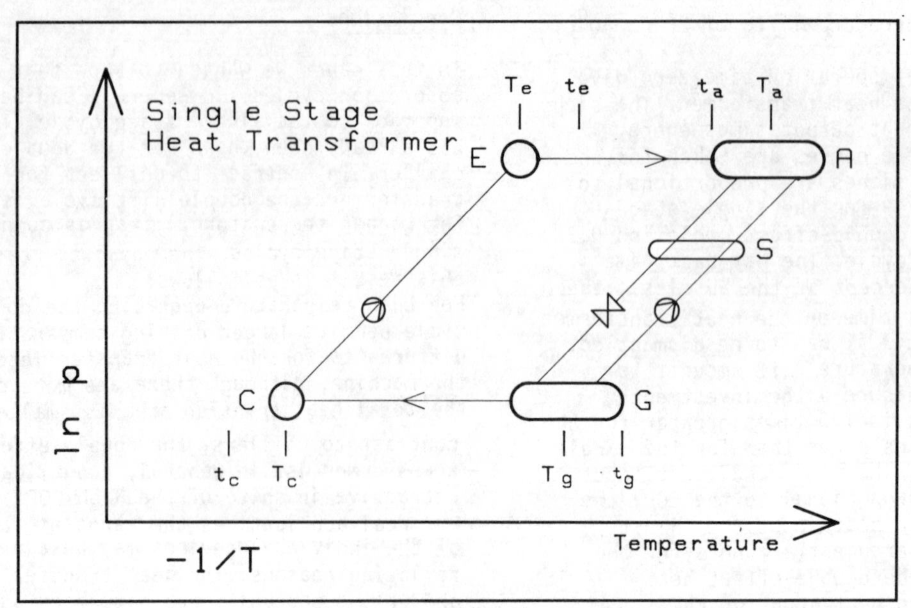

Fig.1: Single Stage Absorption Heat Transformer (A: Absorber, C: Condenser, E: Evaporator, G: Generator, S: Solution Heat Exchanger)

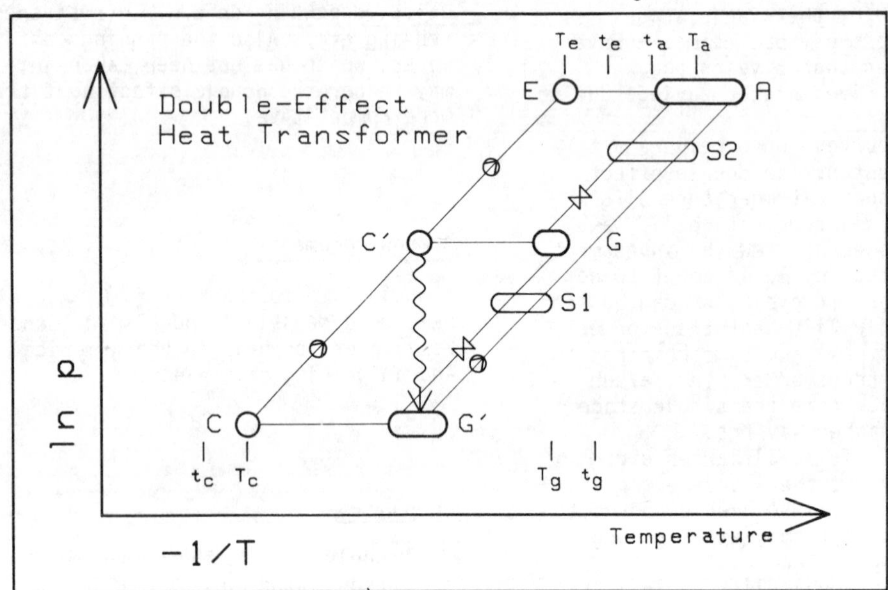

Fig.2: Double-Effect Absorption Heat Transformer (A,E,C,G,S: see figure 1, G': Second Generator, C': Second Condenser)

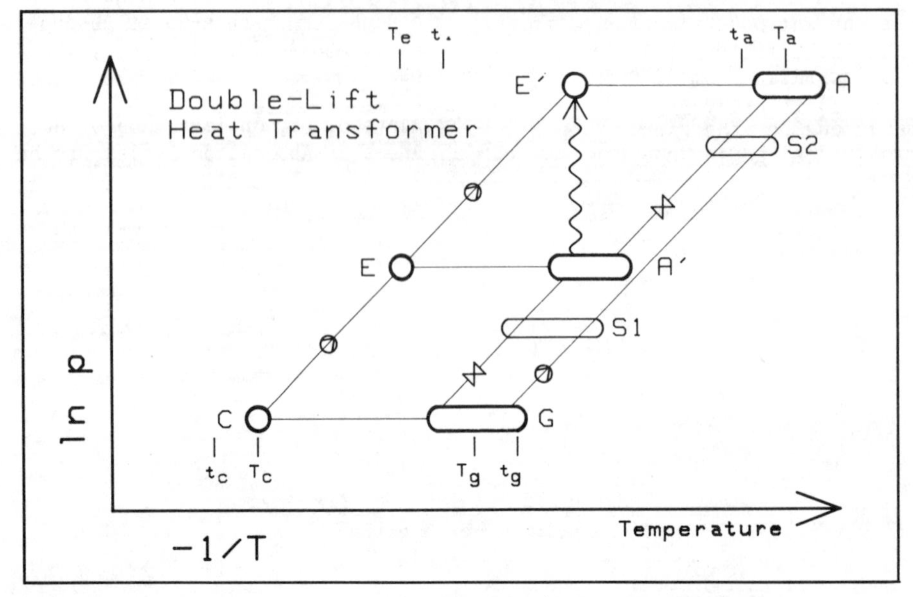

Fig.3: Double-Lift Absorption Heat Transformer (A,E,C,G,S: see figure 1, A': Second Absorber, E': Second Evaporator)

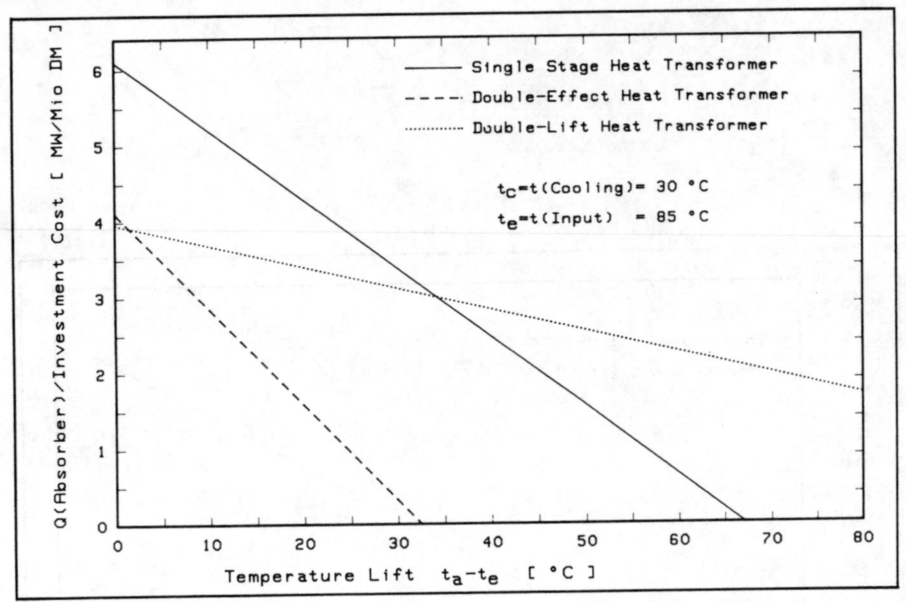

Fig.4: Heat Output per Investment Cost as a Function of the Temperature Lift (numerical calculation)

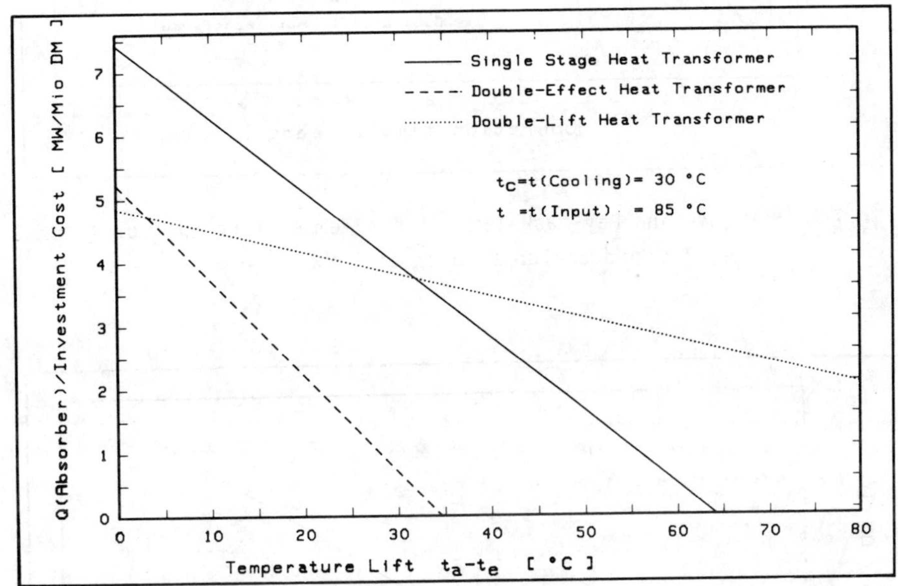

Fig.5: Heat Output per Investment Cost as a Function of the Temperature Lift (analytical calculation)

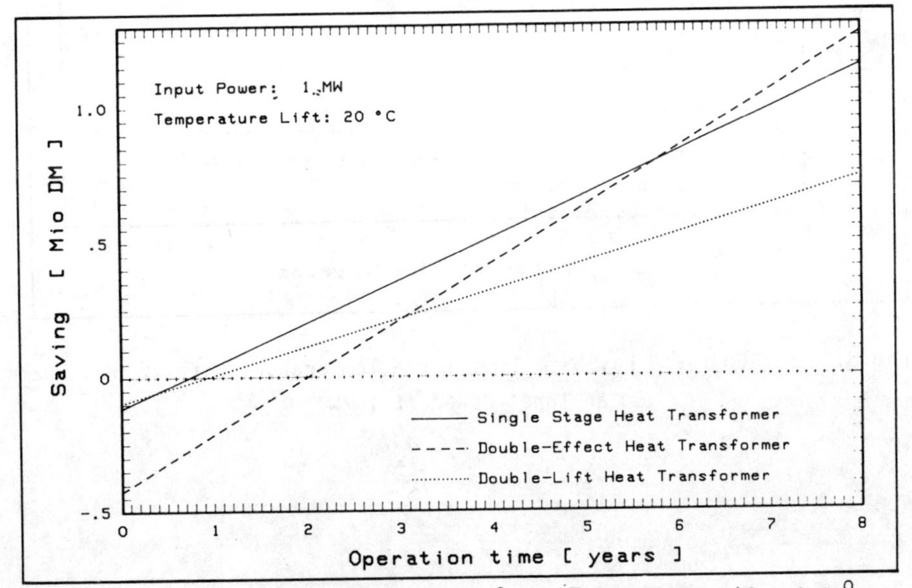

Fig.6: First Cost and Pay-back Time for a Temperature Lift of $20^{0}C$ and an Input Power of 1 MW

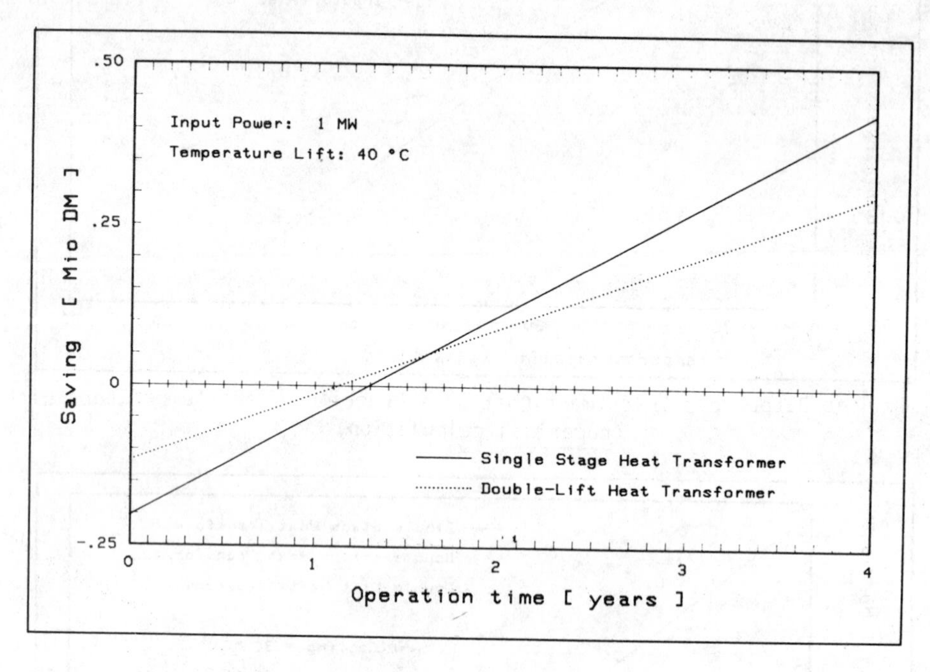

Fig.7: First Cost and Pay-back Time for a Temperature Lift of 40^{0}C and an Input Power of 1 MW

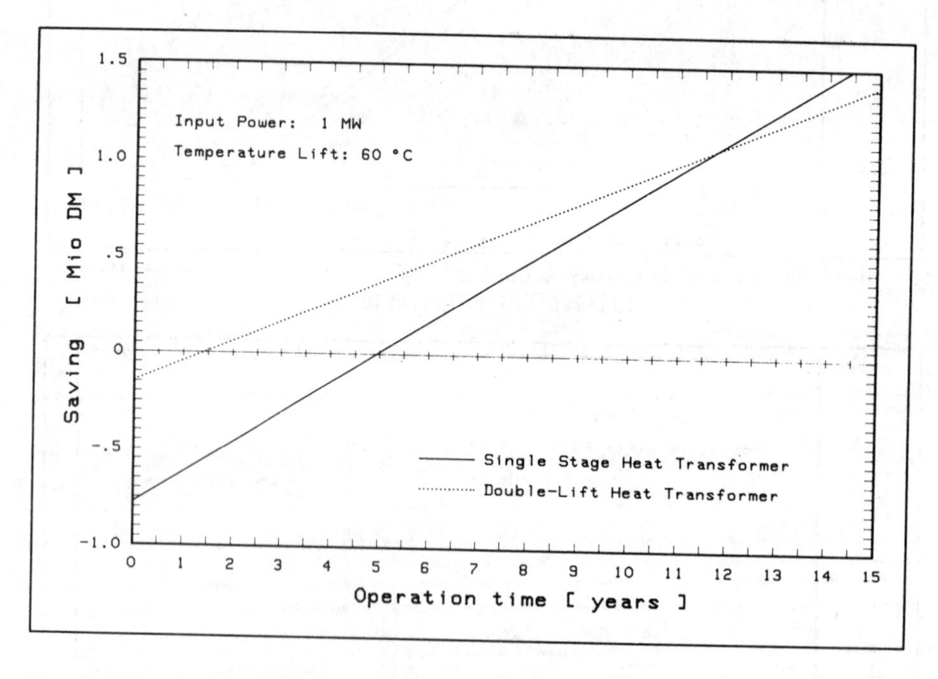

Fig.8: First Cost and Pay-back Time for a Temperature Lift of 60^{0}C and an Input Power of 1 MW

3rd International Symposium on the

Large Scale Applications of Heat Pumps

Oxford, England : 25-27 March 1987

PAPER G3

PART-LOAD BEHAVIOUR OF AN ABSORPTION HEAT TRANSFORMER

P.Riesch, J.Scharfe, F.Ziegler, J.Völkl, G.Alefeld

Physik-Department E 19 der Technischen Universität München, 8046 Garching, W-Germany

SYNOPSIS

A single-stage heat transformer working with water/LiBr has been built and extensively tested. The part-load behaviour was studied between 8 and 2 kW. The COP varies between 0.48 and 0.42. The heat output can be plotted as a linear function of the difference between the evaporator - condenser and absorber - generator temperature. This relation can be used to estimate the investment cost for industrial applications.

Nomenclature

A	Heat-transfer area	[m^2]
a	Part of the total heat-transfer area	
c	Specific heat capacity	[kJ/kg·K]
COP	Coefficient of performance	
k	Overall heat-transfer coefficient	[$kW/m^2 \cdot K$]
l	Heat of solution	[kJ/kmol]
L	Heat losses	[kW]
m	Mass-flow rate	[kg/s]
p	Pressure	[bar]
Q	Heat-flow rate	[kW]
r	Heat of vaporization	[kJ/kmol]
t	External temperature	[°C]
T	Internal temperature	[K]
x	Concentration of refrigerant	[kg/kg]
ΔT	Mean logarithmic temperature difference	[K]

Meaning of the indices:

A	Absorber
G	Generator
C	Condenser
E	Evaporator
S	Solution-heat exchanger
s	Strong absorbent
w	Weak absorbent
'	Ideal insulated
"	Ideal insulated with infinite solution heat-exchange area
0,1,2	Different levels
i	Variable index

INTRODUCTION

Heat transformers are mostly used for the recovery of waste heat from industrial processes like drying or distillation. Approximately one half of the waste heat at medium temperature is boosted to a higher temperature, while the other half of the waste heat is rejected to the ambient. In this paper experimental results are reported for the COP and the temperature lift obtained on a heat transformer of laboratory scale. The machine has been operated at different thermal loads by varying the temperature of cooling water, waste heat and upgraded heat.

DESCRIPTION OF THE CYCLE

The absorption heat transformer uses the effect of the boiling-point elevation of solutions. This property of the working pair can be described by a (log p) - (-1/T) - diagram. In figure 1 the four main components of the heat-transformer cycle (evaporator, absorber, generator and condenser) are plotted into a log p - 1/T - diagram according to the pressure and temperature inside the component /Nie/. In the evaporator the refrigerant (x=1) is vaporized at the temperature T_E by supply of waste heat. In the absorber this vapour is absorbed by the strong absorbent ($x=x_s$) and releases the heat of absorption at a high temperature level around T_A. The diluted absorbent ($x=x_w$) is regenerated in the condenser-generator pair which works reverse to the evaporator-absorber part, but at a lower pressure level. Refrigerant vapour is produced in the generator and flows to the condenser where it is liquefied at T_C. By cooling the condenser down to ambient the heat at temperature T_G required to reconcentrate the weak absorbent again can be provided by the waste heat source. The cycle is closed by pumping the liquid refrigerant back into the evaporator and the strong absorbent into the absorber.
The performance of the machine is

Held at St. Catherine's College Oxford, England. Organised and sponsored by BHRA, The Fluid Engineering Centre, Cranfield, Bedfordshire, MK43 0AJ England.

improved by installing a counterflow heat exchanger between the weak and strong absorbent. This solution heat exchanger reduces the amount of sensible heat transported by the weak solution from the absorber to the generator.

THE EXPERIMENTAL SET-UP

For the experiments water was used as refrigerant and an aqueous solution of lithium-bromide as absorbent. The figures 2 and 3 show the experimental set-up. Evaporator, absorber, generator and condenser are four vertical falling-film heat exchangers which are cooled or heated by a turbulent flow of water as heat-transfer fluid. The heat input to the evaporator and generator or the output from the absorber and condenser respectively is measured by the mass-flow rate and temperature difference between inlet and outlet of the heat-transfer fluid. The location of flowmeters and thermometers (thermocouples) is indicated in figure 2.
The height of the four heat exchangers is 2.2 m with a heat-transfer area of 0.9 m^2 each. The liquid refrigerant drained from the condenser is fed to the evaporator by a small membrane pump. In the same way the strong salt-solution from the generator and the weak solution from the absorber is circulated through the solution-heat exchanger to the head of the absorber column and the generator column respectively. A 13 m long coaxial tube wound to a coil was used as a solution-heat exchanger. Its heat-transfer area amounts to 0.5 m^2 for the inner tube surface, while the outer surface is enlarged with fins. Glass cylinders are used as vacuum vessels for the evaporator, absorber, generator was condenser. The whole apparatus was insulated for the measurements to keep the heat losses small.

COEFFICIENT OF PERFORMANCE

The heat input or output at the four heat exchangers was calculated from the measured mass-flow rate and the temperature difference $|t_i - t_o|$ between inlet and outlet of the heat-transfer fluid. These heat loads ranged from 2000 to 8000 W. The heat losses L' and the pumping energies P are negligible for large-scale plants. In the case of our laboratory-scale unit the measured heat loads must be corrected by the values of the net heat losses $L = L' - P$. Equation (1) gives the balance of energy for the whole set-up:

$$Q_G + Q_E - (Q_A + Q_C) = (L_G' - P_G) + (L_E' - P_E) + (L_A' - P_A) + (L_C' - P_C) \quad (1)$$

Since the condensation temperature T_C was a few degrees below room temperature no condensation could occur on the glass cylinder and furthermore the heat loss at the condenser L_C' can be neglected. The energy input P_C of the pump for the condenser is smaller than 5 W and thus negligible. The energy input P_E, P_A or P_G at the three other pumps amounts to about 60 W each. For the net heat loss at the generator L_G a value of 200 W was measured and the same value was estimated for the net heat loss L_E. The loss at the absorber L_A was measured as 550 W. In figure 4 two curves for the COPs are shown: For the lower curve the COP is calculated with the measured heat input and output, without applying any correction for heat losses. This COP is calculated according to equation (2).

$$COP = \frac{Q_A}{Q_G + Q_E} \quad (2)$$

For the upper curve the measured heat loads are corrected for the net heat losses, resulting in the following COP':

$$COP' = \frac{Q_A + (L_A' - P_A)}{Q_G - (L_G' - P_G) + Q_E - (L_E' - P_E)} \quad (3)$$

Figure 4 shows that even for a laboratory model the high values of COP'= 0.48 can be reached. Furthermore, it can be seen that the COP' decreases only slightly in spite of a load decrease as large as a factor of 4. This result demonstrates the excellent part-load behaviour of absorption machines.

THE QUALITY FACTOR OF THE HEAT-TRANSFORMER CYCLE

We define the quality factor of the heat-transformer cycle as the measured COP' devided by the COP_C of a completely reversible Carnot machine. The COP_C is defined as

$$COP_C = \frac{(T_1 - T_0)}{T_1} \cdot \frac{T_2}{(T_2 - T_0)} \quad (4).$$

First we evaluate a $COP_{C,E}$ which is determined by the temperatures of the external heat source and heat sink. For T_0 the mean value of the inlet and outlet temperature of the cooling water at the condenser has to be taken. T_1 and T_2 are defined analogously. This Carnot coefficient $COP_{C,E}$ is represented by the middle curve in figure 5. The corresponding quality factor is the second curve from above. With increasing heat output the quality factor $COP'/COP_{C,E}$ decreases as a consequence of increasing irreversibilities due to rising heat loads on the four exchange units.
The second curve from below represents the reversible $COP_{C,I}$ calculated with internal process temperatures. Again for the absorber and generator the middle temperature must be taken. The corresponding quality factor $COP'/COP_{C,I}$ (first curve from above) characterizes the irreversibilities without heat transfer. The high quality factor of

about 90 % indicates that the irreversibilities inherent to the process are very small. No other thermally operated machine has such a high quality factor.

PART-LOAD BEHAVIOUR

Figure 6 shows typical temperatures of the tested apparatus for heat outputs of 2, 5 and 8 kW. If the waste-heat temperature $t_1 = 75^oC$ and the temperature of the useful heat $t_2 = 110^oC$ were kept constant, the heat output Q_A can be changed only by varying the cooling-water temperature t_0. According to figure 6 the change of the cooling-water temperature causes an increase or decrease of the driving temperatures for the heat transfer at the four exchange units. The heat output Q_A at the absorber was measured for different temperatures of the external heat-transfer fluids. Furukawa et al. /Fur/ derived a linear relation between the heat output Q_A and a certain difference between the inlet temperatures at the four heat-transfer fluids similar to equation (5). Furukawa et al. also verified this relation by measurements. Following the same ideas a similar relation, yet with the heat-transfer coefficients and working-pair properties explicitly taken into account, will be derived in the next chapter. This theoretical approach suggests to plot the heat output Q_A against a total temperature difference ΔT_T as defined by equation (5) (see figure 7).

$$\Delta T_T := (t_E - t_C) - R\,(t_A - t_G) \qquad (5)$$

The temperatures t in equation (5) are the mean values of the inlet and outlet temperature of the heat carrier. The dimensionless factor R can be determined by the properties of the working-fluid pair water/LiBr and amounts to 0.86 in our experiments. Figure 7, in which the actually measured values for the heat output (devided by the total heat-transfer area of the four exchange units) are plotted, shows that the experimental points fall indeed on a straight line. The meaning of the slope and intercept will be discussed in the next chapter.

DISCUSSION OF THE PART-LOAD BEHAVIOUR

From the log p - 1/T -diagram in figure 6 the following equation 6 can be derived:

$$(T_C - T_E) - R\cdot(T_G - T_A) = 0 \qquad (6)$$

$$\text{with } R := \frac{r+1}{r}\cdot\frac{T_E\cdot T_C}{T_A\cdot T_G}$$

$\frac{r+1}{r}$ = ratio of the heat of absorption to the heat of vaporization of pure water,

T = internal process temperatures.

For T_A and T_G the mean values of the equilibrium temperatures of weak and strong solution at the absorber or generator respectively are taken. The value of R varies only from 0.87 to 0.85 for a heat output from 2 to 8 kW. Consequently for R the constant value of 0.86 will be used.

With equation (5) as definition of the total temperature difference and equation (6) we obtain:

$$(T_C-t_C) + (t_E-T_E) + R(t_G-T_G) + R(T_A-t_A) = \Delta T_T \qquad (7)$$

The heat-transfer equation for each of the four exchange units is given by:

$$Q = k \cdot A \cdot \Delta T \qquad (8)$$

where k = overall heat-transfer coefficient [W/m^2 · K]
A = overall heat-transfer area [m^2]
ΔT = mean logarithmic temperature difference [K]

In figure 8 the relation between the temperature difference $|t-T|$, as used in equation (7), and the mean logarihmic temperature difference ΔT, as used in equation (8), is shown. The parameter s is equal to $s_1 - s_2$ for counter-flow heat exchange (s_1 = temperature spread of the heat carrier, $s_2 = 0$ for condenser and evaporator, s_2 = solution width for absorber and generator). To characterize the individual situation for the four exchange units the dimensionless factor z defined as

$$z = \frac{\Delta T}{|t-T|} \qquad (9)$$

will be used.
According to figure 8, for $\Delta T > s$ the factor z is close to one. In our measurements this was the case for the absorber and generator. For the condenser or evaporator the difference s is exactly the temperature rise $Q/m\cdot c$ of the heat transfer fluid (m = mass flow of water, c = specific heat capacity). The flow rate m and therefore $\Delta T/s = m\cdot c\ /\ k\cdot A$ was constant for all measurements. For the evaporator $\Delta T/s$ was found to be 0.26 which leads to $z_E = 0.5$ (figure 8). The analogous values for the condenser are 0.2 and $z_C = 0.38$.
By introducing equations (8) and (9) into (7) the following relation is obtained:

$$\Delta T_T = \frac{1}{A_T}\cdot\sum_{i=C,E,G,A} Q_i\ C_i \qquad (10)$$

$$\text{with } C_i = \frac{R_i}{z_i\ k_i\ a_i} \qquad (11)$$

where $R_i = 1$ for i = C,E and
$R_i = R$ for i = G,A

A_T = sum of transfer areas in the four main components

$a_i := A_i/A_T$ (= 0.25 for the tested heat transformer)

The thermal loads Q_i on the four main components have a certain ratio to each other which does depend only on the COP of the machine. Ziegler and Alefeld have shown that in the case of a heat transformer the ratio Q_i/Q_A is 1 for the absorber or generator and (1-COP)/COP for the condenser or evaporator /Zie/. With help of equations (12) the heats Q_i'' of an ideal machine having no heat losses and an infinite area of the solution heat exchanger are related to the actually measured heat quantities Q_i by taking into account the heat losses L and the irreversible heat flux Q_S from absorber to generator:

$$\begin{aligned} Q_C'' &= Q_C \\ Q_E'' &= Q_E - L_E \\ Q_G'' &= Q_G + Q_S - L_G \\ Q_A'' &= Q_A + Q_S + L_A \end{aligned} \qquad (12)$$

For the working pair water/LiBr the heat quantities Q_C'', Q_E'', Q_G'' and Q_A'' are equal within a few percent /Zie/, i.e.

$$Q_C'' = Q_E'' = Q_G'' = Q_A'' \qquad (13).$$

By inserting equation (12) and (13) into (11) the following relation is obtained:

$$\Delta T_T = \frac{Q_A}{A_T} \cdot \sum_i C_i + \frac{Q_S}{A_T}(C_C + C_E) + \frac{L_T}{A_T} \qquad (14)$$

with $L_T := L_A \cdot C_C + (L_A + L_E) \cdot C_E + (L_A + L_G) \cdot C_G$

According to equation (14) the slope of the straight line in fig. 7 is given by:

$$1 \Big/ \sum_i \frac{R_i}{z_i\, k_i\, a_i} \qquad (15)$$

The intercept at the abscissa is determined by the general heat losses (e.g. radiation) and by the solution heat-exchanger losses which is the dominant term for large machines.

Under the assumption of a heat-transfer coefficient for film evaporation of 4500 W/m^2K, for condensation of 9000 W/m^2K, for desorption of 1100 W/m^2K and for absorption of 700 W/m^2K, we estimated the overall heat-transfer coefficients k_C = 4000 W/m^2K, k_E = 3000 W/m^2K, k_G = 950 W/m^2K and k_A = 650 W/m^2K. With the R_i, z_i, a_i and k_i values as given above we find a slope of 70 W/m^2K from expression (15) which fits within the experimental errors to the measured slope in figure 7. The intercept at the ΔT_T axis amounts to 2.6 K (see figure 7). If we use the measured values for the losses L_A = 550 W and $L_E = L_G$ = 200 W the term L_T/A_T accounts to 1.7 K. This leaves 0.9 K resulting in a solution-circle loss of Q_S = 600 W . This value corresponds to an effectivness of the solution heat exchanger of 85%, which means that 15% of the maximal transferable heat is transported from the absorber to the generator.

INVESTMENT COST

For a given heat-exchange area A_T the part-load relation (14) describes the heat output Q_A for varying external temperature conditions. For a fixed temperature of the heat source (t_E, t_G) and the heat sink (t_C) we see that the heat output Q_A decreases linearly with an increasing output temperature t_A. Thus, for a given driving temperature difference ($t_G - t_C$) the heat output Q_A can be determined from the temperature lift ($t_A - t_E$) and vice versa.
Further, relation (14) can also be used to estimate the heat-transfer area A_T for a given application. The external temperatures t are fixed by the temperatures of the available waste heat or cooling water respectively and the required output temperature. Thus, the value of ΔT_T is given and the total amount of the heat-transfer area A_T in the main components can be read off from figure 7 for the designed heat output Q_A; about 80% of the total heat-exchanger cost can be determined in this way.
In order to make the best use of the invested area A_T the individual areas A_i in the four components should be related in a way that the slope (14) has its maximum value. This optimum depends on the given heat-exchanger parameters (z_i, k_i) and working-pair parameter (R_i). An optimization of this kind is shown by Kern /Ker/.

CONCLUSIONS

The measured COP of the heat transformer approached values of around 0.45 even under part-load conditions. The inherent irreversibilities of the heat-transformer cycle with aqueous lithium-bromide solution are small compared with the irreversibilities in the heat transfer from the four exchange units to the heat carriers.

A linear relationship between the output of useful heat and the external temperature conditions was found experimentally in consistency with a theory. This theory contains factors of the working-fluid pairs and the heat-exchanger properties. The relation can be used for the calculation of the part-load behaviour or the estimation of the required heat-transfer areas in the four exchange units.

REFERENCES

/Fur/ T. Furukawa, M. Furudera, T. Kitamura: "Study on Characteristic Temperatures of Absorption Heat Pumps". Proc. 20th Japan. Heat Transfer Conf., 1983, p. 508-510 (in Japanese language).

/Ker/ W. Kern: Economic Criteria for Application of Single-Stage or Double-Stage Absorption Heat Transformers, 3rd Int. Symp. on Large Scale Applications of Heat Pumps, Oxford, 1987 (this issue).

/Nie/ W. Niebergall: Sorptionskältemaschinen, Band 7/Handbuch der Kältetechnik, Heidelberg, 1981.

/Zie/ F. Ziegler, G. Alefeld: The Coefficient of Performance of Multistage Absorption Cycles. Int. Journal of Refrigeration, 1987 (to be published).

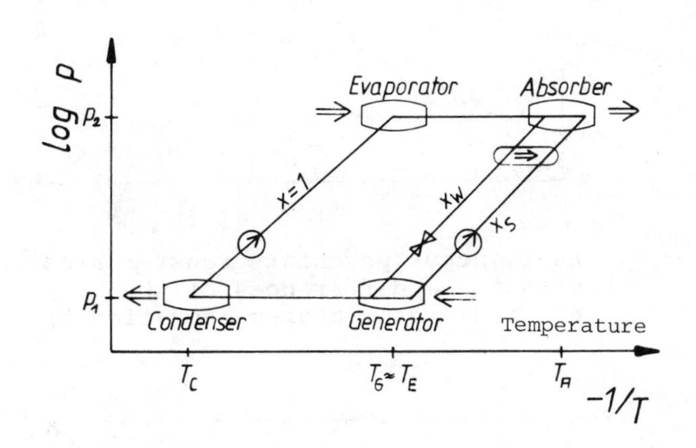

1 The heat-transformer cycle in a (log p) - (-1/T) - diagram

3 Photograph of the experimental set-up

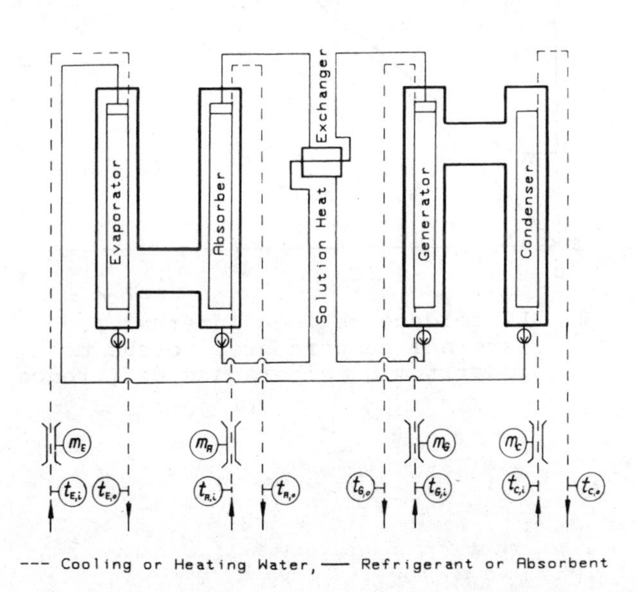

2 Schematic diagram of the experimental set-up

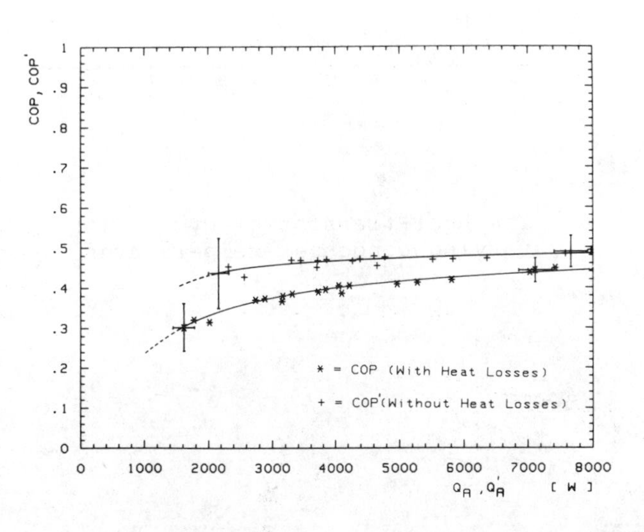

4 Coefficients of performance versus output of useful heat

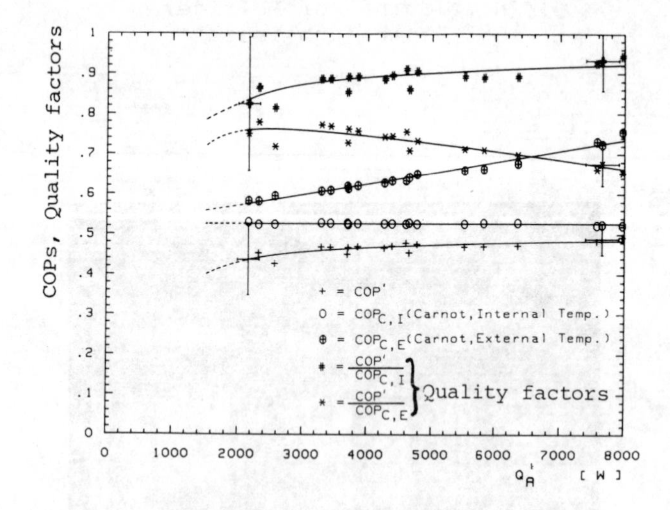

5 COPs and quality factors versus output of useful heat

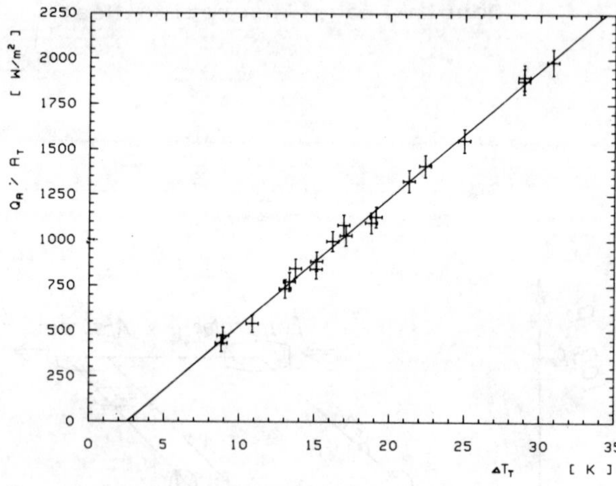

7 Heat output per heat-transfer area versus the differences of the external temperatures (equation 5)

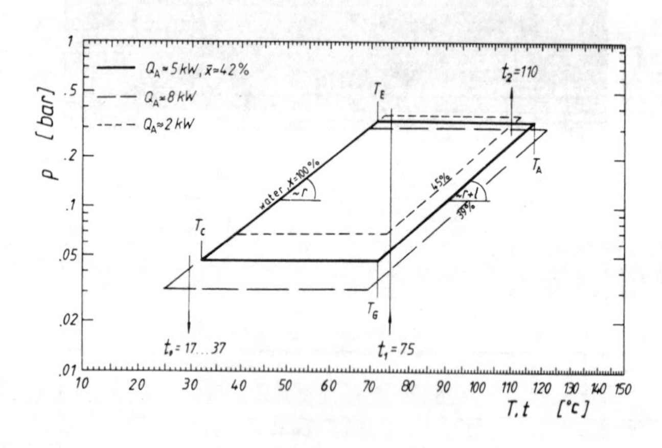

6 The heat-transformer cycle with varying condenser temperatures

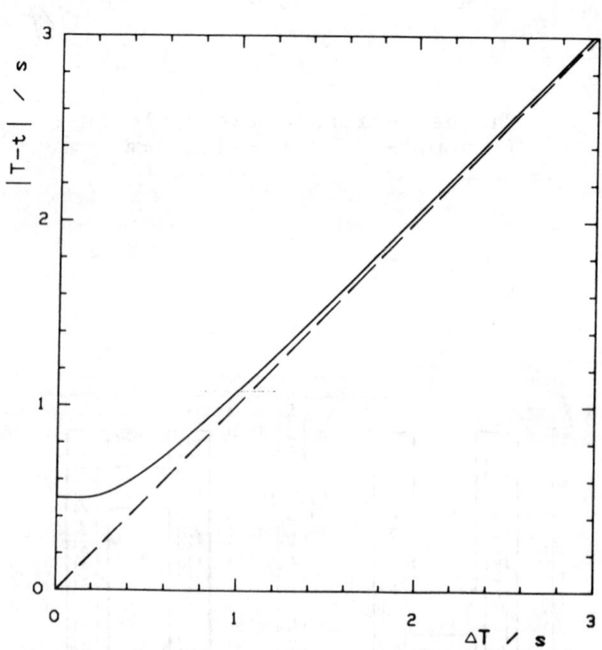

8 Difference between internal and external temperatures versus mean logarithmic temperature difference

3rd International Symposium on the

Large Scale Applications of Heat Pumps

Oxford, England : 25-27 March 1987

PAPER G4

Solid-gas chemical heat pumps in the range 150-500°C. - Research method and expected performances.

M. LEBRUN
S. MAURAN
B. SPINNER

CNRS-IMP (Institut de Science et Génie des Materiaux et Procédés), U.P.32, Université, F-6625, PERPIGNAN

Synopsis :

The type of chemical heat pump developed by the authors, uses gas-solid reactions and is able to function at high temperatures up to about 500°C. The solide-gas pairs selected should firstly be suitable for thermodynamics reasons : they should also have favorable heat and mass transfer proporties.
According to these results, a choice between three types of reactor must be done : simple fixed bed reactor, thin layers fixed bed or fluidised bed reactor.
Then, a computer simulation model which we have developed enables us to predict the performance of discontinuous or pseudo-continuous processes. This model utilizes results obtained on laboratory demonstration plants.

Introduction

The chemical heat pumps (C.H.P.) and the chemical heat transformers (C.H.T.) are based on a reversible reaction between a gas and a solid :

$$\langle MX, nG\rangle + y(G) = \langle MX, (x + y)G\rangle + y\,\Delta H$$

The pump may either consist of two reactors containing different reactants exchanging the same gas, or of a reactor and a condenser/evaporator.

A recent conference /1/ took stock of the research of such processes, where the aimed objectives were essentially applications at temperatures less than 150°C.
The need to extend the temperature range upwards towards 500°C for industrials applications was emphasized by ZEGHERS /2/.

The problems to be solved in addition to those connected with the thermodynamic feasibility of the process are :

- the production of a thermal power per unit volume in excess of 1000 kW/m^3.

- an acceptable capital cost for an industriel unit, including the installation cost. To achieve this the output should probably exceed about 1MW.

In such conditions, the introduction of the C.H.P. or H.P. process becomes very interesting for the thermal energy management as GOURLIA /3/ and recently DURAN and GROSSMAN /4/ have demonstrated.

In addition, the chemical heat pump or the chemical thermal transformer based on gas-solid reactions have some important characteristics arising from the principles on which they operate. These are :

- thermal energy storage without degradation, without limit in time ;

- the possibility of changing the power output by a simple modification of the thermodynamic operating conditions.

On the opposite, the chemical cycles repetition needs the control of the eventual reactivity decreasing ; at relatively low temperature ($t < 150$°C), no changes in performance have been observed on repeated cycling ; at higher temperature, it was studied for the CaO - H_2O cycle /5/ and NiO - SO_3 /6/ ones for which solutions were found.

In conclusion we whish to stress (1) that the cost factor is essentially bound to the magnitude of the interfacial area for heat and mass tranfer which is available per unit mass of reactant (2)

Held at St. Catherine's College Oxford, England. Organised and sponsored by BHRA, The Fluid Engineering Centre, Cranfield, Bedfordshire, MK43 0AJ England.

that a C.H.P. must function in a continuous manner in an industrial application (it is necessary to repeat that the process base is discontinuous mode on functionning).
We dicided on a research programme based on the simultaneous progression of three points of research :

a) selection of solid-gas reactions which were suitable on thermodynamic, energetic, financial and safety grounds.

b) the characterization of the performances of these pairs in terms of the power out put per unit volume, or per unit mass (and so of energetic density) combined with a study of the heat and mass transfer and of the chemical kinetics.

c) modelisation and simulation of the process, characterization of the input and ouput variables of the system which permit the definition of the command factors of the process.

Selection of the solid-gas pairs

The temperature range ($t < 500$°C) and the pressure ($0.1 < P < \approx 20$bars) is well suited to the use of ammonia reacting with chlorides and sulphates.

Desirable criteria which the pair should meet for the implementation of the chemical heat pumps or heat transformer are :

- great energetic density with regard to mass and (or) volume ;
- great power output per unit volume or per unit mass ;
- chemical stability ;
- absence of corrosion, toxicity, inflammability problems ;
- satisfactory financial and ecological aspects.

Further factors to be considered in relation to the temperature level are :

- changes in reactivity on cycling (fritting and stoichiometry)
- reactants stabillity under reaction gas atmosphere;
- presence and size of a pseudo-equilibrium field.

A first selection of solid reactants which can react with ammonia was made using standard chemical texts :

- chloride : $NiCl_2$, $MgCl_2$, $ZnCl_2$, $MnCl_2$;
- sulphate : $NiSO_4$, $ZnSO_4$, $MgSO_4$, $Fe_2(SO_4)_3$.

The difficulty of dehydrating $MgCl_2$ (it forms an oxychloride at 172°C), the fact that $Zn Cl_2$ melts at 318°C, and the poor reactivity of $MgSO_4$ with NH_3 leading to a very poor energetic density /7/ led us to reject these three salts from our experimental study.

Furthermore after experimentation, $NiSO_4$ and $Fe_2(SO_4)_3$ were eliminated for reasons of reactivity deficiency.

The reactivity of other salts was followed by thermogravimetry under controlled pressure in the scale beyond 1 bar. The results are summarized in table 1.

Table 1 : Experimental data of salt-NH_3 pairs.

Salt	$NiCl_2$		$MnCl_2$			$ZnSO_4$		
Equilibrium	2/1	1/0	2/1	1/0.5	0.5/0	2/1	1/0.5	0.5/0
ΔH ($kJ.mole^{-1}$)	88.2	97.8	83.2	85.7	94.1	67.7	68.6	90.3
Temperature (°C) * of Equilibrium and Pseudo-Equilibrium	280 ±25	368 ±5	230 ±24	310 ±15	345 ±15	250 ±32	340 ±20	419 ±3
Temperature (°C) * of irreversible decomposition	395		637			460		

* measured at P = 1 bar

A reactivity decline of $NiCl_2$, $MnCl_2$ and $ZnSO_4$ can be observed during cycling. It becomes stabilized after 4 or 5 cycles. It is due to a fritting (the scanning elctron miscroscope replies show this fact). The initial reactivity was recovered by reforming the higher ammine, for which the expansion is very important (in order 10 in volume with regard to $NiCl_2$).
In the case of $NiCl_2$, a kinetic study was done for different absorption temperatures under controlled pressure. These feats were carried out using small quantities of reactant (100 mg) without important heat and mass transfer limitations.

The table 2 summarizes the experimental conditions and the measured powers :

Table 2

ΔT equilibrium (°C) at P = 0.57 bar	63	43	25
Power (kW/kg) at X = 0.8	11,6	9,1	3,7

These values must be compared to those obtained with measurements done at the laboratory on a reactor with an exchanger, containing about 100 g of $CaCl_2$, reacting with methylamin, with on inert filling material which improved the heat and mass transfers. Using this equipment, the following values were obtained (Table 3) :

Table 3

ΔT equilibrium (°C) at P = 2 bars	46.6	26.6
Power (kW/kg) at X = 0.8	4,9	2,33

Problems related to the heat and mass transfer, chemical kinetics and the interplay between them.

In previous laboratory works, Mazet and Coll /7, 8/ have characterized and emphazied the main role of this coupling on the functionning, sizing and management of the C.H.P. or C.H.T.
These studies show the prime importance of the operating conditions on the performance of this reactant.

A possible solution consists in minimizing the heat and mass transfer limitations by choosing a fluidized bed reactor.

Some experimental measurements will be done to verify that attrition of the grains is not a blocking process for the C.H.P. development.

In a fixed bed reactor a great improvement in heat and mass transfer rates has been obtained by adding an inert filling material in the porous reactive medium /9/. This improvement strongly depends on the inert material percentage and on the porosity of the medium.
An optimum, resulting from a compromise between heat and mass tranfers, was found in the case of the $CaCl_2$ and CH_3NH_2 reaction.

The establishment of this optimum for other of the pairs selected requires the determination, under the operating conditions of these reactants, of the quantities characterizing these transfer, namely the effective thermal conductivity λ_e and the permeability B_0.
This experimental determination is based on the same methodology.

In the case of unidimensional flow in a steady state, FOURIER and DARCY's laws (modified for a compressible fluid) allow λ_e and B_0 to be calculated.

$$\lambda e = \frac{-\phi_c}{\text{grad } T} \qquad (1)$$

$$B_0 = \frac{-\eta \, \phi_m}{P_m \cdot \text{grad } P} \qquad (2)$$

with ϕ_c, ϕ_m, grad T, grad P the heat and "mass" (in $M.T.^{-3}$) flows and the temperature and pressure gradients,
η the dynamic viscosity,
P_m the average pressure.

However it must be ensured that the heat and mass transfers are not coupled either with themselves or with the chemical reaction. This fact is essential if correct coefficents are to be obtained.

Considering the heat transfer example, (the case of mass transfer is exactly analogous).
If we want determine the effective thermal conductivity of the porous media without mass transfer limitation, we attempt to achieve an

uniform pressure within the bed (by using a high porosity or increasing the percentage of inert filling material).

Then the measurement consists in determining the temperature at two different levels for a fixed heat flow.
In these conditions a part of the porous medium is in a state of chemical imbalance.
The reactions between two solids and one gas that we study are monovariant. At equilibrium these is therefore a one-to-one relation between the temperature and the pressure. So the lack of chemical balance induces an absorption (or desorption) of gas that produces an enthalpy effect strongly disturbing the measurement of λe.

It is necessary to find an experimental way of eliminating this transfer-chemical reaction coupling.

Since the problem is basically the same for the two types of transfer we shall analyse only the case of mass transfer and then show how we prevented this coupling from taking place.

i) Limits of Darcy's equation

In terms of the diffusivity K, equation (2) can be written :

$$K = \frac{\phi_m \cdot Z}{\Delta P} = \frac{B_0 P_m}{\eta} \qquad (3)$$

with $\phi_m = \dfrac{QiPi}{\Omega}$

Qi the volumetric flow rate of the gas (in $L^3 T^{-1}$) at the i-coordinate at which the pressure is Pi, Ω the normal section to the stream.
In practice this relation does not exactly represent the behaviour of a real gasstream flowing across a porous medium.
Experimentaly, we observe two types of flow (fig1).

These are slip flow and Kundsen's flow, the relations for which in the K vs P_m graph are respectively :

$$K = \frac{B_0}{\eta} P_m + K_0 \qquad \text{(curve 2)}$$

$$K_{kn} = C^{te} \qquad \text{(curve 3)}$$

By plotting the diffusivity K versus the average P_m we can determine :

- The permeability B_0 of the medium. This is the main objective since it is this parameter that accounts for the mass transfer.
- The K_0 and K_{kn} coefficients from which we can estimate the diameter of the pores /10/ ; a parameter that occurs in heterogeneous kinetics models /11/.

ii) Apparatus and procedure.

The experimental device (fig. 2) allows two distinct pressures P1 and P2. to be assigned at the boundaries of the porous medium.
The bed of the reactive salt is under study is formed by a cylindrical core, the sample having a height of $Z = 30$ cm and a section of $\Omega = 19{,}6$ cm^2.
Its temperature Tc is fixed by a P.I.D. regulation.
We consider that it is also the flowing gas temperature. The pressure P1 up-stream of the bed is fixed by an ajustable expansion-valve in the range 0.5 - 8 bars against the liquid NH_3 tank R1.
The downstream pressure P2 is fixed with a regulation of PI type by an electrovalve connecting up the reactor and the liquid NH_3 tank R2 at low temperature (T = -50°C).
In these conditions, the accuracy of the ΔP (= P1 - P2) measurement achieves 10^{-3} bar.

The mass flux Q_m crossing the bed is measured in continuous after condensing the NH_3 into the tank R2 through an iron rod, which constitutes the core of a shifting-gauge, rigidly locked with a float.
From this mass flow we calculate ϕ_m considering NH_3 as a perfect gas.

Every experiment consists in measuaring Q_m for fixed bed temperature Tc, as a function of the upstream and downstream pressures P1 and P2.
By varying P1 and P2, we can plot K versus p and hence deduce B_0 (see fig 2), the NH_3 viscosity being known in the desired temperature range 150-500°C.
As for the experimental procedure, it must ensure an uncoupling as clear as it is possible between the mass transfer, the heat transfer and the chemical kinetics. The mass-heat transfer uncoupling is realized by holding the salt and the gas which crosses it, at constant temperature Tc.

In other respects, we can measure the permeability of a type $\langle MX \cdot nG \rangle$ salt by a gas G without disturbance due to the chemical reaction
$\langle MX \cdot nG \rangle + p(G) = \langle MX \cdot (n+p)G \rangle$ (fig3)
Consider this equilibrium in the LnP vs 1/T

diagram. We can lay down a pressure gradient between P1 and P2 without having a chemical reaction if the salt is thermodynamically stable in this pressure range. This is the case between the two equilibrium lines 1 et 2 for the ‹MX. nG› salt. The same holds true for the ‹MX . (n+p) G› salt between 2 and 3 lines.

Hence to study the permeability of the ‹MX . nG› salt by a gas G, we plot K versus P_m ; each point of the graph being a unit of the experimental run : (P_{11}, P_{21}) ; (P_{11}, P_{22}) ;.....; (P_{11}, P_{2n}) ; (P_{12}, P_{2n}) ;.....; (P_{1n}, P_{2n})

The same procedure can be applied for the ‹MX . (n+p)G› salt. After determining the permeabilities of these two salts we can value by linear interpolation, in first approach, the permeability of the reactive mixture (1-X) ‹MX . nG› + X ‹MX . (n+p)G› at any reaction rate X.

The approach that we have developed for the mass transfer can be applied also to the study of heat transfer. On the basis of these experimental studies it is possible to determine the best reactant conditions to employ in a fixed bed. In order to achieve this and determine the best cycle time, a model for the process is required.

<u>Modelisation and Simulation of the thermochemical heat pump process.</u>

A numerical simulation of the behaviour of a reactor during an absorption-desorption reaction was devised in order to perform the sizing of the solid-gas reactor of a chemical heat pump prototype.
This simulation is based on modelisation of heat and mass transfers occuring on the various thermochemical heat pump components.
Aimed at sizing, its purpose is not the description of the phenomenons occuring on microscopic scale of the matter, taking into account the discontinuity of the porous reactive medium and the composition and temperature drops which it involves, but rather to predict the macroscopic behaviour of the reactive medium as a continuous phase.
The model used is based on the hypothesis that homogeneous composition and temperature are maintained accros the reactive medium while the reaction occurs. Geometry of the reactor is therefore not a factor, and the model can apply to any reactor shape compatible with this hypothesis.

The overall behaviour of the medium is described by a semi-empirical relation comprising a term for activation, a term to account for the departure from equilibrium conditions and a term for the degree of advancement of the reaction as follows :

$$\frac{dX}{dt} = f(X)\, a \exp(-b/T) \, Ln\,(P/P_{eq})$$

This kinetic model is able to represent exactly the existence of several simultaneous or consecutive chemical reactions. It contains as many equations of this type as there are chemical reactions. In the case of the $CaCl_2$ - CH_3NH_2 pair where in our temperature and pressure processing conditions two reactions occur, two such kinetic equations have been coupled.

The heat transfer modelisation takes into account all the heat exchanges which occur in the various thermochemical heat pump components. Thermal balances have been made in particular on the solid porous reactive medium, the liquid and gaseous phases of the reactive gas, each thermal fluid and the main metallic parts of the machine.

The first step of the process simulation consists in the modelisation of an ideal installation where the condenser-evaporator is much oversized with respect to the gas-solid reactor capacity.
At this stage, the model must be fited to experimental results and the determination of the optimal conditions must be done. A series of experiments on laboratory demonstration plant using only a few hundred grams of solid has enabled us to optimise several key variables (the equilibrium gap, packing density of the salt and the inert filling material percentage) and to determine several parameters of the model.
The following studies were done on optimum conditions with regard to parameters values.

From this simulation base, the reactor is coupled with a real condenser-evaporator. Once the separated simulation of absorption and desorption phases achieved, this work leads to consider the realization of the process simulation on fully cycle operating on discontinuous or pseudocontinuous modes.

Figures 4-5 illustrate the simulation results on the power profiles during a typical cycle.
Figure 4 shows results for a discontinuous process with only one solid-gas reactor. Figure 5 shows results for a pseudo-continuous process using two identical solid-gas reactors which are in phase opposition. The units shown on these graphs are

arbitrary but are the same on both. We can notice that there is an important difference between the results for these two processing modes.

The emploiement of two opposite solid-gas reactors allows more regular profiles to be obtained and this leads to more constant performance.

The simulation described above enables the **performance of both discontinuous and pseudo-continuous thermochemical** heat pumps **to be predicted not only for the** $CaCl_2$ - CH_3NH_2 pair which was used in this study but on many other reactive pairs over a variety of temperature and pressure ranges.

Conclusion

The research which we have carried out enables us to comment farther on several of the problems posed in the introduction :

- power criterion :

by comparing the few experimental results actualy obtained on the process functionning in the range 150-500°C with those obtained on lower temperature we can assert that average powers from X = 0 to X = 0.8 to 0.9 of about several kW per kg, and even a few tens of kw per kg of reactive salt could be obtained for temperature differences between the chemical equilibrium and the "heat flow" fluid of 30 to 60°C ; the voluminal powers are for the same conditions 1 to 4 MW/m^3.

- cost criterion :

the permeability measurements, coupled with the heat transfer ones - not described in detail here enable the optimum operating conditions (porosity, inert filling material percentage) and the power output to be predicted. These also allow us a more compact reactor as possible to conceive.

- industrial funtionning criterion :

The simulation described above enables us to assert that it is possible to pass from a discontinuous process to a pseudo-continuous system.These simulations confirm results obtained from a 20 kW prototype functionning at temperatures less than 100°C.

Acknowledgements :

This work is a part of a CNRS-PIRSEM and AFME CRA (Coordinated Research Action) headed : "Chemical Heat Pump".

References :

/1/. Proceedings of international workshop on heat transformation and storage. Joint Research Centre, Commission of the European Communities, Ispra (Italy), 9-11 october 1985.

/2/. ZEGHERS P., C.E.C. Heat pump research and development ; Ibid, p. 435.

/3/. GOÜRLIA J.P. in "Energétique Industrielle" from LE GOFF P., vol. 3, ch. VIII ; Techn. and Doc., Paris, 1982.

/4/. DURAN M.A. and GROSSMANN I.E.; AIChE J., vol 32, n°1, 1986, p. 123.

/5/. ROSEMAY J.K., BAVERLE G.L. and SPRINGER T.H., J. Energy, vol.3, n°6, 1979, p. 321.

/6/. BERJOAN R. and EL GHANDOUR N., réf. 1, p. 209.

/7/. MAZET N., PAYRE D., MAURAN S., CROZAT G. and SPINNER B., Proc. O.R.C.H.P. Technology, VDI Seminar, Zürich (C.H.), 10-12 sept.1984, p. 945-959.

/8/. MAZET N., SPINNER B. and ARNAUD G., Bull. Soc. Chim. Fr., n° 6 , 1985, p. 1047.

/9/. COSTE C., MAURAN S. and CROZAT G., Brevet 84-40/2/2.

/10/.CARMAN P.C. in "Ecoulement des gas à travers les milieux poreux" ; Presse universitaire de France, Paris 1961.

/11/.BATHIA S.K. "Analysis of distributed pore closure in gas-solid reactions" AIChE J., vol.31, n° 4, April 1985, p. 642-648.

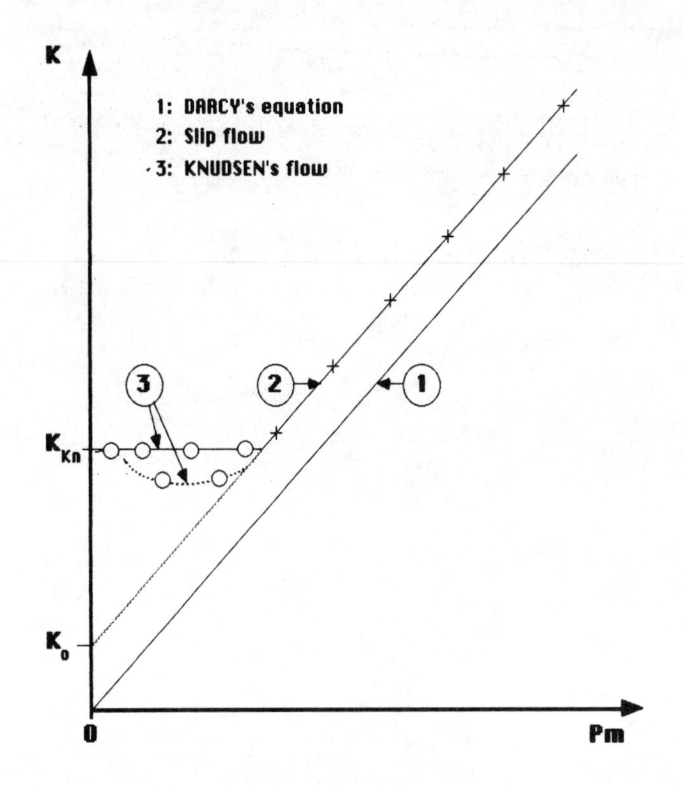

Figure 1 : Diffusivity K of a gas flowing through a porous medium. In some cases K passes through a minimum (dotted curve 3)

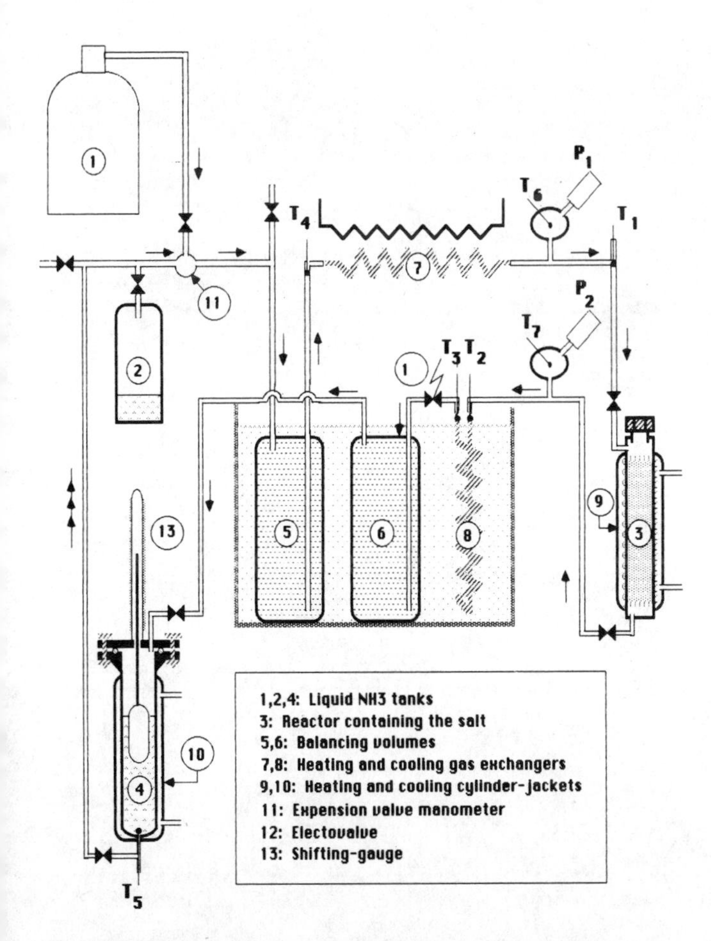

Figure 2 : Diagram of the experimental device. The gas flow direction is show by -->. When R1 is empty we blow off the liquid NH_3 in the->>>way.

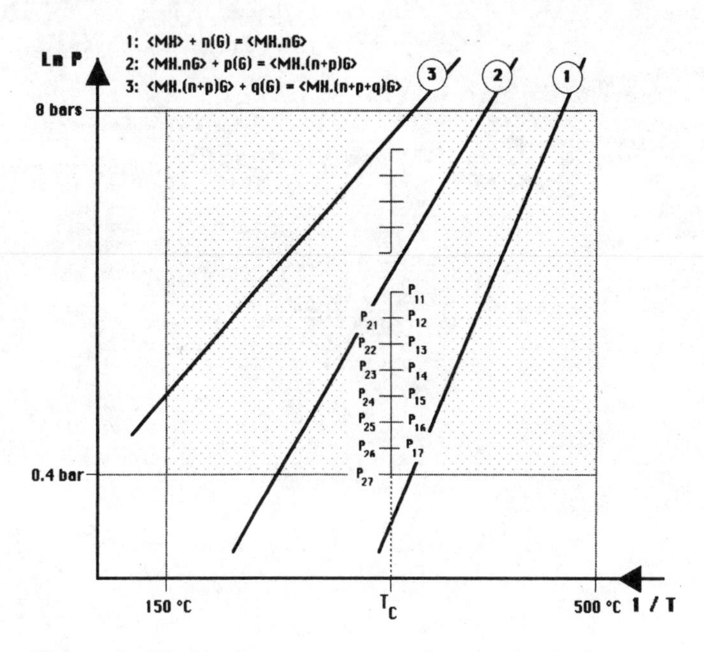

Figure 3 : Measuring experimental procedure of the permeability uncoupled with chemical reaction.

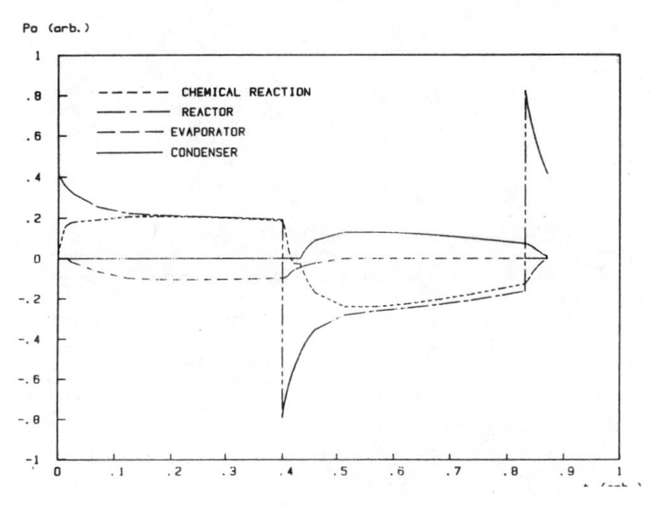

Figure 4 : Power generated by a discontinuous processing.

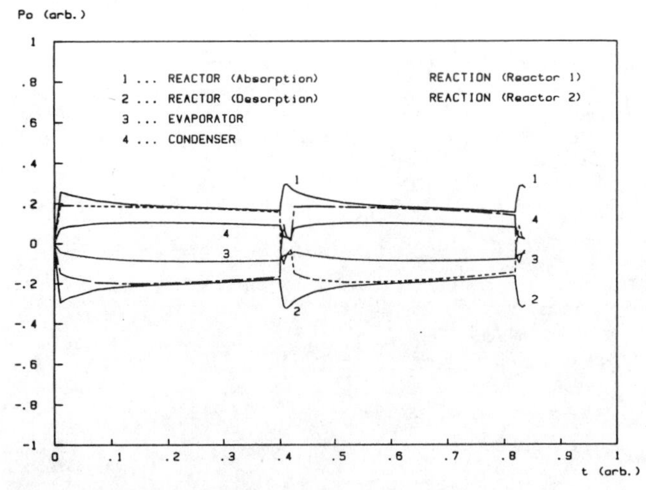

Figure 5 : Power generated by a pseudo-continuous processing.

3rd International Symposium on the

Large Scale Applications of Heat Pumps

Oxford, England : 25-27 March 1987

PAPER H1

HEAT PUMP RESEARCH CARRIED OUT BY THE EUROPEAN COMMUNITIES

P. Zegers
Commission of the European Communities, Brussels, Belgium

ABSTRACT

CEC heat pump research is executed in a four year Non-Nuclear Energy R and D Programme running from 1985 to 1988; at present 22 CEC funded heat pump projects are being carried out by different research organizations in Europe. This programme has two main objectives :

- Reduce the cost and improve efficiency of heat pumps for domestic and commercial heating applications in order to achieve economic feasibility and bring about a breakthrough on this market will consumes 26 % of the primary energy in the E.C.
- Develop industrial heat pumps which can produce heat up to 300°C.

Research to achieve the first objective for compressor heat pumps is aiming at the use of fluid mixtures, development of more efficient and cheap compressors, of oil free compressors, better control and development of efficient low pollution combustion engines. Work on absorption heat pumps is mainly focussed on fluid pairs and on the development of cheap and efficient heat exchangers.
Research on high temperature industrial heat pumps is carried out along several lines. For compressor heat pumps, fluid mixtures for high temperatures (180°C) are being investigated. For high temperature absorption heat pumps, work is focussed on new fluid pairs and on the study, construction and testing of different cycles (absorption heat pumps, heat transformers and other cycles). Finally, the possibility to develop solid / fluid absorption heat pumps up to 300°C (solid/fluid combinations, reactor, heat exchangers) is systematically explored.

INTRODUCTION

The European Community is funding heat pump research in its Non-Nuclear Energy R and D Programme. 22 R and D projects are being carried out by different organizations in Europe on compression and absorption heat pumps in the framework of this programme which runs from 1985 to the end of 1988. In total 6 million ECU (*) is being spent on this research of which the Commission is paying 60 %.

---

* 1 ECU (European Currency Unit) = 1,1 $

Held at St. Catherine's College Oxford, England. Organised and sponsored by
BHRA, The Fluid Engineering Centre, Cranfield, Bedfordshire, MK43 0AJ England.

IMPACT OF HEAT PUMPS ON ENERGY SAVINGS, OIL SUBSTITUTION AND POLLUTION ABATEMENT

R and D on heat pumps forms part of a broader energy policy of the Commission which aims at :

- Energy saving
- Substitution of oil by coal, nuclear or renewable energy
- Pollution abatement.

Energy saving

For the provision of energy in the European Community about 30 billion ECU per year is spent on energy investment, this is about 35 % of the total industrial investments. In addition over 100 billion ECU per year is spent on fuels. Energy savings of only a few percent will lead to the saving of billions of ECU.
The primary energy consumption in the European Community in 1985 was 940 million tonne oil equivalent. This energy was distributed over the different demand sectors as shown in Table 1.

Table 1. Primary energy use in the E.C.

Buildings 41 %		Industry 41 %			Transport 18 %
Heating	Power	Power	Non energy use	Process heat	
26 %	15 %	13 %	7 %	21 %	18 %

Heat pumps may contribute to energy savings in two sectors : heating of buildings and industrial process heat.
Heat pumps can provide heat up to temperatures around 120°C. For heating in buildings heat is needed at temperatures of around 70 - 90 °C, which can easily be provided by heat pumps. The potential market is thus the whole building stock and if one assumes that heat pumps use 30 - 50 % less energy than conventional heating systems, the potential energy savings will lie between 8 and 13 % of the overall primary energy use.
Heat pumps in industry mainly serve as a tool to transform waste heat into heat at a higher temperature level where it may be used. Process heat in industry is required in a temperature range from 100°C to 1500°C and the amount of heat in industry as a function of temperature is roughly as indicated in Fig. 1. The use of heat pumps for industrial process heat is limited to applications which use temperatures below 120 °C. This limits the potential market to about 10 % of the overall industrial process heat consumption. The potential energy savings lie around 1 % of the overall primary energy consumption. If heat pumps could be developed which produce heat at temperatures up to 400 °C the energy saving potential could be increased to around 3 - 4 % (see Fig. 1). Development of high temperature heat pumps is therefore an important objective of this programme.

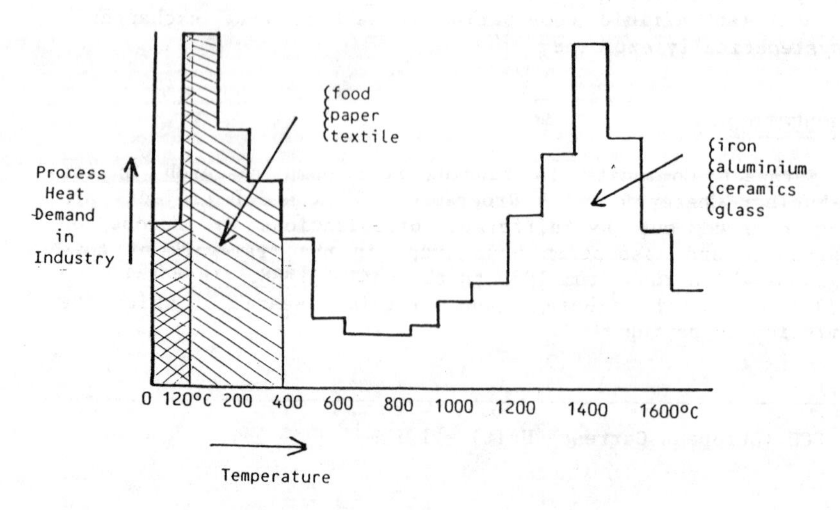

Fig. 1 : Typical process heat demand curve as a function of temperature (for Germany)

As for the size of the heat pumps one may expect that for the case of a large scale introduction of heat pumps about 50 % of this heat will be provided by small heat pumps (typically 10 kW) for individual heating of dwellings. Large heat pumps of over 100 kW will be needed for the remaining 50 % (heating of offices, industrial buildings, apartment buildings, hospitals, schools etc.).

Substitution of oil

The use of oil in the EC presently amounts to 42 % of the primary energy use (down from 63 % in 1973). It is the objective of the Commission to further reduce this percentage for two reasons :
- In the long term oil will be in short supply
- Around 80 % of the oil is imported; mainly from politically unstable areas

Table 2. Oil consumption in the different demand sectors

	Heating in Buildings	Electricity	Industrial Process Heat	Transport
Primary en. use	26 %	28 %	21 %	18 %
Oil use	55 %	16 %	37 %	85 %

Of the different demand sectors (see Table 2) the transport sector has the highest percentage of oil consumption. It is however very difficult to replace oil in this sector by other fuels in the short and medium term. In the long term electrical vehicles may possibly take a share of the market.
Other sectors which have a large oil consumption are : heating of buildings and industrial process heat. In both sectors heat pumps and in particular compressor heat pumps could contribute to substitution of oil as they provide heat by using electricity which can be produced from nuclear energy, coal or renewable energy. Absorption heat pumps could replace oil by gas, coal or wood.

Pollution

The conversion and use of energy has always been an important cause of environmental pollution; e.g. exhaust gases of cars, of power plants of domestic heating systems and of industrial boilers. Recently environmental pollution has taken a dramatic dimension as it is more and more clear that a large part of the forests in Europe are threatened by acid rain caused by such a pollution. The cost for society of environmental abatement is presently estimated to be 3 - 5 % of the GDP and is likely to increase due to an increasingly severe legislation.
Potentially heat pumps could make a large contribution to pollution abatement, in particular in densily populated areas. Heat pumps produce heat 30 - 50 % more efficient than conventional systems and give per unit of heat produced less pollution. If used on a large scale this heat pump could make a large contribution to the pollution abatement; about 28 % of the primary energy consumption could be provided by heat pumps.
In addition the major part of the heat pumps will be electrically driven heat pumps which would replace coal, oil or gas fired conventional systems by a completely clean system. The additional power required would result in a somewhat increased pollution at the electricity plant but this pollution could be more easily taken care of.
For fossil fuel fired absorption heat pumps the reduction of the pollution level would mainly be brought about by lower energy use, which could be 50 %.

Another aspect of heat pumps could have a negative impact on the environment. In particular, some fluorochemicals used in heat pumps can attack the ozon layer. In view of the present worry on keeping the ozon layer intact a ban on the use of these refrigerants in the USA and Europe may very well be possible. This would have a strong impact on heat pump R & D and the consequences should be investigated.

OBJECTIVES FOR EC HEAT PUMP R AND D

Work carried out in the E.C. heat pump programme is aiming at two objectives :

- Reduction of the cost and improvement of the efficiency of compressor heat pumps by 50 %
  Although heat pumps save energy and reduce pollution levels theay are in Europe economically only marginally attractive. At present in Europe around 200.000 heat pumps have been installed and a breakthrough can certainly not be expected in the near future. The main reason is that heat pumps are still too expensive and the performance is too low. However, both for cost and performance there is scope for improvement. Large scale production could lead to cost reductions of 30 - 50 %; research could also lead to cost reductions. In particular for compressor heat pumps the practical seasonally averaged COP values are a factor three lower than theoretically can be expected. R and D should therefore be able to increase performance levels by 50 %.
- The second objective is the development of high temperature heat pumps for industrial applications. The potential for application of heat pumps in industry could strongly increase if the temperature range of heat pumps could be extended for present values of around 120°C to 300°C. To that end new refrigerants, working fluid paris, solid gas pairs and new cycles are being investigated.

RESEARCH PROGRAMME

Heat pump research in the European Community is carried out in its third Non-Nuclear Energy R and D Programme (1985 - 1988). At present 22 projects have been approved for a total cost of 6 million ECU of which the Commission pays 50 %.

Compressor heat pumps

Compressor heat pumps are commercially available but their economic feasibility is marginal. Moreover their seasonally averaged efficiencies (from primary fuels) of 90 - 100 % are considerably lower than for other heat pump types such as absorption heat pumps or internal combustion engine driven heat pumps. On the other hand, electrical heat pumps could make a major contribution to pollution abatement and have the additional advantages to allow the replacement of precious premium fuels such as oil and gas by coal or nuclear energy. R and D on compressor heat pumps was therefore continued, the more so as there is still scope for considerable (50 %) improvement of the efficiency and a reduction in cost.
The second Non-Nuclear Energy R and D Programme (1979 - 1983) established that the compressor heat pump could be improved in the following way :

- Improved compressor design leading to higher efficiency and lower cost.
- Dissolution of lubrication oil of the compressor in the refrigerant leads to a decreased performance; ways should be found to avoid this.
- Use of non-azeotropric mixtures can lead to increased performance and lower cost.
- Fluctuation in the refrigerant flow should be avoided.

Four projects are aiming at the improvement of compressor heat pumps :
GENERAL SUPPLY, Greece is developing a simple oil free highly efficient and cheap compressor. It is a rotating compressor with a maximum speed of 15 000 rpm, the volume swept per rotation is 2 x 100 cm^3.

UNIVERSITY OF ULSTER, UK is studying a number of ways to improve the compressor heat pump. This group has studied the influence of dissolution of compressor lubricant oil in the refrigerant and found that this could decrease the performance by as much as 30%. Two options exist to avoid this problem. The first one is the development of a lubricant oil free compressor; this way is followed by GENERAL SUPPLY and FICHT. A second possibility is to develop lubricants which are dissolving less in the refrigerant; as a first step this project is studying dissolution properties of a wide range of lubricants. Experiments with simple rolling compressors are also being carried out and first experiments indicate that they give a better performance than reciprocating or screw compressors. Other ways to improve the compressor heat pump are the development of methods to control evaporator superheat with a microcomputer based expansion system and the development of a cycle in which an active process is used for flashing the refrigerant to evaporating pressure and which according to Granryd should lead to significant performance gains.
R and D on the use of non-azeotropic mixtures in the previous E.C. Non-Nuclear Energy Programme (1979 - 1983) showed that these mixtures could improve the performance by 10 - 20 % and reduce the cost. In this programme ELF and IFP France are developing non-azeotropic mixtures for industrial heat pumps which produce heat above 110°C.
It is known that for on/off operation of a heat pump the duration of the operation interval has a strong influence on the overall performance and it is generally assumed that operation of at least 15 - 20 minutes is required to optimize heat pump performance. Shorter time intervals could reduce the performance by 10 - 15 %. On the other hand it has also been established that very short operation intervals can improve operation performance of heat pumps as compared to continuous operation. This is being investigated by FORDSMAND, Denmark. First experiments showed improvements of 20 %, with cycle times of 20 - 60 seconds and with a ratio of on and off time operation of 0,25.
Contrary to electrically driven heat pumps, internal combustion engine driven heat pumps allow the recovery of waste heat of the engine, which is added to the heat produced by the heat pump. This leads to considerably higher efficiencies. Also the heat output can be varied more easily by regulating the speed of the motor and switching off one or possibly more cylinders of the compressor. The heat output can thus be matched to the heat requirements of the house which leads to further energy savings.
Two projects are concerned with internal combustion engine (ICE) driven compressor heat pumps.
The first project by FICHTEL AND SACHS, Germany deals with a 20 kW ICE driven heat pump developed in the second E.C. Non-Nuclear Energy Programme. A barrier for introduction on the market was the high pollution caused by the engine. In this project a lean burn gas engine and a low polluting Diesel engine are being developed, suitable for operation with a 20 kW heat pump. The following objectives have been set :

	Oil	Gas
NO_x	800 (2650) mg/m³	400 (2500) mg/m³
CO	600 (650)	310 (1250)
CmHn	510 (680)	65 (360)
Soot	0,1 - 0,2 (1,0 - 14) in Bosch number	

The data in brackets give the present values.

The second project by FICHT, Germany aims at the construction of a combustion engine - compressor unit where the crankslot has two piston rods one for the piston of the combustion engine and one for the piston of the compressor; in this way transmission losses are minimized. The conical rotary slide engine is a lean burn engine with very low CO and NO_x levels, far below the levels legally required. Due to the low number of components the cost is expected to be low and the reliability high.

Absorption heat pumps

Absorption heat pumps are expected to have a much higher efficiency (PER = 1.3 - 1.5) than electric heat pumps and they may use a variety of fuels such as coal, oil, gas, wood or even waste heat. Absorption heat pumps do cause more pollution than electric heat pumps which are intrinsically clean. However as compared to conventional oil or coal combustion, absorption heat pumps cause a much lower pollution in particular due to the much higher energy conversion efficiency.

A large part of the research work in this programme on absorption heat pumps is aiming at the development of high temperature heat pumps for applications in industry.

For absorption heat pumps chemical decomposition of the working fluid pairs is often the bottleneck for achieving a high temperature heat production. Research on working fluids therefore has a high priority. In the second Non-Nuclear Energy R and D Programme (1979 - 1983) a systematic search for new working fluid pairs has been carried out, which resulted in a number of promising pairs which will be further studied by the UNIVERSITY OF ESSEN, Germany in the ongoing programme. Stability, compatibility with materials used in heat pumps and also heat and mass transfer problems will be investigated.

For high temperature operation the absorption heat pump is at a disadvantage as the temperature of the produced heat at the absorber and condenser, is considerably lower than the temperature of the generator which has the highest temperature in the circuit; here decomposition is likely to occur first. This is not the case for a heat transformer where medium temperature heat is given to the generator and evaporator to produce high temperature heat at the absorber. In a heat transformer heat is thus produced in the highest temperature point of the circuit. This makes the heat transformer intrinsically more suitable for high temperature operation than absorption heat pumps. It has the additional advantage that, apart from waste heat, no extra energy input is needed.

Heat transformers are therefore being developed by GEA, Germany in particular for operation at high temperatures. Working fluid pairs identified by the UNIVERSITY OF ESSEN, suitable for high temperature operation, will be tested in a heat transformer producing 20 kW heat at 170°C from 50 kW waste heat at 100°C. In addition also an absorption heat pump will be built, with the new working fluids. Different combinations of the heat transformer and the absorption heat pump will be tested and of the most promising combination a 100 kW installation will be built.

A similar project will be carried out by the UNIVERSITY OF MÜNCHEN, Germany. Here a combination of a heat transformer for the low temperature part and an absorption heat pump for the high temperature part is envisaged. As a working pair $LiBr/H_2O$ will be used. This combination is interesting as it combines the advantages of providing sufficient heat at high temperatures with the ability to also use all waste heat from the process at lower temperatures. The absorption heat pump alone generally produces sufficient heat but can not recover all the waste heat of an industrial process; on the other hand a heat transformer can generally not cover the heat demand but can use all the waste heat. There is no risk for decomposition of LiF as this component is stable up to high temperatures.

A third two-stage heat pump type is constructed by BATTELLE, Germany. This is a two-stage $LiBr/H_2O$ absorption heat pump with which 100 kW of heat at 130°C will be produced from waste heat at 105°C.

A periodically operating absorption heat pump is being developed by the UNIVERSITY OF AACHEN, Germany. This heat pump consists of two reservoirs where liquids are separated, but where vapors can freely move from one reservoir to the another. They are used as a generator and condensor for half a period and as an absorber and evaporator for the other half period. During the generator/condensor phase, heat is given to the generator and vapour is condensed in the condensor where heat is extracted. During the evaporator/absorber phase low grade heat evaporates the refrigerant which is absorbed in the absorber where released heat is extracted. The cycle time is 1 - 1,1/2 hour. An advantage of this heat pump type is that no solution pump is required. The working fluids are CH_3OH/H_2O-LiBr. In the second programme a 10 kW unit has been built which produced heat at 50°C (with the generator at 120°C and the heat source at 0-10°) with a PER value of 1,23. In the ongoing programme, research is aiming at improving this absorption heat pump and at the construction and testing of a second pilot plant.

A second periodic absorption heat pump is developed by the CNRS, France. The main feature of this heat pump is a new concept for the generator and condensor and the possibility for storage of the working fluids. A 30 kW unit is expected to be in operation in 1988.
Reduction of cost is an important objective in this research programme also for absorption heat pumps. As a large part of the heat pump is related to heat exchangers a cost reduction and improved performance of heat exchangers can make a large contribution to economic feasibility of absorption heat pumps, heat transformers or other cycles. To that end, DUINTJER, Netherlands is developing a platefin heat exchanger which has the promise of good performance and cheap mass production. This type of heat exchangers will be tested in heat pumps for heat and mass transfer and flow resistance. If satisfactory the manufacturing of the heat exchangers will be developed.
An interesting option to develop high temperature heat pumps is the use of solid and gas as working media. It is believed that, with suitable solids and gases, heat pump operation up to 300°C is possible. The fact that such absorption heat pumps always operate periodically could be a disadvantage for industrial applications; however solutions for this problem may be found. Other problems are the development of suitable heat exchangers and reactors. This topic is studied by three heat pump projects.
A group of one Italian (CNR) and four French (CNRS, UTC, CETIAT, BLM) organizations is carrying out a systematic search for solid/gas combinations (with a emphasis on zeolite/water and active coal/methanol) suitable for operation up to 250°C. Other important R and D topics are the design of the reactors and heat exchangers in the different components of the heat pump, the optimization of the components, and the operation of cycles which consist of different stages.
The TECHNICAL UNIVERSITY OF DENMARK will systematically investigate NH_3 (refrigerant) and different metal halides (absorbants) in a number of cycles. In this project an attempt will be made to design a system which operates in a quasi-continuous way in order to circumvent the main draw back of solid absorption heat pumps : periodic operation. Two small systems will be constructed : a air to air heat pump for energy recovery from a drying process and a 2,3 kW two-stage heat transformer producing heat up to 300°C.
INPG, France investigates the possibility to develop a solid absorption heat pump with NH_3 (refrigerant) and graphite containing metallic salts. It is believed that this system, which operates via insertion of metallic salts in graphite, will have a much better performance than metal halides and pure NH_3.

ACKNOWLEDGEMENT

The author wishes to thank Ir. J.A. Knobbout for his help and valuable advice, as expert of the Commission.

3rd International Symposium on the
Large Scale Applications of Heat Pumps
Oxford, England : 25-27 March 1987

PAPER H2

A REVIEW OF R & D ACTIVITIES ON INDUSTRIAL HEAT PUMPS BY A CANADIAN ELECTRIC UTILITY

J.T. Strack

The author is in charge of the Process-Applications Unit at Ontario Hydro Research Division, Toronto, Canada

SYNOPSIS

An electric utility can have an important catalytic effect on the transfer of new technologies to its customers. To aid this transfer process, Ontario Hydro has established unique development and demonstration facilities for selected electro-technologies at its Research Division. This paper reviews a variety of projects conducted by Ontario Hydro Research Division aimed at helping industrial customers benefit from the use of heat pump technology. Included are descriptions of field and laboratory projects dealing with wood and textile dehumidification dryers, and a steam-raising heat pump system.

1.0 INTRODUCTION

Ontario Hydro has throughout its history played an important role in the introduction of new electrotechnologies to its customers. This aspect of utility business is no less important today than in the past. Much of the current attention is directed at introducing electrotechnologies into the manufacturing sector.

The Research Division of Ontario Hydro is responsible for providing technical support to utility programs, and has established unique development and demonstration capabilities in a number of electrotechnology areas including compression heat pumps. This paper describes three projects, that have been completed by the Research Division, to help customers take advantage of heat pump technology.

2.0 THE RESEARCH DIVISION ROLE

Providing technical support for a technology involves a number of activities including:

1. Field monitoring;
2. Application optimization;
3. Feasibility assessment; and
4. Technology development.

Field monitoring is necessary to gather performance information on a specific installation of a technology. Often industry does not install enough instrumentation to report the performance of a new piece of equipment, and if a case study is to be published, a third party will be required to install monitoring equipment. The Research Division has developed specialized field monitoring philosophies and equipment that are used to gather information efficiently from field installations.

The second activity area deals with the application of existing technology in new applications. Utility personnel responsible for electrotechnology development can often provide a broad perspective and technical resources not available to the average consultant and equipment supplier that specialize in a particular industry. One example involves heat pump dehumidifier suppliers in North America. These manufacturers currently supply heat pumps to the wood products industries, but recognize new market potential in other manufacturing sectors such as the textile, and food and beverage industries. One of the difficulties of applying existing equipment in new applications is performance optimization. The Research Division can often provide the necessary technical assistance to help bridge the gap between equipment supplier capabilities and customer needs.

Other applications of electrotechnologies, generally of a larger scale and newer in nature, require detailed engineering assessment before the customer will make a financial commitment. "Risk-sharing" by the electric utility through support of these early studies is often the ingredient that is needed in order to initiate these projects. In addition, utility participation in a feasibility study is perceived by many customers to provide a more objective view of an application than one conducted by an equipment supplier.

The last activity area involves technology development. A certain fraction of utility resources are used to explore technological development that will provide benefits to customers in the medium and long term (5 to 20 years) time frame. This is a perspective not available to most manufacturers. The results of this effort are developments in technology that would not occur, or at least, are moved forward in time. The improvements are normally communicated to manufacturers which incorporate the concepts into their products and deliver them to the ultimate users. One example of this in Ontario Hydro involves the development of an improved air-source heat pump for northern climates (Ref 1). This project was co-funded by Ontario Hydro and the Canadian Electrical Association.

To date utility support of industrial heat pump technology has concentrated on the first three activities. Results of three typical projects are summarized in the following sections of the paper.

3.0 DESCRIPTION OF THREE PROJECTS

A brief description of three projects is given. The first is a field monitoring project that measured the performance of a dehumidifier in an accepted application - wood drying. The second example describes analysis that was conducted to optimize dehumidifier performance in a new application - textile drying. The third deals with an investigation into the feasibility of using a combined closed-cycle and open-cycle heat pump to recover and upgrade heat from an exhaust air stream in a food processing plant.

Held at St. Catherine's College Oxford, England. Organised and sponsored by BHRA, The Fluid Engineering Centre, Cranfield, Bedfordshire, MK43 0AJ England.

3.1 FIELD MONITORING OF A WOOD DRYER

In the early 1980's furniture makers in Canada were just starting to consider using electric dehumidification systems in their drying kilns. In order to be able to provide customers with quantitative information on this efficient technology, a commercial kiln was monitored for six drying cycles over a six month period (Ref 2). A furniture manufacturer, located about 100 km west of Toronto, agreed to allow Ontario Hydro Research to monitor the performance of one of four kilns operating on his site. The dehumidification kilns, pictured in Figure 1, were installed by the owner in 1981 as part of a plant expansion to replace previously operated steam-heated kilns.

Heat Pump System Description: A schematic of the heat pump system is shown in Figure 2. The system was manufactured and installed by a local supplier. It employs a liquid-to-liquid heat pump and two water-to-air heat exchangers, connected by water loops to provide dehumidification. The system has a typical water removal rate of about 27 kg/h. The kiln uses a 42 kW electric heater for initial heat-up and to maintain the kiln at operating temperature throughout the drying schedule.

Monitoring System: In order to measure the performance of the kiln, instrumentation was installed to record all major parameters of the kiln as shown in Figure 3. Measured were heat pump electric consumption, heat pump thermal output, heater electric consumption, electric consumption by pumps, fans, and controls, kiln dry-bulb temperature, and kiln wet-bulb temperature. This information was recorded, as 15-minute averages, on magnetic tape for subsequent analysis.

Results: The results from a typical dehumidification drying cycle are summarized and compared to a standard steam-heated cycle in Table 1.

The main advantage of heat pump dehumidification over the conventional method is a large savings in energy consumption. The net energy requirements of the subject cycle was 13,343 kWh versus an estimated 31,000 kWh for a comparable steam-heated cycle (Ref 2). This corresponds to energy-savings of about 60%.

Figure 4 illustrates the daily energy consumption during a drying cycle. The plot is broken down into three main quantities: electric heater consumption; heat pump consumption; and auxiliary (fans, pumps, controls) consumption.

The auxiliaries consume energy at a relatively constant rate throughout the drying cycle. The maximum consumption is about 200 kWh/day.

The heat pump consumption is also relatively constant except during the start of the drying cycle (Oct 6 to 12), and at the end of the drying cycle (Oct 28 to Nov 6). Heat pump cycling at the start is attributed to slow drying rates demanded by the kiln operator in order to produce high quality wood. Heat pump cycling at the end of the drying cycle is a result of the reduced ability of the wood to give up water at low moisture contents.

The major contributor to the electricity demand is the 42-kW electric heater. This unit adds a "peaky" component to the total kiln energy consumption. A closer examination of the hourly kiln load revealed that peaks are concentrated during daytime and often occurred after the kiln operator adjusted the kiln controller set-points. Transferring kiln heater operation to off-peak times (night), or reducing the heater capacity during the daytime will benefit the customer by reducing the demand component of the electricity charges. It has been estimated that shifting heater operation from 24 hr/day to night-time only (16 hr/day) will reduce the average cost of electricity from $0.040/kWh to $0.032/kWh.

In addition to energy savings, the operator reports a slight improvement in product quality compared to the previously operated steam-heated kilns. Product quality is important in a furniture operation and contributes to the overall operator satisfaction.

In general dehumidification drying is well received by hardwood operators. A survey of wood kilns conducted in 1983 (Ref 3) found that over 50% of the hardwood kilns in Ontario were using heat pump dehumidifiers, and new construction often employed electric heat pump technology.

3.2 OPTIMIZATION OF A DEHUMIDIFICATION TEXTILE DRYER

In 1985 Ontario Hydro established a product drying-rate measuring facility at the Research Division. The purpose of this facility is to provide data on the product that is not normally available from the customer, but none-the-less necessary if dehumidification system performance is to be optimized. An overview of the measurement and analysis procedure is illustrated in Figure 5. The approach combines experimental measurements of product drying rates, and published dehumidifier water-extraction rates.

The procedure works like this. A sample of customer product is tested in an environmental chamber as shown in Figure 6. Drying rate measurements are taken at various temperature and humidity conditions as the products dries, and the values are stored on a computer diskette. Once the product drying-characteristic is known, it can be combined with the water-extraction- characteristic of various sizes of commercially available dehumidifiers. This leads to the prediction of energy consumption and drying time for different dehumidification systems. System optimization is simply a matter of searching for a minimum in one parameter (eg drying time, energy consumption, capital cost) while keeping the others within acceptable limits.

The first product tested in this facility was a textile - large spools of thread that require drying after a dyeing operation. This project was initiated by utility field staff, with the goal of providing information to the customer on methods of reducing the cost of drying his product.

Current Situation: The drying process is a batch operation that is currently accomplished using one gas-heated and three electric-heated dryers. A typical batch consists of 648 spools of 100% nylon thread. Each spool has a weight of approximately 0.83 kg (dry weight), with an initial moisture content (MC) of 12% (weight, dry basis). The product is dried to a MC of about 2.5% at 82°C to 93°C in 6 to 8 hours.

Product Measurements: The product drying characteristic was measured at three dry-bulb temperatures and various relative humidities. The results are illustrated in Figure 7. Here the measurements have been processed and displayed as "Drying Rate versus Moisture Content". In this case, the product is being dried in the falling-rate section of the drying characteristic where the drying rate is proportional to the moisture content.

Dehumidifier Performance: System performance was predicted for R12 dehumidifiers with compressor sizes ranging from about 2 kW to 11 kW. Predicted drying time and energy consumption are shown in Figure 8. The 2-kW unit will dry 648 spools in about 9.6 hours, while the 11-kW unit will reduce drying time to less than 7 hours, but substantial increases energy consumption.

Dehumidifier Selection: Selection of the "best" size of dehumidifier is a trade-off between drying time, operating cost, and capital cost. Table 2 lists dehumidifier capacities and expected performance values to achieve shortest drying time, lowest energy cost, lowest capital cost, and one that is a compromise between low energy and capital cost, and short drying time. If the drying time of about 9.6 hours can be tolerated, the 2-kW unit is a good choice, being cheaper to buy and operate than the other units. Energy cost savings for this unit are estimated to be about $2,400 per year compared to the electric-resistance heated dryer. The cost of installing the unit is estimated at $8,000 giving a simple payback of 3.3 years.

3.3 FEASIBILITY OF RECOVERING HEAT FROM A DRYER-EXHAUST STREAM

This project initially started as an offshoot of dehumidifier application evaluations. Many industries use continuous drying operations to process their products. These dryers generally operate at high temperatures (120°C to 170°C) and low relative humidities (10% to 30% RH) which makes it impractical to operate dehumidification systems. However, the moist air exhaust streams leaving the dryers are waste heat sources that can be tapped and used to produce process heat by employing electric heat pump and mechanical vapour recompression technologies.

The purpose of this study was to investigate the technical and economic feasibility of using available electric heat pump technology to recover heat from dryer exhaust to produce low-pressure steam. The approach taken was that of a case study using operating conditions from an actual plant in eastern Ontario. The plant processes corn into liquid sweeteners, starch, and meal. The focus of the study was the gluten feed dryer.

Current Plant Conditions: A major waste heat source in the plant is the exhaust from the gluten feed dryer. This exhaust stream has a flow of 18.9 m^3/s at a temperature of 111°C and humidity ratio of 0.216 $kg_{water}/kg_{dry-air}$. The dryer is operated at least 6,720 h per year.

The gluten feed dryer is direct-fired using natural gas and the exhaust contains traces of combustion products which will become corrosive when the stream is cooled below the dew point (about 64°C). In addition, there is a significant level of particulate matter in the exhaust. Any heat recovery device will have to be resistant to corrosion, and easily cleaned.

A potential waste heat sink in this plant is the production of low-pressure (LP) steam at 150 kPa. The plant operates with two large high-pressure natural-gas-fired boilers. LP steam is produced using high-pressure steam through a pressure-reducing valve (PRV) and from flash steam recovery on the high-pressure condensate return system as illustrated in Figure 9. The LP steam is used for process heating and make-up water heating in the deaerator. An average of 2.58 kg/s of LP steam is needed for approximately 8,000 h per year. Of this total, 1.83 kg/s is produced by the boilers and 0.75 kg/s is recovered from the condensate system. The system uses 8.3 l/s of make-up water that enters at 38°C, but is eventually heated to 104°C in the deaerator.

The opportunity that was investigated was the complete displacement of the 1.83 kg/s of LP steam, that is currently produced by the boilers, with heat recovered from the gluten feed dryer exhaust and upgraded using an electric heat pump system.

Heat Pump Alternatives Investigated: Three different heat pump systems were investigated as possible equipment alternatives for this application. These are shown in Figure 10. In all cases the heat pump system provides heating as a combination of make-up water preheating before the deaerator, and the production of LP steam. The equipment and performance variations are summarized in the following paragraphs.

Alternative 1 uses an air-to-water heat exchanger, a water-to-water preheater, a high temperature water-to-water heat pump, and an external steam separator. The system recovers about 3 MW from the exhaust stream and delivers 0.8 MW in the preheater and 3.3 MW as LP steam. The system will require 1.18 MW of electricity to operate and has an overall system coefficient of performance of 3.5.

Alternative 2 uses an air-to-water heat exchanger, a water-to-water preheater, a medium temperature water-to-steam heat pump, and a steam compressor. The system recovers about 3.1 MW from the exhaust stream and delivers 0.8 MW in the preheater and 3.3 MW as LP steam. The system will require 1.02 MW of electricity to operate and has an overall system coefficient of performance of 4.1.

Alternative 3 uses an air-to-water preheater, a medium temperature air-to-steam heat pump, and a steam compressor. This arrangement eliminates one heat exchanger and the water-loop compared to the previous alternatives. The system recovers about 3.3 MW from the exhaust stream and delivers 1.2 MW at the preheater and 2.9 MW as LP steam. The system will require 760 kW of electricity to operate and has an overall system coefficient of performance of 5.4.

Installed Cost of the Heat Pump Systems: The cost of installing the three alternative heat pump systems was estimated by obtaining prices from equipment suppliers for major components and estimating installation cost based on actual site conditions at the "host-plant". A summary of the installed-cost is given in Table 3. The main difference in the cost of the three systems is the equipment cost. Alternative 1 is the most expensive at $3.2 millions, while alternative 3 is the least expensive at $1.7 millions.

Cost-Benefit Analysis: The operation of this heat recovery system will reduce natural gas consumption normally required to generate steam in the boilers. At the time of the study (summer 1986), the cost of producing steam in the plant was about $0.02/kWh ($0.0141/kWh fuel costs @ 71% overall steam-raising efficiency). The dryer operates 6,720 h per year and the average electrical cost (energy and demand charges) for an incremental load is $0.035/kWh. Table 4 summarizes the costs and benefits for the three heat pump alternatives. Alternative 3 is clearly the best, with the lowest capital cost and the highest coefficient of performance. The simple payback for this system is 4.4 years.

Table 5 gives simple payback periods for other steam and electricity costs. This will allow other customers to begin to assess the impact of this type of system on their operations. The table illustrates that heat pump systems have shorter payback times in areas with high energy prices.

4.0 CONCLUSIONS

An electric utility can play an important role in the introduction of electrotechnologies into new applications, by providing technical resources to bridge-the-gap between equipment supplier know-how and customer needs. This is true for industrial heat pumps as well as other technologies.

Hardwood drying with dehumidifiers can reduce energy consumption by 60% compared to steam-heated kilns. This is now the accepted method of drying by over half of the hardwood kiln operators in Ontario.

Dehumidifier manufacturers and suppliers in Canada are experienced in supplying equipment for hardwood kilns, and in the last few years some have made inroads into the softwood industry. New markets exist in other industries, but due to the general lack of information on how products dry, providing dehumidification equipment for these applications (textile, food, paper products etc.) is often a hit-

and-miss proposition. Electric utility development and demonstration facilities can play an important role by providing the technical resources needed to bring the equipment supplier and customer together.

Components can be purchased today that will allow the recovery of heat from moist air exhaust streams and the production of low pressure steam and/or hot water. One such system, capable of delivering 4.1 MW of heat has a coefficient of performance of 5.4. The system reduces annual heating cost by $390,000 and has an installed cost of $1,730,000, which provides a simple payback of 4.4 years.

Low energy prices (both fuel and electricity) in Canada contribute to the relatively long payback periods for heat pump equipment. For example, a 30% increase in the real cost of both fuel and electricity would reduce the payback period for the steam-raising heat pump from 4.4 years to 3.3 years. In the case of dehumidification equipment, improvements in product quality (eg color, shrinkage, dimensional accuracy, etc.) often overshadows the energy-cost issue in making the equipment cost-effective.

5.0 ACKNOWLEDGEMENTS

I would like to acknowledge the contributions of A. Goldenberg in preparing the figures for this paper, as well as the various members of the Process-Applications Unit at Ontario Hydro that conducted the tests and performed the calculations that form the basis of the paper.

6.0 REFERENCES

1. Young, D.J. and Lange, H., "Design of a Northern Climate Heat Pump", Canadian Electrical Association Report CEA 78-86, March, 1979.

2. Ciz, J., "Dehumidification Heat Pump Wood Dryer - Performance Evaluation", Ontario Hydro Research Division Report #84-59-K, April 4, 1984.

3. Strack, J.T., Howell, B.T., "Survey of Wood Drying Kilns in the Province of Ontario", Ontario Hydro Research Division Report #83-33-K, March 22, 1984.

TABLE 1

DEHUMIDIFICATION KILN PERFORMANCE SUMMARY

Drying Cycle Dates:	Oct 6 to Nov 5, 1982
Duration:	30 days
Type of Wood:	Beech, Hard Maple, Ash
Volume of Wood:	50.3 m^3 (21,320 fbm)
Moisture Content: - Start (wt., dry basis) - Finish	60 % 9 %
Mass of Water Removed:	15,467 kg
Dehumidification Kiln Performance	
Elect. Consumption	
- Heat Pump:	4,334 kWh
- Heater:	3,912 kWh
- Auxiliaries:	5,097 kWh
total	13,343 kWh
Heat Pump COP:	2.6 to 5.2
Performance Factor:	0.86 kWh/kg_{water}
Conventional Steam Kiln (Ref 2)	
Typical Performance Factor:	2.0 kWh/kg_{water}
Estimated Consumption for similar volume of wood:	31,000 kWh

TABLE 2

SELECTION OF A HEAT PUMP DEHUMIDIFIER FOR DRYING TEXTILE PACKAGES

	SELECTION CRITERIA		
	LOW ENERGY OR LOW CAPITAL COST	CAPITAL/ ENERGY/ TIME COMPROMISE	SHORT DRYING TIME
DEHUMIDIFIER CAP. (kW_{comp})	2.2	3.8	11.1
APPROX CAPITAL COST (Cdn$,000)	8	10	20
EXPECTED PERFORMANCE: DRYING TIME (h)	9.6	8.7	6.9
ENERGY CONSUMPTION (kWh)	75	86	139
EXISTING DRYER PERFORMANCE: DRYING TIME	6 TO 8 HOURS		
ENERGY CONSUMPTION	150 kWh		
ECONOMICS: ANNUAL ENERGY COST SAVINGS* (Cdn$,000)	2.4	2.0	0.3
SIMPLE PAYBACK (yr)	3.3	5.0	58

* Based on 650 batches per year of 648 spools

TABLE 3

CAPITAL COST ESTIMATES
FOR STEAM-RAISING HEAT PUMPS
(Cdn$,000)

ITEM	ALTERNATIVE 1	2	3
EQUIPMENT			
Heat Exchanger	560	560	n/a
Preheater	35	35	220
Heat Pump	2014	900	630
Steam Compressor	n/a	303	303
Piping	80	80	63
Elect & Controls	85	85	85
Building (installed)	104	104	104
TOTAL EQUIPMENT COST	2878	2067	1405
INSTALLATION			
Heat Exchanger	21	21	n/a
Preheater	3	3	21*
Heat Pump	20	8.4	8.4
Steam Compressor	n/a	8.4	8.4
Piping	42	42	42
Elect & Controls	17.5	17.5	17.5
Shipping & Handling	7	7	7
Contingency	75	75	75
TOTAL INSTALLATION	185.5	182.3	179.3
ENGINEERING	150	150	150
TOTAL CAPITAL COST	3214	2399	1734

* Includes installation of hp evaporator

TABLE 4

COST/BENEFIT SUMMARY
FOR STEAM-RAISING HEAT PUMPS

ITEM		ALTERNATIVE 1	2	3
Heat Output	(MW)	4.1	4.1	4.1
Annual Heat Output	(GWh)	27.6	27.6	27.6
Elect Power Req't	(MW)	1.18	1.02	0.76
Annual Elect Usage	(GWh)	7.93	6.94	5.10
Coeff. of Performance: System		3.5	4.1	5.4
HP (& Steam Comp.)		2.8	3.4	3.8
Annual Fuel Savings	(k$)	569	569	569
Annual Elect Costs	(k$)	278	243	179
NET ANNUAL SAVINGS	(k$)	291	326	390
CAPITAL COST	(M$)	3.21	2.40	1.73
SIMPLE PAYBACK	(years)	11.0	7.4	4.4

TABLE 5

THE EFFECT OF ENERGY PRICES
ON THE PAYBACK PERIOD FOR A STEAM-RAISING HEAT PUMP
(YEARS)

COST OF STEAM ($/GJ)	(¢/kWh)	AVERAGE ELECTRICITY PRICE (¢/kWh) 1.0	2.0	3.0	4.0	5.0	6.0	7.0	8.0	9.0	10.0
1.39	0.5	20	48	NA	NA	NA	NA	NA	NA	NA	NA
2.78	1.0	7.7	9.9	14	24	82	NA	NA	NA	NA	NA
4.17	1.5	4.8	5.5	6.6	8.2	11	16	30	288	NA	NA
5.56	2.0	3.5	3.8	4.3	5.0	5.8	7.0	8.9	12	19	41
6.94	2.5	2.7	2.9	3.2	3.6	4.0	4.5	5.2	6.1	7.5	9.6
8.33	3.0	2.2	2.4	2.6	2.8	3.0	3.3	3.7	4.1	4.7	5.4
9.72	3.5	1.9	2.0	2.1	2.3	2.4	2.6	2.8	3.1	3.4	3.8
11.11	4.0	1.6	1.7	1.8	1.9	2.0	2.2	2.3	2.5	2.7	2.9
12.50	4.5	1.5	1.5	1.6	1.7	1.8	1.8	2.0	2.1	2.2	2.4
13.89	5.0	1.3	1.4	1.4	1.5	1.5	1.6	1.7	1.8	1.9	2.0

Figure 1: A Commercial Dehumidification Hardwood Kiln at a Furniture Plant

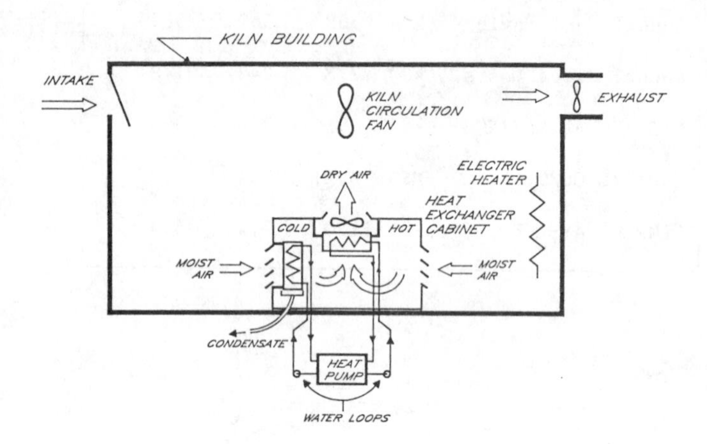

Figure 2: Schematic of the Dehumidification Kiln

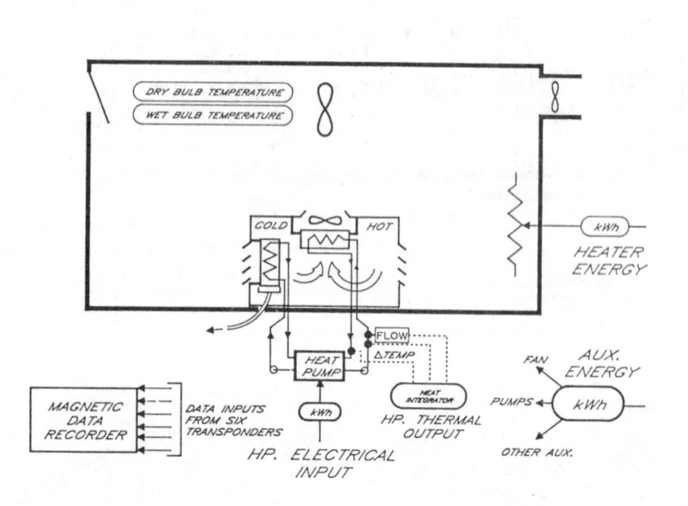

Figure 3: Instrumentation Used to Measure the Performance of the Dehumidifier

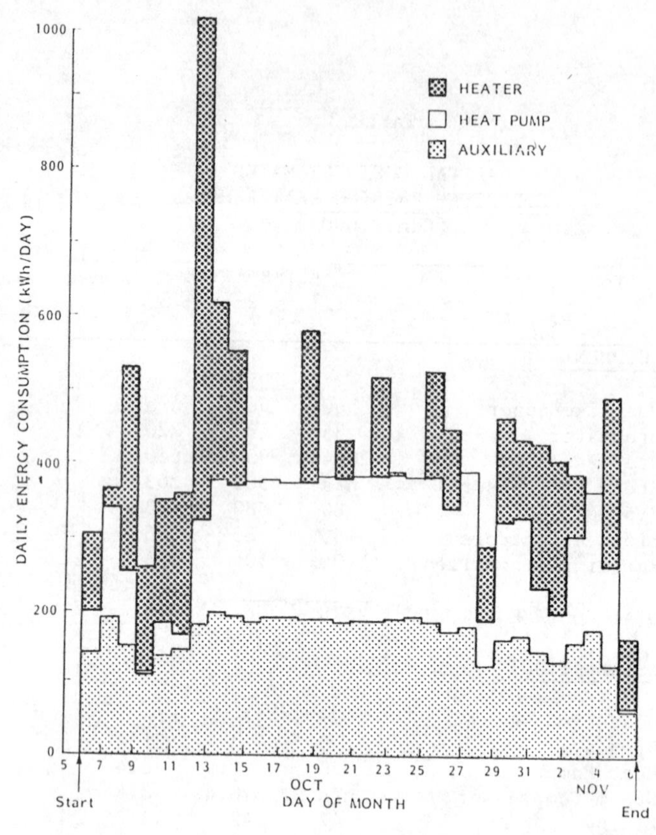

Figure 4: Energy Consumption by the Wood-Kiln for a Typical Drying Cycle

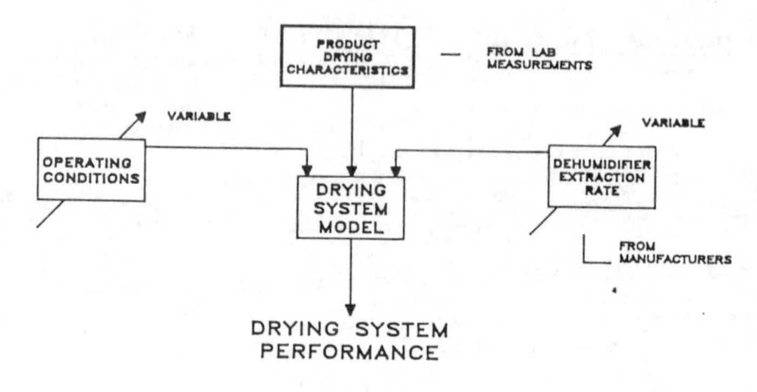

Figure 5: Block Diagram of the Ontario Hydro Method of Predicting Dehumidification Dryer Performance

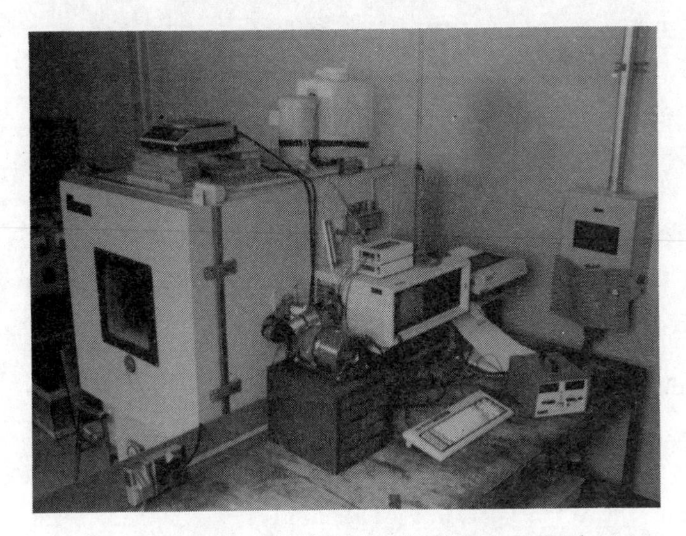

Figure 6: Testing Facilities Used to Measure Product Drying Rates

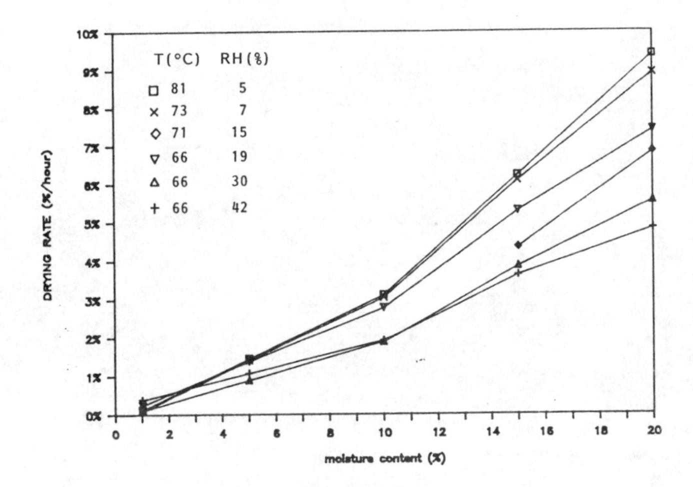

Figure 7: The Drying Characteristic for Spools of 100% Nylon Thread

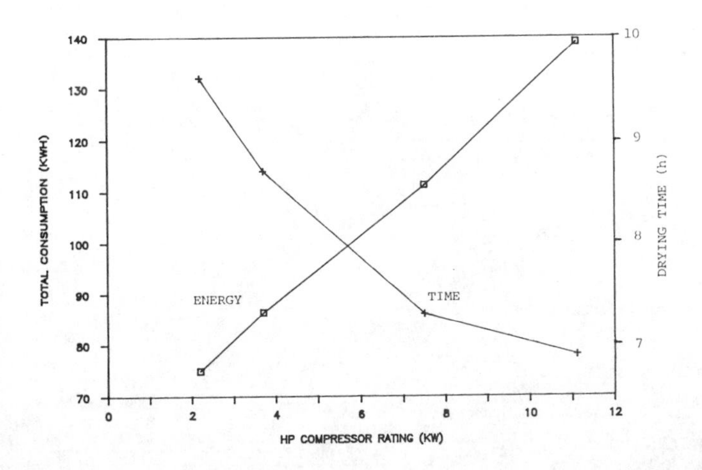

Figure 8: Energy Consumption and Drying Time for Dehumidification Systems Drying 648 Packages of Nylon Thread

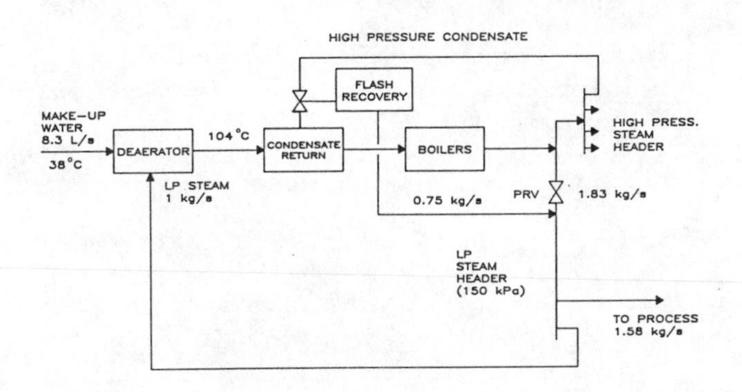

Figure 9: Heating Loads in a Corn Processing Plant in Ontario

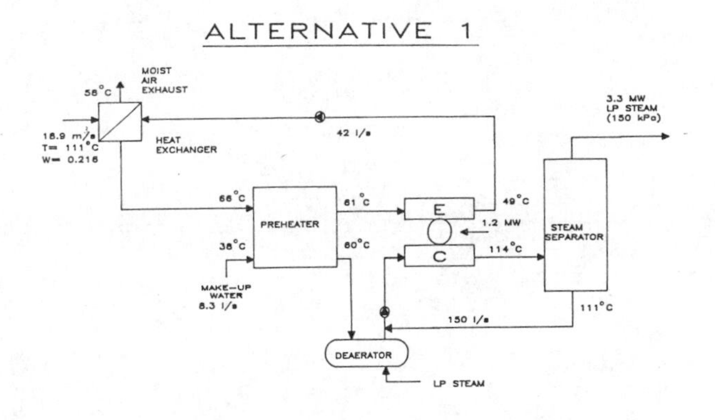

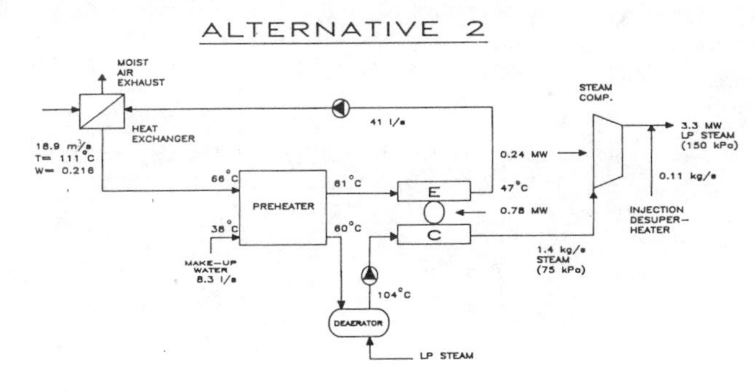

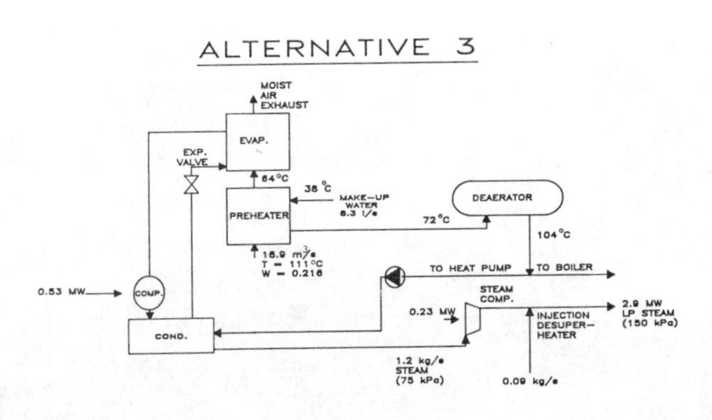

Figure 10: Steam-Raising Heat Pump Alternatives for the Corn Processing Plant

3rd International Symposium on the
Large Scale Applications of Heat Pumps
Oxford, England : 25-27 March 1987

PAPER J2

THE DYNAMIC THERMAL MODELLING AND CAPACITY CONTROL OF THE ABSORPTION CYCLE HEAT PUMP

C.P. Underwood
Newcastle upon Tyne Polytechnic

SYNOPSIS

A dynamic thermal model has been developed, tested and applied in the design of a capacity control system for the Lithium Bromide/Water absorption cycle heat pump.

The model has been developed using a Lumped Parameter methodology and solved by computer. Transient disturbances in cycle input conditions have been investigated at typical running conditions by the computer model and experimentally and the two sets of results compared. The input transients considered have been source and sink water temperatures and the generator heat input rate representing typical cycle disturbances in an industrial environment.

In general good agreement has been obtained between model and experimental results with an average coefficient of determination over all results of 0.55.

The model was used in the development of a capacity control system which was tested experimentally. A proportional plus integral algorithm was used employing a proportional gain of 111% $^{\circ}C^{-1}$ and an integral time of 10 seconds. This was implemented digitally using a sampling interval of 15 seconds and was found to give a stable response in condenser water outlet temperature whilst maintaining the cycle COP during modest load changes.

INTRODUCTION

The study of transient performance and capacity control for the absorption cycle heat pump was based on a 10 kW experimental rig designed and constructed by Treece (1). The purpose of the study was to identify a suitable modelling methodology for investigating cycle input disturbances such as might occur in an industrial environment and to use this methodology in the design of a suitable capacity control system.

It was assumed that any capacity control strategy would be based upon achieving a constant sink water leaving temperature whilst undergoing load changes manifesting themselves as variations in sink water entering temperature. This assumption was in-keeping with many process applications as well as mass market orientated space heating applications and, particularly, boiler replacement/retrofit situations.

The control system was perceived as having three aims to satisfy. Firstly the establishment of stable operation during load changes. Secondly, the capability of satisfying a wide range of loads and, thirdly, the achievement of constant optimum coefficient of performance (COP) during load changes.

SYMBOLS USED

Hgsso — Generator strong solution enthalpy ($kJkg^{-1}$)

Hgswi — Generator weak solution entering enthalpy ($kJkg^{-1}$)

Hpssi — Heat exchanger strong solution entering enthalpy ($kJkg^{-1}$)

Hgro — Condenser entering refrigerant enthalpy ($kJkg^{-1}$)

Hgr — Generator leaving refrigerant enthalpy ($kJkg^{-1}$)

Xgsso — Generator strong solution concentration (%)

Xgswi — Generator weak solution entering enthalpy (%)

Xpssi — Heat exchanger strong solution entering enthalpy (%)

Qg — Generator heat input rate (kW)

ϕsw — Weak solution flow rate ($kg\ s^{-1}$)

ϕss — Strong solution flow rate ($kg\ s^{-1}$)

ϕgr — Refrigerant flow rate ($kg\ s^{-1}$)

ρgsso — Generator strong solution density ($kg\ m^{-3}$)

ρpssi — Heat exchanger strong solution entering density ($kg\ m^{-3}$)

ρco — Condenser refrigerant vapour density ($kg\ m^{-3}$)

acg — Generator cross sectional area (m^2)

Lgss — Generator solution level (mm)

Pcg — Generator pressure ($kN\ m^{-2}$)

Held at St. Catherine's College Oxford, England. Organised and sponsored by BHRA, The Fluid Engineering Centre, Cranfield, Bedfordshire, MK43 0AJ England.

Pea	Evaporator pressure ($kN\ m^{-2}$)
kss	Strong solution flow constant
Vgps	Generator strong solution line volume (m^3)
Vgr	Generator vapour line volume (m^3)
R^2	Coefficient of linear determination
E	Controller output
Kc	Controller gain
ti	Integral time (seconds)
Eo	Controller output in the steady-state
ε	Control error (^{o}C)

System Description

A general layout of the subject heat pump rig is given in schematic form (Figures 1 and 2).

Vapour is generated at low pressure (typically 3-5 $kN\ m^{-2}$) by two evaporators boiling in parallel with series-connected heating water which is pumped between the evaporators and two electric water heaters. The vapour passes into the absorber where it is dissolved into aqueous lithium bromide. The absorber comprises a sprayed tube bundle in which the vapour dissolves into solution which is pumped over four concentric tube bundles. Solution is recirculated from the absorber "bottoms" and mixes with strong concentration solution from the generator to produce a larger quantity of absorbing solution but at a lower concentration than exists in the bottoms.

Heat is evolved in the absorber due to the diffusion of refrigerant vapour into solution. This heat is carried away by an internal cooling water circuit which passes first through the absorber and then through the condenser. Facilities exist for the absorber and condenser to be cooled independently enabling transients to be investigated in the absorber without disturbing the condenser/generator.

The weak and cool solution is then pumped to the generator through a solution heat exchanger which transfers heat from the hot strong solution flowing from the generator to the cool weak solution flowing from the absorber. The presence of this heat exchanger is important as it relieves the absorber of what would otherwise be a significant sensible heat load which would need to be satisfied before equilibrium absorption of the low pressure vapour can be approached.

The weak solution enters a pool of strong solution in the generator. Pure vapour is separated from the less-volatile solution by boiling. At the generator/condenser pressure, the saturation temperature of water is less than that of the solution. Thus the vapour which boils out of solution is superheated and passes to the condenser for de-superheating and condensing. The heat of condensation is removed by cooling water piped from the absorber. This cooling water is itself cooled by an external cooling water circuit fed from the local cold water main. Refrigerant liquid condensing at the upper cycle pressure can be stored in the refrigerant storage vessel or directly throttled to the lower cycle pressure and passed to the evaporator pool.

A refrigerant heat exchanger was fitted in the evaporator/absorber vapour line to exchange heat between the hot condensate liquid and cool refrigerant vapour. This heat exchanger was physically large to give low pressure drop on the vapour side and, consequently, a large volume of water existed on the condensate side of the heat exchanger. Preliminary testing by the previous investigator revealed that this large volume of water, initially cold on start up, took a considerable time to be consumed by the evaporators and replaced by hot fresh condensate. For this reason, this heat exchanger was bypassed on the condensate side and not used in the experimental work.

The strong solution returning from the bottom of the generator was forced out the top of the generator as a result of the vapour pressure created inside. Its passage through the generator pool took place through coiled tubes thus enabling some measure of heat transfer between the strong hot bottom solution with the cooler weaker solution higher up the pool.

System Modelling

A lumped-parameter methodology was adopted in model building with the assumption that heat losses to surroundings may be neglected and pressure losses experienced by flowing fluids may be ignored.

A block diagram representation of the system was expressed and model equations derived around it. Figure 3 shows the block diagram representation of the generator.

Energy and component continuity balances about the generator system gives:

Energy,

$$\phi sw.Hgswi - \phi ss.Hgsso - \phi gr.Hgr + Qg = acg.\frac{d}{dt}(\rho gsso.Lgss.Hgsso) \quad ...(1)$$

Total continuity,

$$\phi sw - \phi ss - \phi gr = acg.\frac{d}{dt}(\rho gsso.Lgss) \quad ...(2)$$

Component continuity,

$$\phi sw.Xgswi - \phi ss.Xgsso = acg.\frac{d}{dt}(\rho gsso.Lgss.Xgsso) \quad ...(3)$$

A steady-state energy balance has been used to establish vapour evolution rates. A film-boiling regime has been expected from the low thermal capacity generator heating coil, making the steady-state assumption plausible.

Accordingly, $\phi gr = \frac{(\phi sw.Hgswi + Qg - \phi ss.Hgsso)}{Hgr}$ ...(4)

A root law has been assumed for the strong solution flow-rate leaving the generator which is forced by vapour pressure-head across a valve before reaching the absorber.

ie $\phi ss = kss(\rho gsso(Pcg - Pea))^{\frac{1}{2}}$ ...(5)

kss is a flow constant which has been determined from experimental data.

The solution distributing line between the generator and solution heat exchanger was assumed to behave as single lump element. With constant flow-rate, energy and component-continuity balances can be written as follows:

Energy,

$$\phi ss(Hgsso - Hpssi) = Vgps. \frac{d}{dt} (\rho ssi.Hpssi) \quad ...(6)$$

Component continuity,

$$\phi ss(Xgsso - Xpssi) = Vgps. \frac{d}{dt} (\rho pssi.Xpssi) \quad ...(7)$$

Also, an energy balance about the generator/condenser gas line gives:

$$\phi gr(Hgr - Hgro) = Vgr. \frac{d}{dt} (\rho co.Hgro) \quad ...(8)$$

A convergence algorithm has been developed to determine solution temperature from enthalpy and concentration data. In general:

Solution Temperature
= f(Enthalpy, Concentration)
and, Solution Density
= f(Temperature, Concentration)

Thermodynamic property data presented by McNeely (2) and Brunk (3) was used for aqueous lithium bromide and that of Keenan et al (4) used for water vapour.

A similar procedure was adopted for the remaining components and a final model equation set comprising 35 differential and 29 other functional relationships arrived at.

The Computer Model

A computer program was written in FORTRAN 66 for the solution of the model equations and implemented, initially, on an IBM 370 mainframe and, latterly using a FORTRAN 77 compiler, on an AMDAHL 5860 mainframe.

The main program listing contains the following major compartments and solution methodology.

(a) Read and Initialising Sections

The data input or read section introduces all constants and initial conditions necessary for the program to function. The initialising section generates further initial conditions for distributing pipes and similar system elements at the initial steady-state. Title-printing proceeds the read and initialising section.

(b) Dynamic Section

In the dynamic section, a call is made to fluid property subroutines and all derivatives are thence explicitly calculated sequentially. The majority of the processing time dedicated to each simulation is spent in this section and, particularly, in the solution temperature subroutine.

(c) Write Section

The output or write section immediately follows the dynamic section. All output variables are printed at the current time point.

(d) Update Section

The update section is the final function in each loop-pass so retaining the sequential solution of all differential equations. "Time" is incremented and all derivatives integrated with respect to time.

The program loop control then returns to the beginning of the dynamic section unless the current time point has been incremented to a value in excess of the "halt-time".

Choice of an Integration Algorithm

A simple Euler integration algorithm has been adopted for two main reasons.

(a) Although reasonable accuracy is required from the integration routine, it is evident that model equations represent a stiff system of fast and slow acting differential equations (47). For instance, equations involving vapour dynamics (eg equations 10, 29) will act very much more rapidly and, hence, with greater potential instability than will equations involving lithium bromide solution dynamics (eg equations 22 and 31). Since the slower acting dynamics are ultimately of most interest, an integration interval chosen to given stable solution of fast acting equations will ensure accurate solution of the slower acting ones which represent the majority of system zones.

(b) The size and complexity of the problem appeared to warrant a simple processor-efficient integration method particularly if future developments on micro-computing equipment were to be in prospect.

An integration interval of 0.5 seconds was first selected for trial simulation runs. This produced instability and further trials were conducted with intervals of 0.1 and 0.05 seconds, both yielding persistingly unstable results. An inspection of the fast-acting equations in the model identified equation number 10 representing an energy balance in the evaporator/absorber vapour line as giving the highest "potential time constant" clearly resulting from the rarefied vapour zone which this equation represented. Accordingly it was replaced with a steady-state energy balance which resulted in model stability with an integration interval of 0.05 seconds.

Application of the Computer Model and Comparison with Experimental Results

To enable a comparison between model performance and actual running conditions to be made, the laboratory rig developed by Treece (1) was instrumented.

Iron/constantan thermocouples were located in the positions given in Figure 4. They were mineral insulated bonded junction type having low time constant.

Turbine flow meters were located in source and sink water circuits and heating current signal interface cards fitted to both the evaporator and generator thyristors giving 0-10 Vdc signal output proportional to heating current output.

Transients in source water temperature, sink water temperature and generator heat input formed the major input disturbances investigated. This was achieved by running the rig at some steady-state condition, typically with the sink water temperature at 65-70°C and the source water temperature 30-35°C. The steady state condition was then interrupted by the input transient. Figures 5-7 show input transients investigated consisting of a step-wise change in generator heat input rate and approximatley ramp-wise changes in source and sink water temperatures. Transient response results were sampled at intervals of typically 10-20 seconds by a 40 channel data logger (Figure 4).

The model was applied to data representing equivalent running conditions obtained in the initial steady-state operating mode of the laboratory rig. Allowances had to be made in the theoretical model for standing heat losses from the experimental rig to enable the model to be initialised at eqivalent experimental conditions.

Figures 9-14 give time response results for an approximately ramp-wise decay in sink water temperature. Experimental and model results have been co-plotted on common axes.

Corresponding decays were noted in both absorber and condenser leaving water temperature (Figures 10 and 12). Heat transfer from both absorber and condenser experienced an initial rippling during the transient decay in source water temperature but eventually settled at, approximately, their starting values (Figures 9 and 11). Accordingly, a similar transient was obtained in cycle coefficient of performance (COP) given in Figure 13.

A gradual decay in absorber solution temperature (Figure 14) was noted confirming the significant thermal capacity in the solution and the bottoms zone.

The coefficient of determination (R^2) indicating the degree of linear association with experimental and model results plotted against one another was excellent with this test, being better than 0.70.

Figures 15-18 give time response results for a gradual decay in source water temperature.

Significant decay was noted in absorber outlet water temperature and heat transfer and, correspondingly, in COP (Figures 15, 16 and 17). An instantaneous functional relationship was used to describe vapour evolution in the evaporators and these results question the adequacy of this with wide disagreement between model and experimental results. This test also reveals that caution must be exercised when interpreting R^2 values with dynamic data sets undergoing approximately constant rate of change but not in agreement with one another.

Figures 19-22 give time response results for an instantaneous change in generator heat input rate.

A rapid rate of rise in condenser heat output and leaving temperature model results was obtained with a more gradual rise in the corresponding experimental resuls (Figures 19 and 20). Again, the disagreement between results is considered to be due to the use of instanteneous functional relationships used for vapour evolution (equation 4). Future work must focus on the representation of boiling and condensing rates by dynamic equations perhaps based on detailed experimental study. The elevated condenser heat transfer and COP (Figure 21) confirm that the initial operating condition was at less than optimum generator heat input for best performance results.

Development of a Cycle Capacity Control System

The capacity control system for the aqueous LiBr and water absorption heat pump was perceived as having three main purposes.

Firstly smooth, stable operation of the heat pump during load changes at the point of heat rejection was required.

Secondly, the control system should enable the maximum possible range of load to be satisfied within the rating of the heat pump system without the need to shut-down the heat pump completely at light load.

Finally, the coefficient of performance be maintained as close as possible to the optimum operating value during load changes.

Control practice in the well-established field of absorption cycle refrigeration was not found to vary significantly in methodology. Most of the earlier and present day refrigerating machines rely on generator heat input modulation in order to achieve constant evaporator leaving water temperatures (5, 6).

The control of condenser leaving water temperature by modulating generator heat input not only appears to offer a responsive method of control (more so than if evaporator leaving water temperature were the controlled condition) but also enables direct matching between heat delivered to sink and heat received at the generator.

Potential applications for the absorption cycle heat pump in process water heating, bottle washing, space heating, etc, usually require to receive a constant rate of water at constant delivery temperature leaving the return temperature to vary according to load. This supports the condenser leaving water temperature of the heat pump being the controlled condition and, of major importance, would enable the heat pump controlled in this way to be used in boiler replacement applications.

A standard proportional plus integral control algorithm was adopted having the following form:

$$E = Kc.(\varepsilon + \frac{1}{ti} \int_0^\infty \varepsilon.dt) + Eo \qquad ...(9)$$

Modifications were carried out to the computer model employing simple Euler integration of the controller integral term. A similar basis was adopted in the development of a real time program developed for a laboratory based Apple IIE micro-computer to enable practical implementation of the capacity control system.

Figure 23 summarises the resulting feedback control loop which was implemented.

The thyristor used to control the ac power input to the generator heaters was rated at 40 amps but had an overcurrent protector set at 28.5 amps (the maximum current handled by the three heating elements combined). Also, the dc positioning-voltage range of the thyristor was 0-5 Vdc. The digital to analogue converter output range was 0-10 Vdc but its range was restricted to 0-3.6 Vdc to match the overcurrent protection of the thyristor by setting a limit of 36% in the Apple IIE controller output to the digital to analogue converter (DAC).

The DAC output signal was fed to the thyristor discretely upon completion of the necessary processing by the Apple IIE.

Tuning the Control Algorithm

The model was applied using a variety of trial controller settings, responding to a unit change in set point. Results that were obtained indicated that desirable control performance was more heavily dependant on the integral time setting than on the gain setting. Figures 24 and 25 give the response to a unit change in set point using a proportional gain of $1.11^{o}C^{-1}$ and integral times of 66 and 10 seconds. With the higher integral time some small degree of offset is evident. Accordingly the settings of Figure 25 were used in the investigation.

A sampling interval for the digital control loop of 15 seconds was used in experimental testing of the control loop. This was determined by observing the frequency of the data logger analogue output annunciator lamp which flashed upon completion of each processing operation.

Testing of the Capacity Control System

Additional instrumentation was added to the experimental rig prior to capacity control system testing. Absolute pressure transmitters were mounted at the generator head and at the evaporator head. To enable the monitoring of solution liquid level in absorber and generator, differential pressure transmitters were connected to the top and bottom flanges of these vessels (Figure 4). These were calibrated for liquid level measurements using existing level sight glasses.

Figure 8 shows a transient disturbance in sink heat load in terms of absorber water inlet temperature which was used to investigate control system performance. This represents an increase in load of approximately 80% followed by a return to the, approximately, initial condition.

Figures 26-37 give time response results of the cycle under control attempting to maintain a set point condition (condenser leaving water temperature) of $72.5^{o}C$. Generator and condenser heat transfers were noted to follow similar trends (Figures 30, 28).

Poor agreement was found with results involving pressure and solution liquid level (Figures 32, 33, 36 and 37). The liquid level results gave scattered experimental data and flat model data. This was due to the pulsating action of the positive displacement pump which delivered absorber-bottoms solution to the generator.

Good agreement was noted between model results and ultimate experimental data on the controlled variable (Figure 29). Gradual rate of rise of experimental data here indicates model inaccuracy to be associated with "instantaneous" refrigerant evolution in the condenser.

CONCLUSIONS

A dynamic thermal model has been developed, tested and applied to the investigation of a method of cycle capacity control for the aqueous lithium bromide and water absorption cycle heat pump.

The model has been developed using a lumped-parameter methodology and consists of 35 first order ordinary differential equations and 29 supporting functional relationships. The model is solved explicitly using a simple Euler integration algorithm with an integration interval of 50 ms.

It would appear that this represents one of the first dynamic models for the absorption cycle system, whether refrigerator or heat pump.

The model has been applied to two distinct lines of investigation. First, a study was made on the influence of disturbances in low and high grade heat sources on the performance of the system, as well as variations in load at the heat sink. This was achieved by imposing step-wise disturbances in generator heat input rate and approximately ramp-wise disturbances in source water temperature and sink water temperature. Second, a study was made on a method of cycle capacity control for the heat pump using a simple P + I control algorithm to vary generator heat input capacity to achieve a constant condenser water outlet temperature.

To enable a comparison of model performance with the real world to be made, a 10 kW scaled laboratory rig which had been developed previously was used to obtain experimental results.

Operation of the experimental rig and model in the study of disturbances in cycle inputs produced the following:

(a) Ramp changes in source water temperature affect, predominantly, the absorber. An increase in source water temperature resulted in an elevation in absorber outlet water and solution temperature and a small increase in absorber heat output. A communicated increase in condenser water outlet temperature occurred with marginal overall change in heat output. The increase in absorber heat transfer resulted in a small elevation in COP. Opposing trends were found with a decrease in source water temperature.

(b) Response to changes in sink water inlet temperature were found, in general, to have little overall effect on absorber and condenser heat transfer after some initial oscillation. Cycle COP was found to follow a similar trend as the combined heat outputs.

(c) Step changes in generator heat input rates were found to influence, predominantly, the condenser heat output. An increase in generator heat input produced an increase in condenser heat output and a small increase in absorber heat output. Little change in COP occurred since these changes were largely self-cancelling. An opposing trend was found for a reduction in generator heat input.

(d) In general good agreement between model and experimental results was obtained particularly with source and sink water temperature and heat transfer parameters. An overall average coefficient of determination of 0.55 was obtained for all results obtained.

The principle causes for inconsistency in model and experimental agreement was threefold.

(i) Difficulty in achieving perfect steady-state conditions in the experimental rig and the unavoidable incompleteness in the measureable data set. This required some parameters, particularly flow rates, to be calculated from measured results to give model initial conditions.

(ii) Discontinuity in the strong solution return flow rate during experiments due to vapour carry over and flashing effects.

(iii) Simplistic heat transfer data used in model development, together with the use of idealised flow laws for strong solution and refrigerant flow modelling, and the dynamic approximation in using steady-state models for refrigerant boiling and condensing rates.

The model was developed to include a capacity control loop using a simple P + I algorithm. A proportional gain of 111% range/^{o}C error and an integral time of 10 seconds was established and a sampling interval of 15 seconds determined experimentally for the implementation of the control algorithm by local micro-computer.

The control system was tested experimentally. Under conditions of enforced load change which was achieved by varying the sink water inlet temperature, a good continuous agreement was generally found between condenser water outlet temperature and set point. Rapid response in generator load variation and, hence, condenser heat output was noted in achieving the set point condition.

At conditions of light load with the condenser/generator fully unloaded, an offset was noted in the condenser leaving water temperature caused by the unchanged heat output by the absorber. Consequently in a practical system, capacity control of the absorber would need to be exercised if a fully modulating system is to be realised. Although this was not investigated, one method of achieving this would be to vary source water temperature or flow rate in sequence with the generator capacity unloading so that optimum COP is maintained. Since the ratio of absorber to condenser output was, typically, 1:1 absorber capacity control would need to be considered when loads of <50% of rated capacity are anticipated.

In general, COP results were found to be reasonably unimpaired during load changes by this method of control.

REFERENCES

1. Treece, R.J.
A Theoretical and Experimental Investigation of the Performance of Aqueous Lithium Bromide Absorption Cycle Heat Pumps.
PhD Thesis (CNAA), 1983.

2. McNeely, L.A.
Thermodynamic Properties of Aqueous Solutions of Lithium Bromide.
In ASHRAE Transactions, 85(1), 1979, pp 413-434.

3. Brunk, M.F.
Thermodynamic and Physical Properties of the Lithium Bromide Solution as a base for the Process Simulation of Absorption Refrigerating Installations.
In Ki Klima-Kalte-Heizung, 10, 1982, pp 365-372.

4. Keenan, J.H., Keyes, F.G., Hill, P.G., Moore, J.G.
Thermodynamic Properties of Water, Including Vapour, Liquid and Solid Phases.
John Wiley.

5. Bannikov, M.T., Fridman, R.N.
Refrigerating Equipment for the Scientific and Production Association.
In Jnl of Chem and Petro Engineering, 20(9), 1984, pp 425-427.

6. Archie, J.L.
Absorption Refrigerating Systems - 1971.
In ASHRAE Journal, September 1971, pp 41-45.

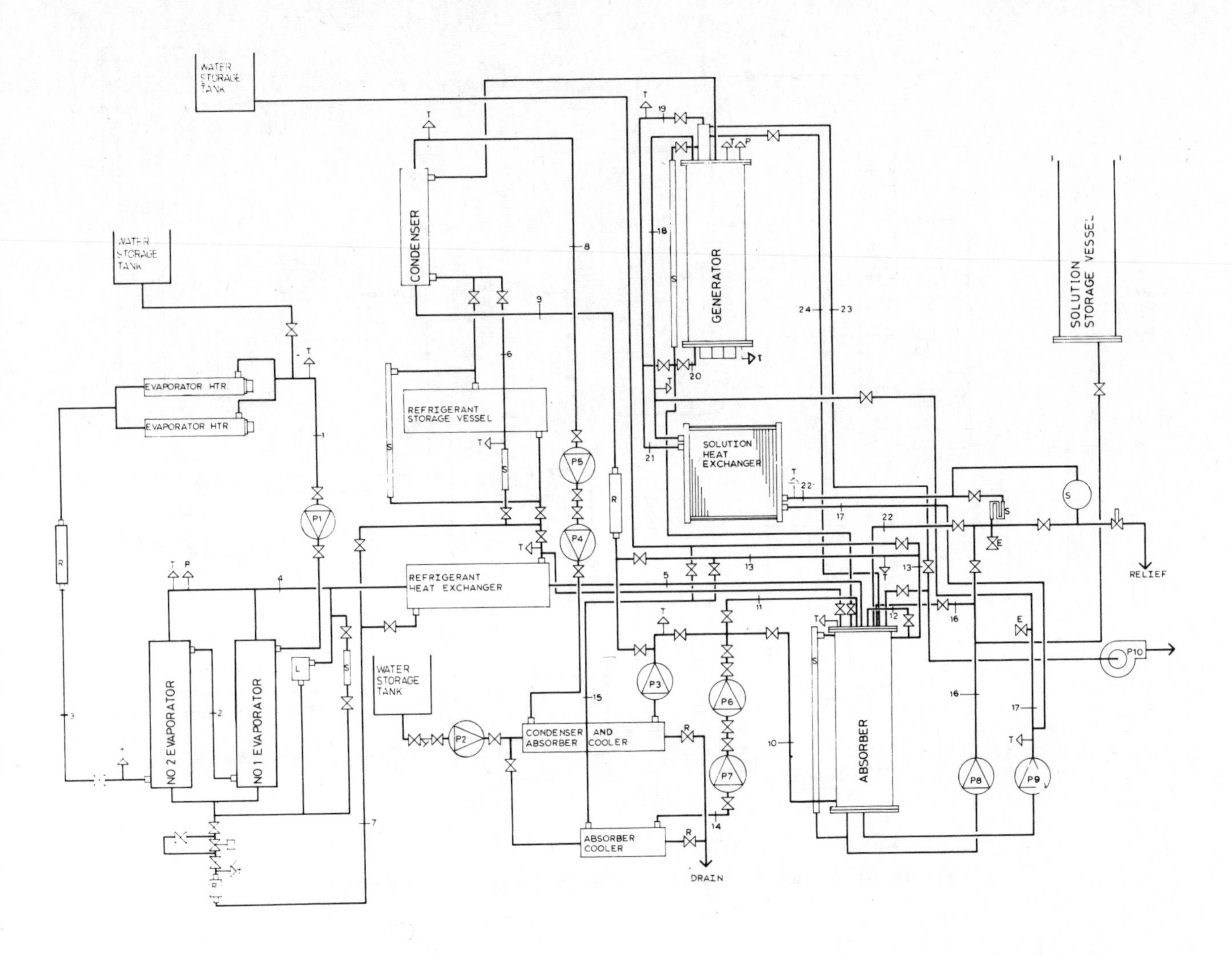

Figure 1: Schematic layout of 10 kW laboratory rig (read in conjunction with Figure 2)

Symbol	Description	Pump	Circuit
⋈	Valve - general purpose	P1	Evaporator source water
⋈	Pressure relief valve	P2	External cooling water
⋈	Metering valve	P3	Internal cooling water
E ⋈	Sampling point	P4	Internal cooling water
⋈	Strainer	P5	Internal cooling water
S	Sight glass	P6	Internal cooling water
R	Rotameter	P7	Internal cooling water
R ⋈	Needle regulating valve	P8	Solution recirculation
↑P	Pressure measurement	P9	Solution delivery
↑T	Temperature measurement	P10	Vacuum pump

Figure 2: Table of symbols used in Figure 1

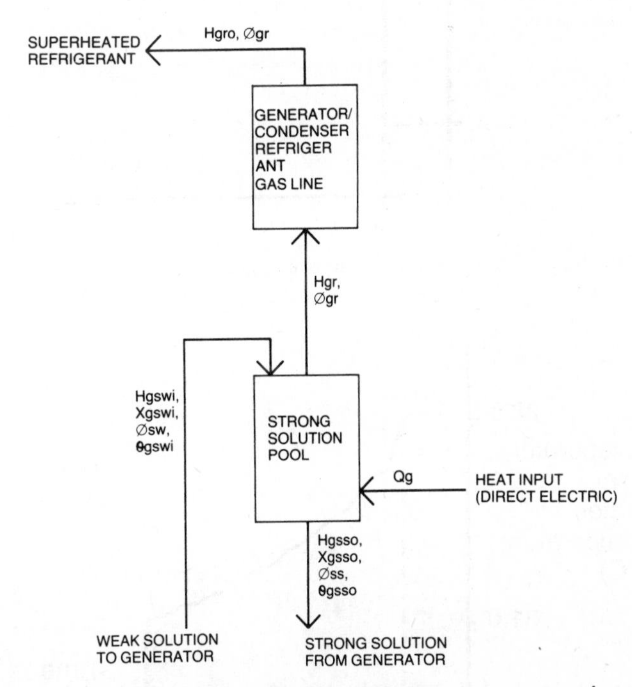

Figure 3: Simplified block diagram representation of the generator

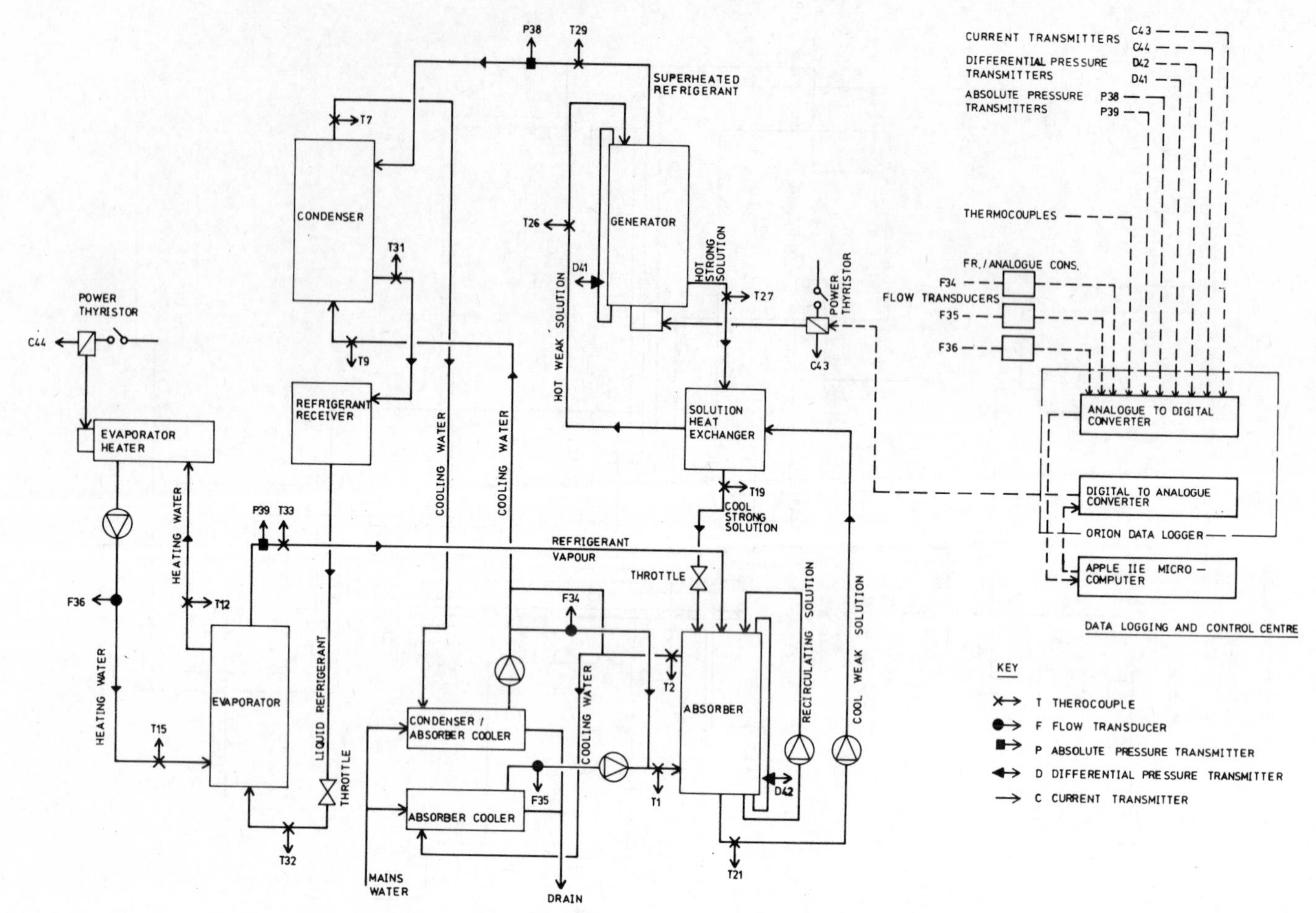

Figure 4: Schematic layout of the instrumented laboratory rig

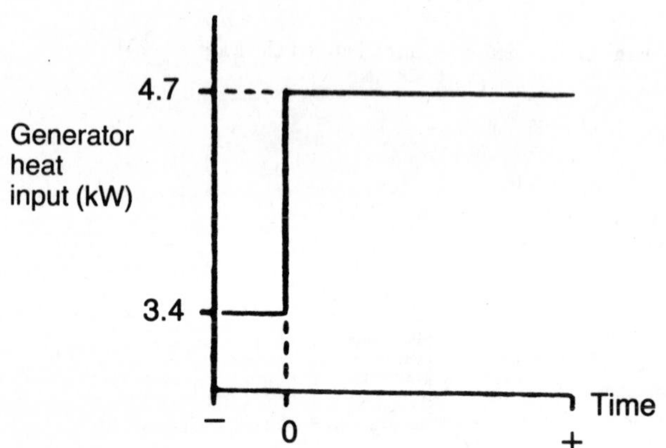

Figure 5: Input disturbance in generator heat rate

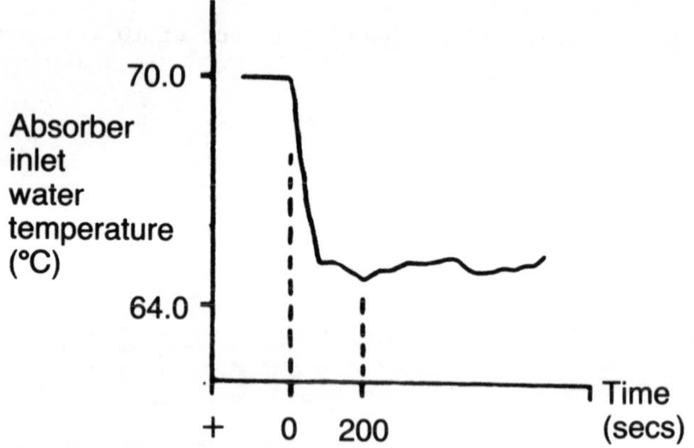

Figure 7: Input disturbance in absorber water inlet temperature

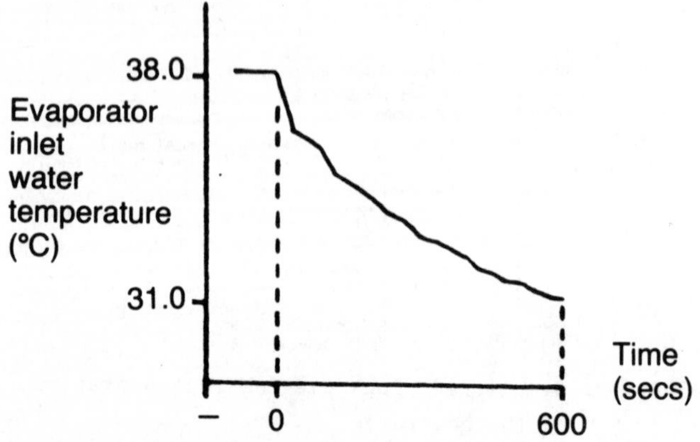

Figure 6: Input disturbance in evaporator water inlet temperature

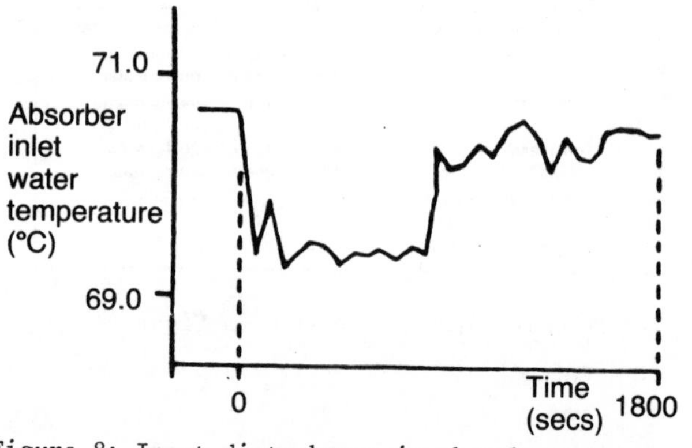

Figure 8: Input disturbance in absorber water inlet temperature (UNDER CONTROL)

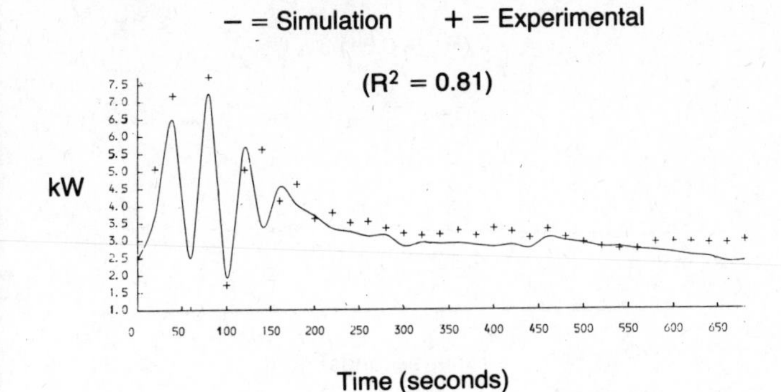

Figure 9: Absorber heat transfer response to a disturbance in inlet water temperature (Figure 7)

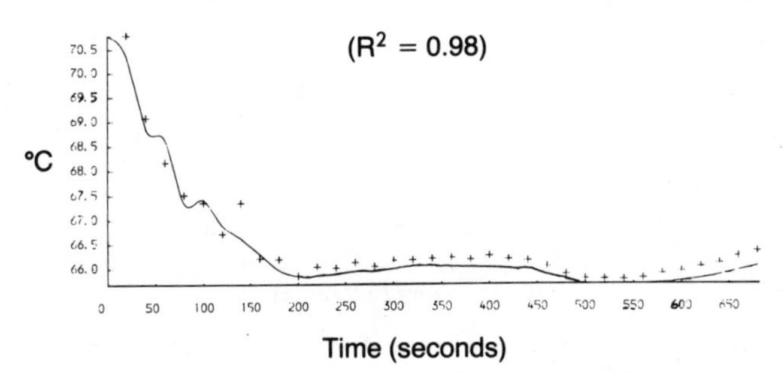

Figure 10: Absorber outlet water temperature response to a disturbance in inlet water temperature (Figure 7)

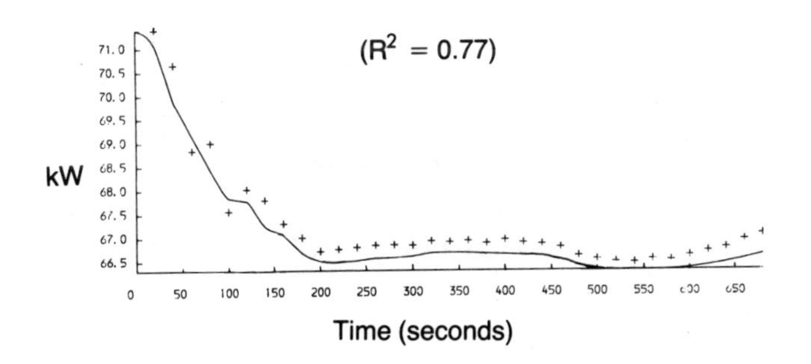

Figure 11: Condenser heat transfer response to a disturbance in inlet water temperature (Figure 7)

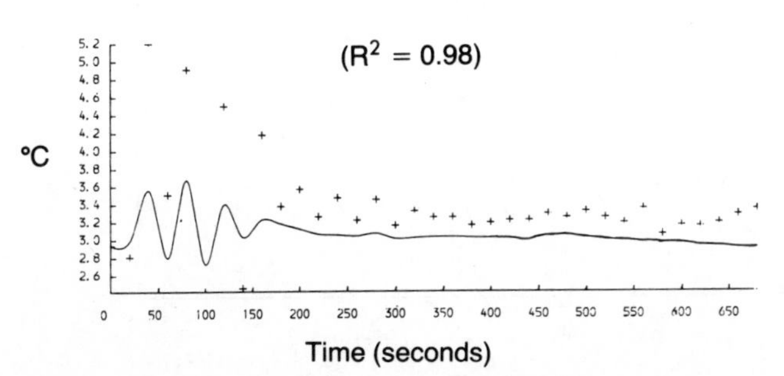

Figure 12: Condenser outlet water temperature response to a disturbance in inlet water temperature (Figure7)

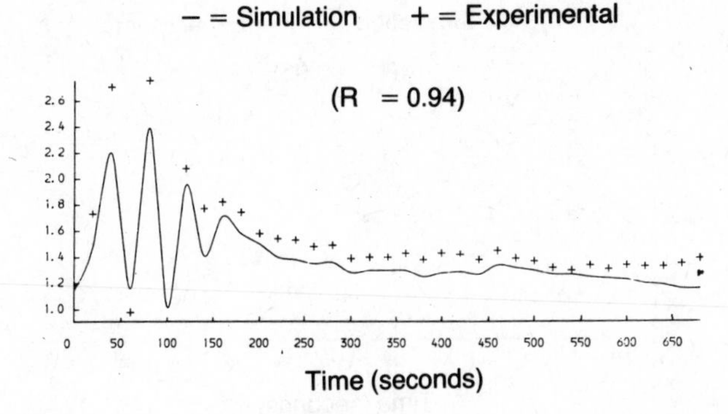

Figure 13: Response of coefficient of performance following a disturbance in inlet water temperature to the absorber (Figure 7)

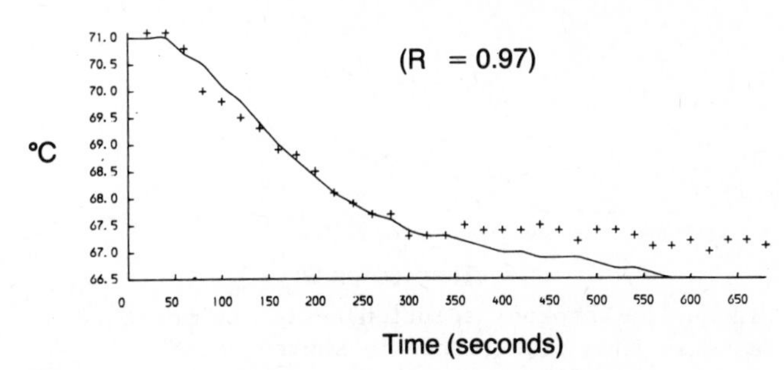

Figure 14: Absorber solution outlet temperature response to a disturbance in inlet water temperature (Figure 7)

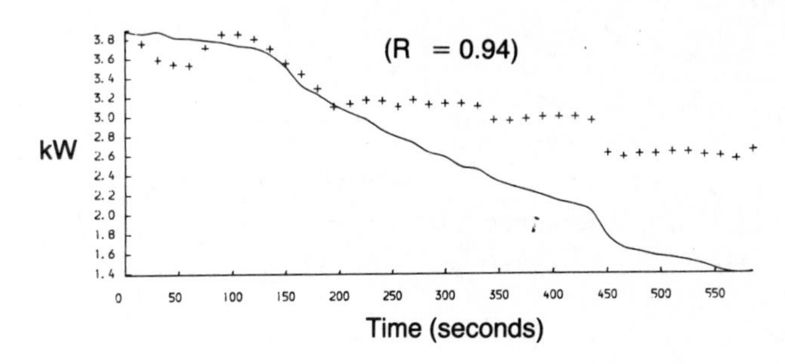

Figure 15: Absorber heat transfer response to a disturbance in source water temperature (Figure 6)

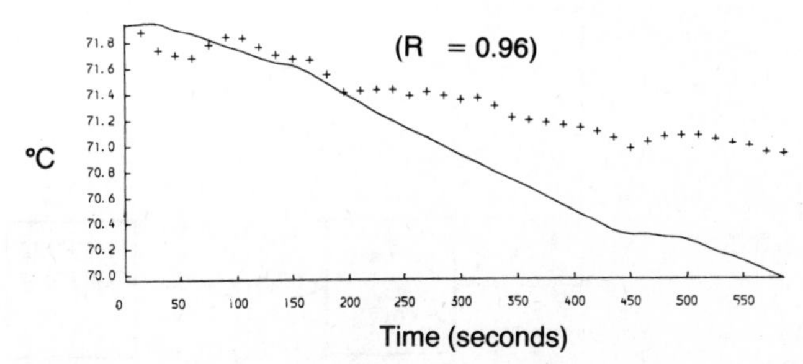

Figure 16: Absorber outlet water temperature response to a distrubance in source water temperature (Figure 6)

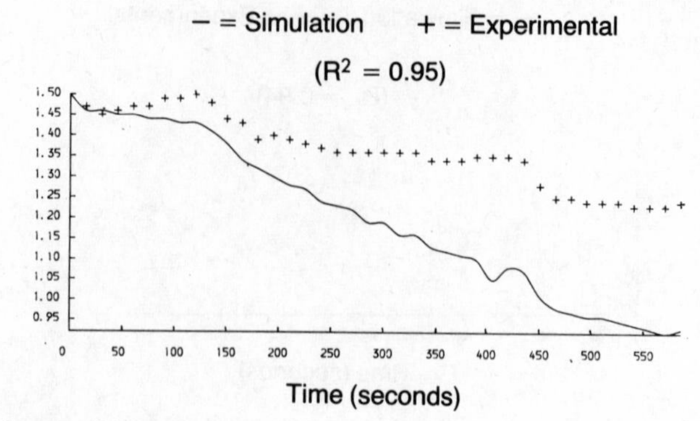

Figure 17: Response of coefficient of performance following a disturbance in source water temperature (Figure 6)

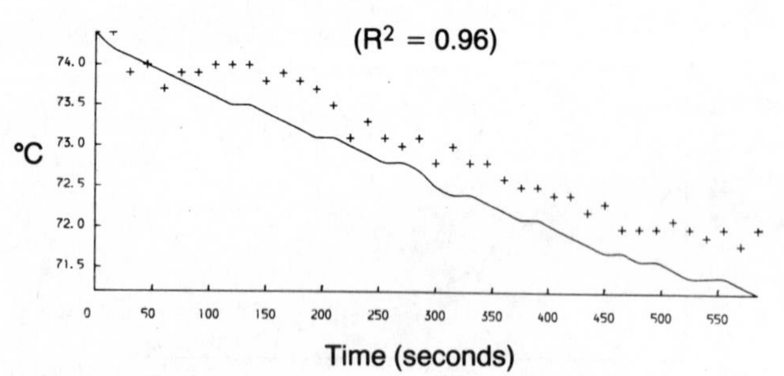

Figure 18: Absorber solution outlet temperature response to a disturbance in source water temperature (Figure 6)

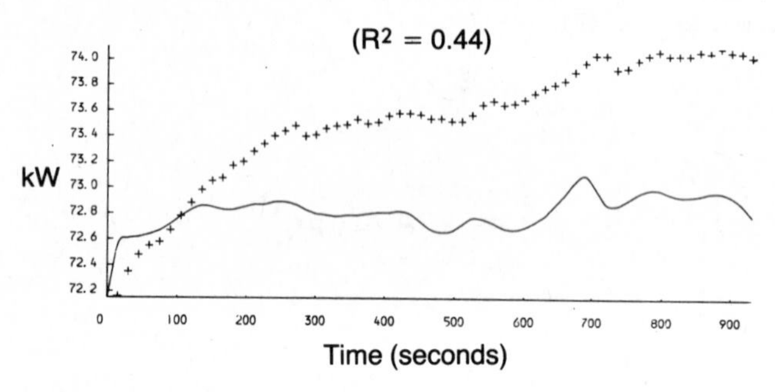

Figure 19: Condenser heat transfer response to a disturbance in generator heat input rate (Figure 5)

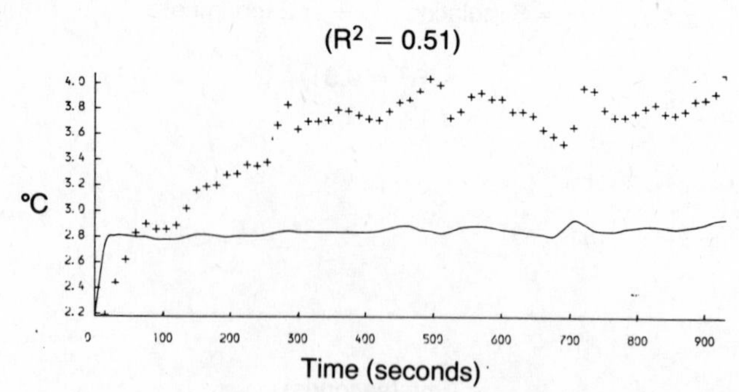

Figure 20: Condenser outlet water temperature response to a disturbance in generator heat input rate (Figure 5)

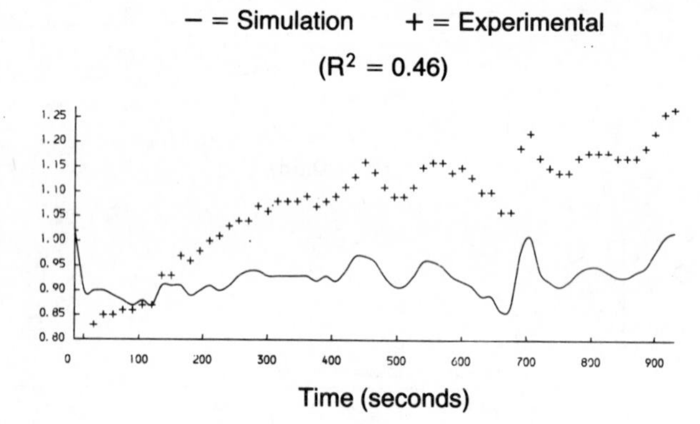

Figure 21: Response of coefficient of performance to a disturbance in generator heat input rate (Figure 5)

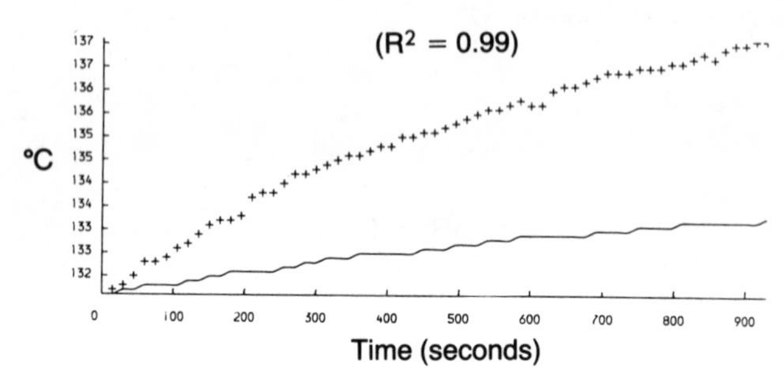

Figure 22: Generator solution temperature response to a disturbance in generator heat input rate (Figure 7)

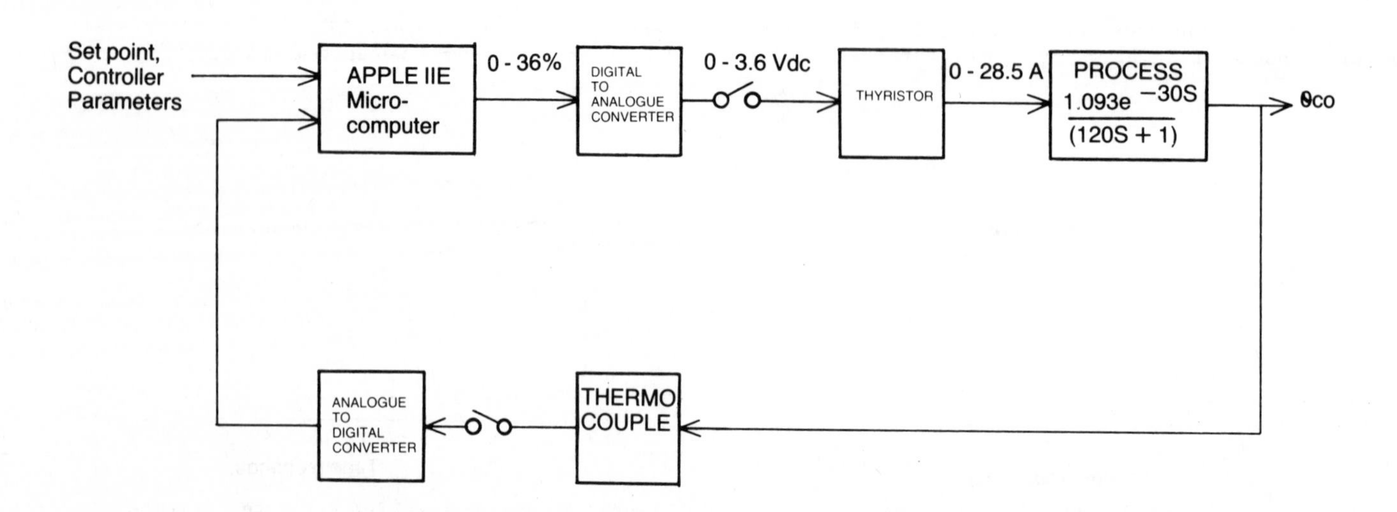

Figure 23: Capacity control system block diagram

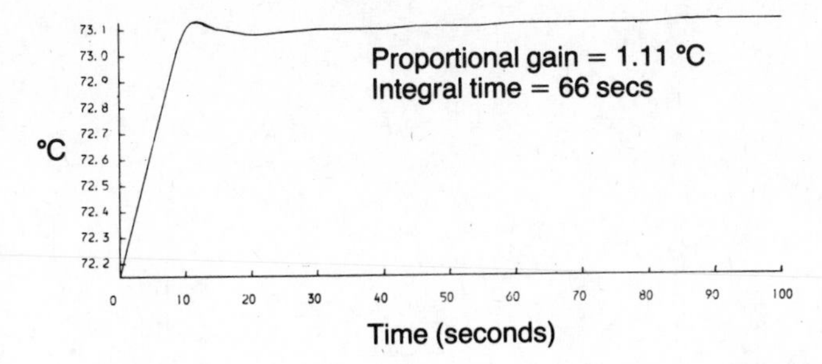

Figure 24: Simulated response in condenser water outlet temperature to a unit change in set point (under control)

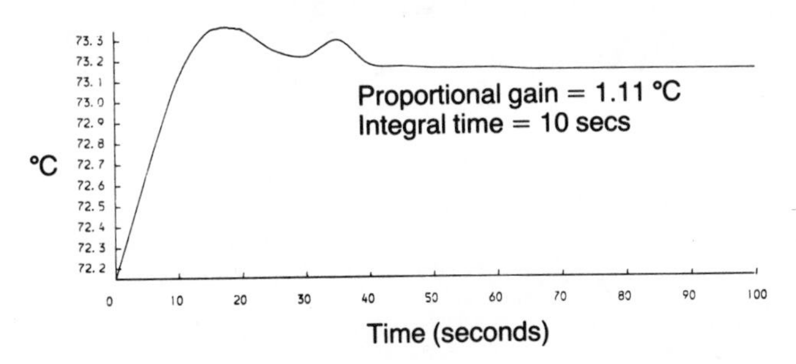

Figure 25: Simulated response in condenser water outlet temperature to a unit change in set point (under control)

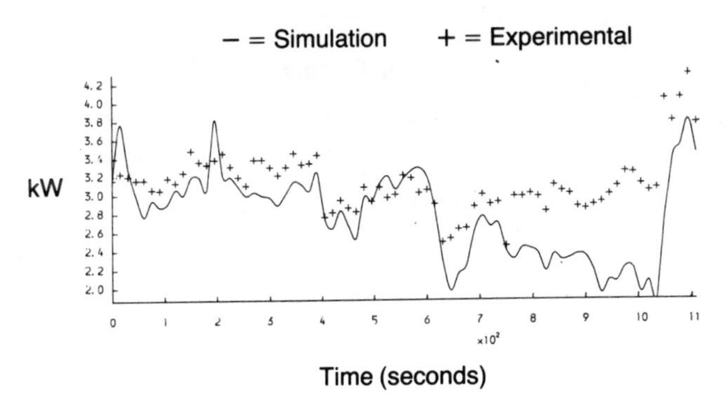

Figure 26: Absorber heat transfer response to a load disturbance (Figure 8) under control

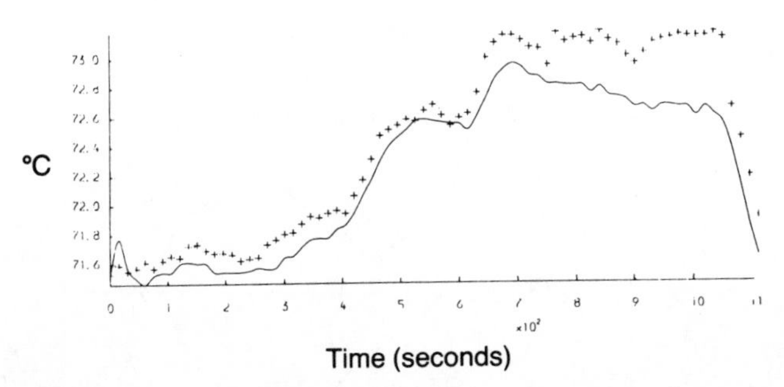

Figure 27 Absorber water outlet temperature response to a load disturbance (Figure 8) under control

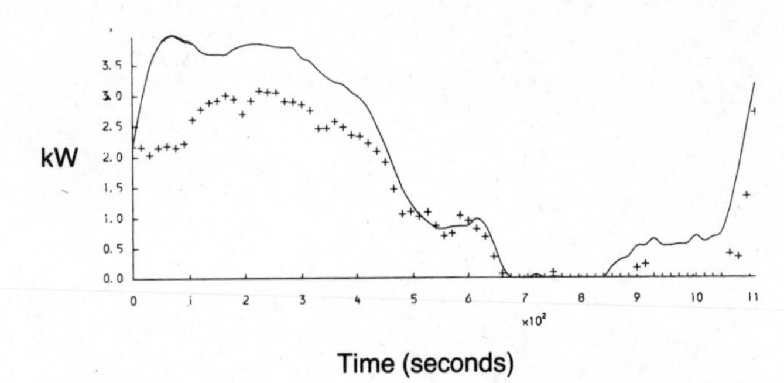

Figure 28: Condenser heat transfer response to a load disturbance (Figure 8) under control

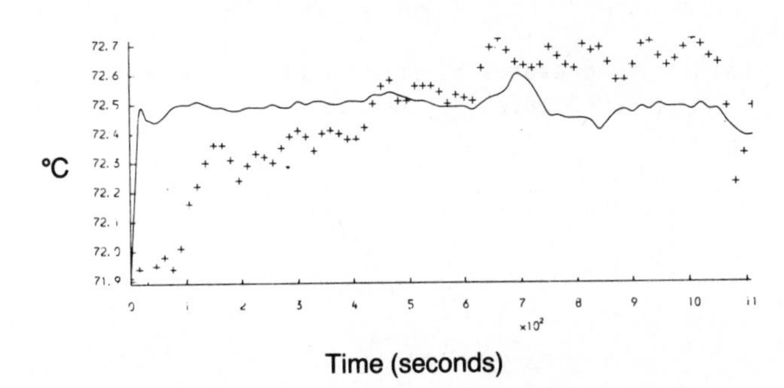

Figure 29: Condenser water outlet temperature response to a load disturbance (Figure 8) under control

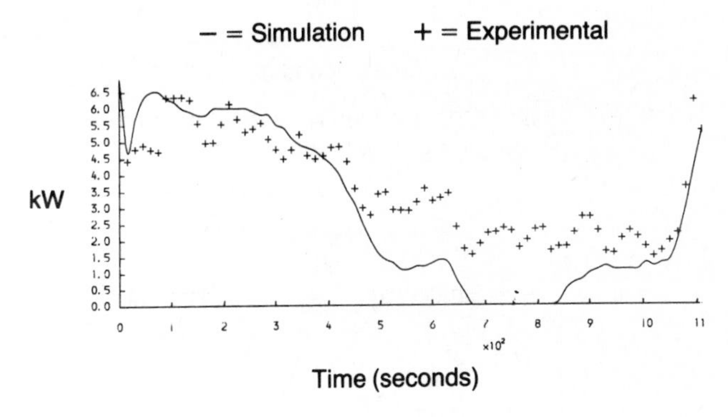

Figure 30: Generator heat input response to load disturbance (Figure 8) under control

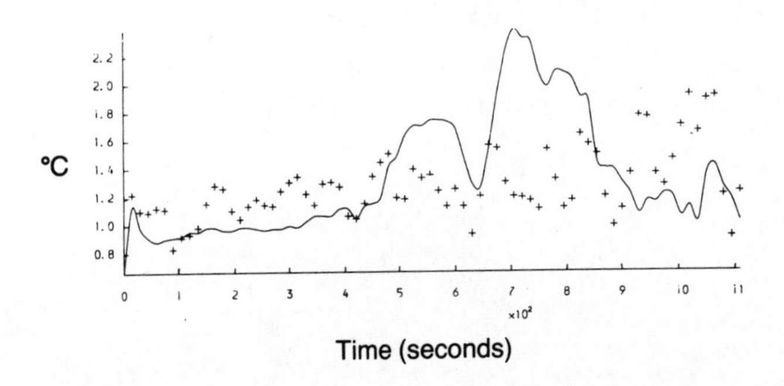

Figure 31: Response of coefficient of performance to a load disturbance (Figure 8) under control

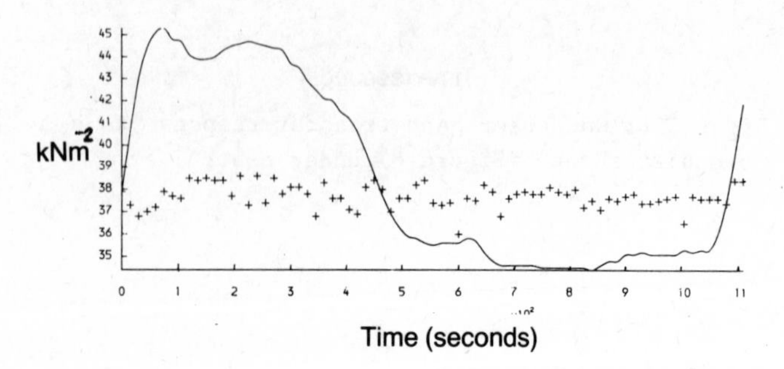

Figure 32: Generator pressure response to a load disturbance (Figure 8) under control

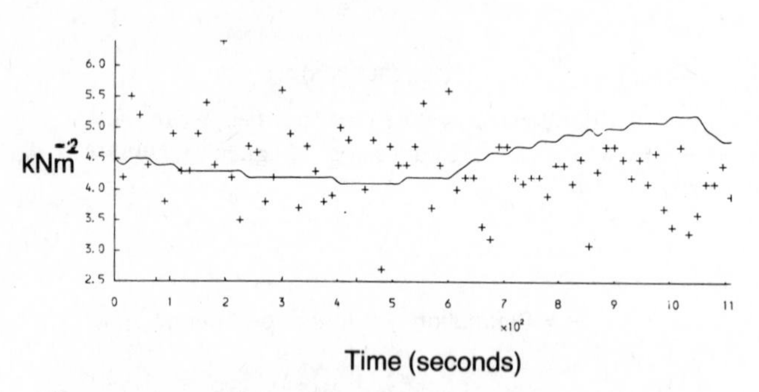

Figure 33: Evaporator pressure response to a load disturbance (Figure 8) under control

— = Simulation + = Experimental

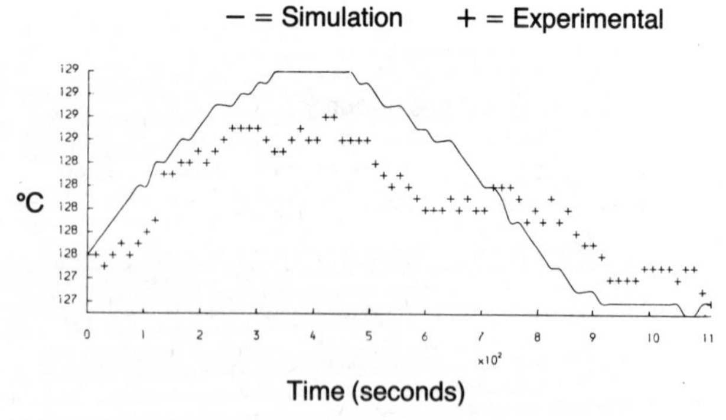

Figure 34: Generator solution outlet temperature response to a load disturbance (Figure 8) under control

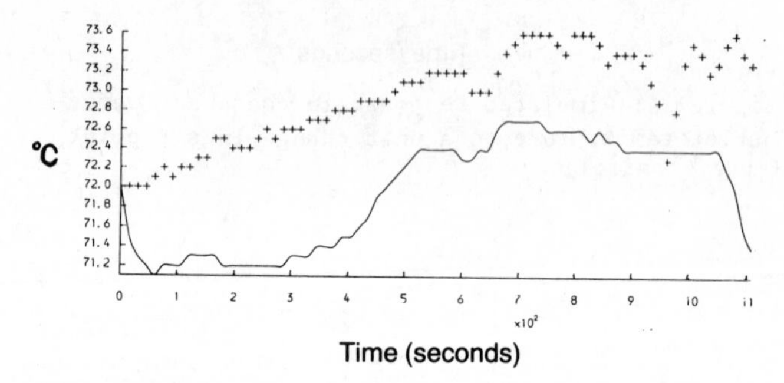

Figure 35: Absorber solution outlet temperature response to a load disturbance (Figure 8) under control

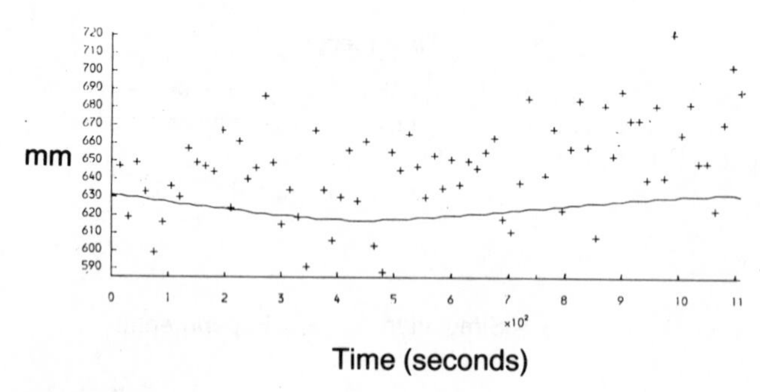

Figure 36: Generator solution liquid level response to a load disturbance (Figure 8) under control

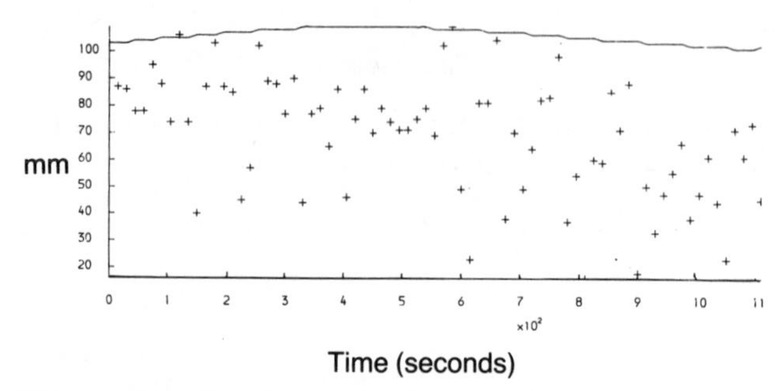

Figure 37: Absorber solution liquid lelvel response to a load disturbance (Figure 8) under control

3rd International Symposium on the

Large Scale Applications of Heat Pumps

Oxford, England : 25-27 March 1987

PAPER J1

THE DEVELOPMENT OF ABSORPTION CYCLE HEAT PUMPS APPLIED TO INDUSTRIAL PROCESS HEAT RECOVERY

M.A. Osei-Bonsu, R.J. Treece
Newcastle upon Tyne Polytechnic

SYNOPSIS

A theoretical and experimental study of the performance of absorption cycle heat pumps applied to industrial processes has been undertaken.

An absorption cycle heat pump test rig, for use with Lithium Bromide and water was designed, constructed and tested.

At the running conditions used experimentally, at temperature levels appropriate to industrial process heat recovery, average coefficients of performance over the range 1.2 to 1.4 were obtained.

INTRODUCTION

The objectives and scope of the study carried out at the Newcastle upon Tyne Polytechnic were to investigate theoretical factors influencing the thermodynamic behaviour of absorption cycle heat pumps and to demonstrate them experimentally.

Heat pumps using the absorption cycle require heat to be admitted at a high temperature to drive the process, but need only a small solution pump to circulate fluid. Absorption heat pumps may therefore be directly fuel fired.

The application of absorption cycle heat pumps to domestic heating is currently being given detailed consideration by several workers. The technical difficulties related to this application centre around the high temperature lift required between the low grade heat sources and intermediate demand temperature.

The application of heat pumps to the industrial sector has certain attractive features compared to domestic application.

It is possible to identify applications where the temperature lift required between the low temperature source and intermediate demand temperature is more modest than that required for domestic application.

The size of the heat pumps required in industrial applications may be several megawatts of heat output offering the prospect of economies of scale compared to domestic units.

However the availability of waste heat and the demand for upgraded heat may occur at different times within industrial processes, and they are often poorly matched in size.

This document describes an investigation to identify and resolve areas of uncertainty associated with the mechanical and thermodynamic performance of absorption cycle heat pumps applied to industrial processes.

Aqueous Lithium Bromide and water was the fluid pair chosen for detailed consideration because of its theoretical suitability at the temperature levels appropriate to industrial heat recovery and its proven application to large scale duties in the field of air conditioning.

The programme of work undertaken involved the mathematical modelling of the heat pump system, followed by the construction and operation of a pilot scale (10 kW of heat output) test rig.

Cycle Analysis

To enable comparison and evaluation of different absorption cycles it is necessary to be able to determine thermodynamic states at points around the system and hence system operating pressures, flow rates, temperatures and other information required for equipment design.

This analysis is accomplished by applying the first law of thermodynamics and continuity of mass, together with a knowledge of the thermodynamic properties of the fluid pair.

The method will be described for the cycle of Figure 1.

The vapour leaving the evaporator (State 11) is assumed to be saturated and at a temperature closely related to that of the low grade heat source, hence the low side pressure is determined. The liquid leaving the condenser (State 8) is assumed saturated at a temperature closely equal to that of heat rejection, hence the high side pressure is determined. The liquid leaving the absorber is assumed to be saturated at the intermediate heat rejection temperature, plus a small temperature difference, hence its concentration may be calculated. States 4 and 7 are assumed at the same temperature, equal to the temperature of high grade heat admission. The concentration of the strong solution may therefore be calculated on the basis that State 4 is saturated liquid at the high pressure. From fluid thermodynamic property data the enthalpies of the fluid at all state points may now be derived, and the relative mass flow rates of strong and weak solutions calculated. Both heat exchangers may be analysed based on a limiting approach temperature. The heat transfer rates across each component may now be calculated, and the required circulation rates scaled to the desired heating effect. This

Held at St. Catherine's College Oxford, England. Organised and sponsored by BHRA, The Fluid Engineering Centre, Cranfield, Bedfordshire, MK43 0AJ England.

analysis assumes that the heat and mass transfer surface is designed such that the desired fluid process may occur, that velocity and gravity effects may be neglected, that solution pump work may be neglected, that no heat transfer occurs with the surroundings (other than that intended to occur within the described system components), and that pressure drops caused by fluid flow may be neglected, except across the two system throttles.

The assumption that vapour and liquid leaving the generator are at equal temperatures (T4 and T7) depends upon the detailed heat and mass transfer design of the generator. In the absence of experimental data, the assumption that T4 and T7 are equal represents a first approximation.

The assumption that liquid leaving the absorber (State 1) is saturated and at a temperature closely related to that of the cooling liquid is very much dependent upon the heat and mass transfer design of the absorber, and the model can only be improved by the use of experimentally obtained data.

A computer model was developed which followed the analytical procedure described.

Fluid thermodynamic property data has been derived from the polynomial equations presented by McNeely (1). These equations described the equilibrium properties of Aqueous Lithium Bromide and water, and also the enthalpy of solutions of varying concentration. These equations have been applied at concentrations up to the solidus line, as described by Ellington (2). Ellington presents a phase diagram for Aqueous Lithium Bromide and this relationship has been put into numerical form and incorporated into the computer model of the fluids' thermodynamic properties.

The thermodynamic properties of water (3) have been represented using the Barner Adler equation of state and the Thek Stiel vapour pressure equation (4). Enthalpy departure functions have been derived and applied to produce enthalpy data after the manner described by Reid (5). Thus the pressure, volume, temperature, density and enthalpy of saturated and superheated vapour, and saturated and subcooled liquid may be determined on the basis of the thermodynamic state involved.

Consideration of the logical boundary conditions required to define the cycle being analysed, shows that after setting the low and the intermediate temperature levels which are controlled by the application, an element of choice exists in the selection of the high grade heat admission temperature. This temperature may be set at a range of levels and the system remain thermodynamically possible.

Application of Computer Model

The computer model has been applied at both conventional air conditioning temperatures, and at temperatures appropriate for industrial heat pump applications.

The graphs of Figure 2 represent the performance of a typical air conditioning cycle, applied over a range of strong solution temperatures leaving the generator. The temperatures used as the other boundary conditions are those in the analysis presented by ASHRAE (6) and at the appropriate generator temperature ASHRAE results are predicted identically by the computer programme.

The design point chosen by ASHRAE is indicated on Figure 2 and the merits of this choice are fairly clear. A high COP is combined with a relatively low solution flow rate, and remains some 12°C below a temperature that may introduce a risk of crystal formation during strong solution cooling in heat exchanger one.

Figure 3 represents computer predictions of the performance of an absorption cycle heat pump applied at industrial temperature levels. The industrial temperature levels in question are of a demand for heat at 80°C combined with a source of heat at 25°C. A range of high grade heat admission temperatures are considered.

In order that the impact of crystal formation on system performance could be observed the procedure of restricting the heat transfer in heat exchanger one was suppressed.

The system illustrated by Figure 3 can be seen to be severely restricted by the risk of crystal formation in the strong solution stream on cooling in heat exchanger one. Even at very close concentration levels, crystal formation can occur. The calculations predict that the loss of performance caused by crystal formation is reduced at higher generator temperatures because the reduced flow rates produce a corresponding reduction in sensible heat loss, as a proportion of the overall heat transfer. Operating at the right hand side of the diagram offers the advantage of smaller flow rates and consequently reduced heat transfer surface. However operation in this region is complicated by the need for the absorber to dilute the solution before significant cooling can be allowed to occur.

Construction of Rig

Refer to Figure 4. Those parts of the rig in contact with the salt solution were manufactured from Stainless Steel AISI type number 316 and glass. Gaskets and seals in contact with the salt solution were made from Polytetrafluoroethylene or asbestos. Copper, bronze and mild steel were used in the construction of items not in contact with the corrosive solution.

The generator consists of a vertical cylinder 1.1 m high, flanged top and bottom. It operates flooded to a depth of 0.7 to 0.8 m. Three electrical heating elements enter a distance of 0.5 m through the bottom flange. Vapour boiled off from the salt solution leaves through a 40 mm diameter pipe and enters a vertical shell and tube condenser. Weak salt solution enters the generator by a distributor through a top flange. Strong solution is drawn from the bottom of the pool vertically up through an internal coil and out through the generator top flange.
The condenser is cooled by water flowing in a common circuit with the absorber. The condenser is of shell and tube construction with the condensation process taking place on the shell side.

The absorber is cooled by a closed water circuit, which also cools the condenser. This closed circuit is itself cooled by once through water going to drain by means of a shell and tube heat exchanger. The cooled closed circuit water leaving this exchanger enters the absorber first, and then the condenser before returning to the exchanger.

The absorber vessel consists of a vertical cylinder 1.2 m high, flanged top and bottom. Strong solution enters the top flange of the absorber and may be caused to flow over a stainless steel coil suspended within the absorber chamber.

The internal stainless steel coil is water cooled, and an external salt solution pump enables salt drawn from the pool in the bottom of the absorber vessel to be recirculated over the internal coils.

Low pressure water vapour enters the absorber from the evaporators through a 75 mm diameter pipe into the absorber top flange. Weak salt solution is drawn from the pool at the bottom of the absorber and pumped, using a diaphragm type metering pump, into the generator.

A stainless steel plate heat exchanger is situated between the absorber and generator. Heat is exchanged between the strong and weak solution streams in counter flow.

The evaporators are shell and tube heat exchangers and operate flooded. Water is evaporated from the tubes by electrically heated water circulating on the shell side.

Those elements of the equipment at temperatures over 90°C were covered with a 75mm thick layer of fibreglass insulation and others with a 50 mm thick layer.

Instrumentation

Refer to Figure 5.

All salt solution and water circuit temperatures are measured by Iron/Constantan J type thermocouples. Those thermocouples in the water circuits, because they are being used in the calculation of heat transfer rates are calibrated to 0.1°C. Water circuit flow rates are measured by rotameters, weak solution flow rate is measured by the calibrated metering pump. An ammeter and voltmeter are used to calculate the generator heat input. Pressures are measured using Bourdon tube pressure gauges. Liquid levels are determined using sight glasses on generator, absorber, evaporator water circuits. Absorber cooling water on temperature is controlled manually using a needle valve. Salt solution concentrations were obtained by sampling and titration.

Experimental Results

The test rig was operated over a range of conditions, results being taken at steady states.

Experimental coefficients of performance were obtained by dividing the heat recovered in the absorber and condenser water circuit by the electrical heat input to the generator. No correction was made for the power absorbed by the circulating pumps.

The coefficients of performance obtained from the experimental results at equal condensing, absorbing and evaporating temperatures have been averaged, and the computer model applied at the same conditions. Figure 6 shows the data plotted against the differences between condenser/absorber and evaporated temperature.

The experimental results showed a discrepancy in heat balance. On average 700 watts less heat was recovered by the absorber and condenser cooling water than was taken up by the generator and evaporator.

Detailed estimates of strong heat loss from the surface of the test rig together with the loss associated with the continuous running of the vacuum pump produce a loss at typical running conditions of 730 watts.

The level of insulation applied to the heat pump was at the limit of practicality and further additions would have produced only marginal improvements. The problem of stray heat loss was compounded by the large size of the vessels in relation to the quantity of heat being transferred. This was a consequence of the experimental nature of the equipment.

The higher COP prediction of the computer model compared to the experimental results evident on Figure 6 is predominantly a consequence of the stray heat losses.

Factors affecting the Performance of the Heat Pump

The equipment was operated with a condenser/absorber coolant temperature of 60°C and an evaporator temperature of 30°C, more frequently than other running conditions in order that the influence of other parameters could be identified.

At each setting of condenser, absorber, and evaporator temperatures the generator temperature was increased progressively giving a range of strong solution concentrations and generator outlet temperatures. Different rates of weak solution flow rate were used, giving rise to a range of recirculation rates in the absorber.

During the operation of the heat pump adjustments must be made to the rate of heating in the generator, and the rate of solution flow from the absorber to cause the equipment to come to the desired steady state condition. Because of the volumes of salt solution involved some considerable time (about 30 minutes) was needed for this, depending on salt solution flow rate.

Equilibrium conditions take much longer to be reached at low solution flow rates, and the gross amount of heat transfer taking place is generally less at low solution flow rates. The likelihood of failing to bring the equipment to equilibrium is greater at low solution flow rates, and low heat outputs.

Figure 7 shows variation of COP with the heat output from the heat pump test rig, for the running condition in question.

The experimental COP shows an increasing trend with increasing heat output. It is thought that this reflects the reducing significance of stray heat loss at increased heat output.

Figure 8 shows the variation of COP with the difference in salt solution concentrations.

At high concentration differences the corresponding circulation rates of salt solution per unit of heat output are reduced. Therefore because the size of heat exchanger one is unchanged, the losses associated with the circulation of salt solution should reduce.

However, the results given in Figure 8 are for different rates of heat output, and no trend of COP with concentration difference is evident.

Figure 9 presents the values of strong solution concentration plotted against generator bottom temperature, all for essentially constant values of condensing pressure. On the same figure, two equilibrium lines are shown representing the relationship between salt solution concentration and temperature at pressures corresponding firstly to a water vapour pressure at 60°C and secondly to the anticipated pressure at the bottom of the pool of solution in the generator.

It can be seen that a strong relationship exists between the solution concentration and temperature at constant condensing pressure.

The solution leaving the generator appears not to be at saturated conditions, but sub-cooled by 5° or 6°C.

The approach to equilibrium of the solution within the absorber is influenced by several factors. During operation of the equipment the recirculation pump was left running at full capacity.

The temperature of the cooling water flowing in the coil of the absorber, and the pressure of the evaporator/absorber system influence the weak solution concentration. The rate at which strong solution is being supplied, and weak solution withdrawn from the absorber is important because it controls the residence time of the solution.

At conditions where the evaporator temperature, absorber temperature, and rate of solution flow through the absorber are closely equal, the resultant weak solution concentration has been plotted against the entering strong solution concentration. Refer to Figure 10.

For each of the above results the equilibirum temperature of the weak solution leaving the absorber has been calculated. The difference between this and the absorber cooling water temperature has been plotted against the strong solution concentration. Refer to Figure 11. This temperature difference represents the degree of deviation from equlibrium conditions of the absorber solution pool.

This temperature difference is necessary in order that the processes of heat and mass transfer within the absorber take place.

The increase in strong solution concentration entering the absorber increases the concentration difference causing mass transfer, larger nett concentration changes are produced, this is illustrated by Figure 10. These larger concentration changes involve increased absorber heat transfer rates.

Crystallization

For each running condition (ie condenser/absorber temperature and evaporator temperature) the generator bottom temperature was increased to produce a stronger solution flowing from the generator to absorber. This increase in bottom temperature was limited by the onset of crystal formation in the strong solution line between heat exchanger one and the absorber.

A glass tube situated in the line between the heat exchanger one and absorber enabled the formation of crystals to be observed. When crystals of solid Lithium Bromide were detected in the glassware the running condition was noted, and the equipment shut down.

Figure 12 presents the operating temperature at the bottom of the generator for each of the running conditions at which crystals were detected. The temperature recorded by the thermocouple in the line where the crystals were observed is also given, together with the concentration of the strong solution where known.

The concentrations derived from these samples do not accurately reflect the average concentration present in the equipment at the point of crystal formation, because solid particles of Lithium Bromide have already been lost from the sample within the equipment.

The need to be capable of observing the strong solution as it passed from the plate heat exchanger to the absorber caused increased heat loss from the solution. Hence the crystallization limit was reached earlier than in fully insulated equipment.

CONCLUSIONS

A theoretical and experimental study of the performance of absorption cycle heat pumps applied to industrial processes has been undertaken.

A computer programme capable of performing a steady state first law analysis of an absorption cycle heat pump was produced. The programme was designed for use with absorption cycle units using Lithium Bromide and water as a working fluid, operating at conditions appropriate to industrial process heat recovery. The data produced by the computer studies was used to assist in the design of a pilot scale (10 kW of heat output) test rig.

An absorption cycle heat pump test rig, for use with Lithium Bromide and water, was designed, constructed and tested.

At the running conditions used experimentally, at temperature levels appropriate to industrial process heat recovery, average coefficients of performance over the range 1.2 to 1.4 were obtained.

The experimental results show a consistently smaller quantity of heat output recovered at the intermediate temperature compared to the low and high temperature heat inputs. This discrepancy is considered to arise from the convection and radiation losses from the surface of the equipment, and a parasitic vapour loss from the absorber caused by the continual running of the vacuum pump.

The stray heat loss depresses the values of the experimentally determined coefficients of performance.

The operating range of the heat pump was restricted by the onset of crystallization in the solution stream flowing from the generator to the absorber. The maximum temperature difference attained between the absorber and the evaporator was 40°C. The design of the equipment enabling crystallization to be detected caused the limit to be reached earlier than would otherwise have been the case.

ACKNOWLEDGEMENTS

The support of the Science and Engineering Research Council and International Research and Development Company Limited towards this project is acknowledged and appreciated.

REFERENCES

1. McNeeley, L.A.
   Thermodynamic Properties of Aqueous Solutions of Lithium Bromide. In ASHRAE Transactions 85, Part 1, 1979.

2. Institute of Gas Technology Research Bulletin. Report 14.
   Ellington, R.T., Knust, G., Reck, R.E., Reed, J.F.
   The Absorption Cooling Process, 1957.

3. Barner, H.E., Adler, S.B.
   Three-Parameter Formulation of the Joffe Equation of State. In Industrial Engineering Chemistry Fundamentals. 9 (4) 1970. pp 521-529.

4. Thek, R.E., Stiel, L.I.
   A New Reduced Vapour Pressure Equation. In American Institute of Chemical Engineers Journal 12 (3). May 1966. pp 599-602.

5. Reid, R C., Prausnitz, J.M., Sherwood, T.K. The Properties of Gases and Liquids. Third Edition. McGraw Hill. 1977.

6. ASHRAE. Handbook of Fundamentals. ASHRAE. 1981.

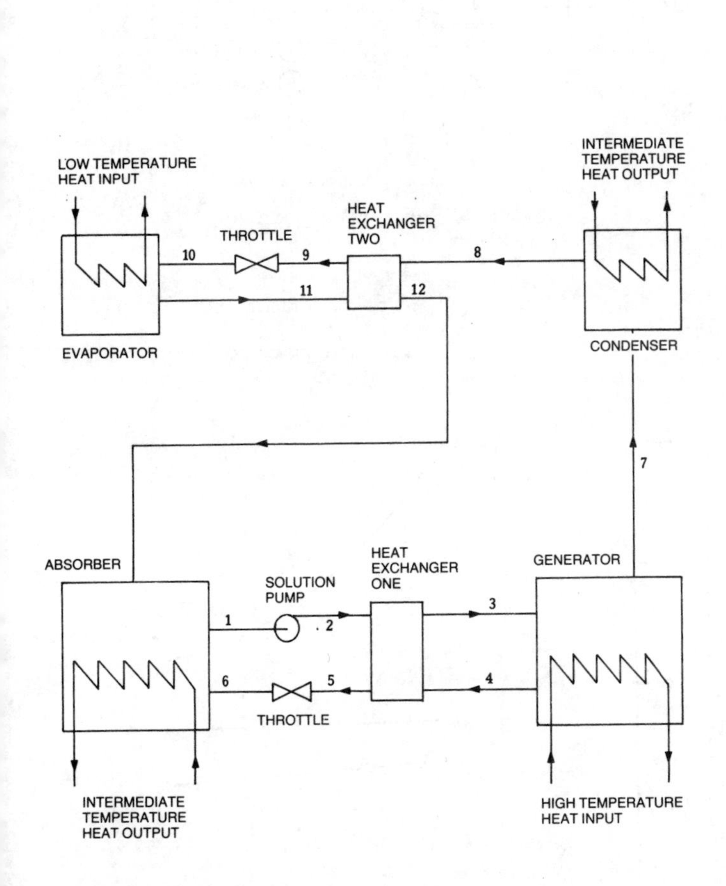

Figure 1
Schematic diagram of an absorption cycle heat pump, using aqueous lithium bromide and water as a working fluid pair

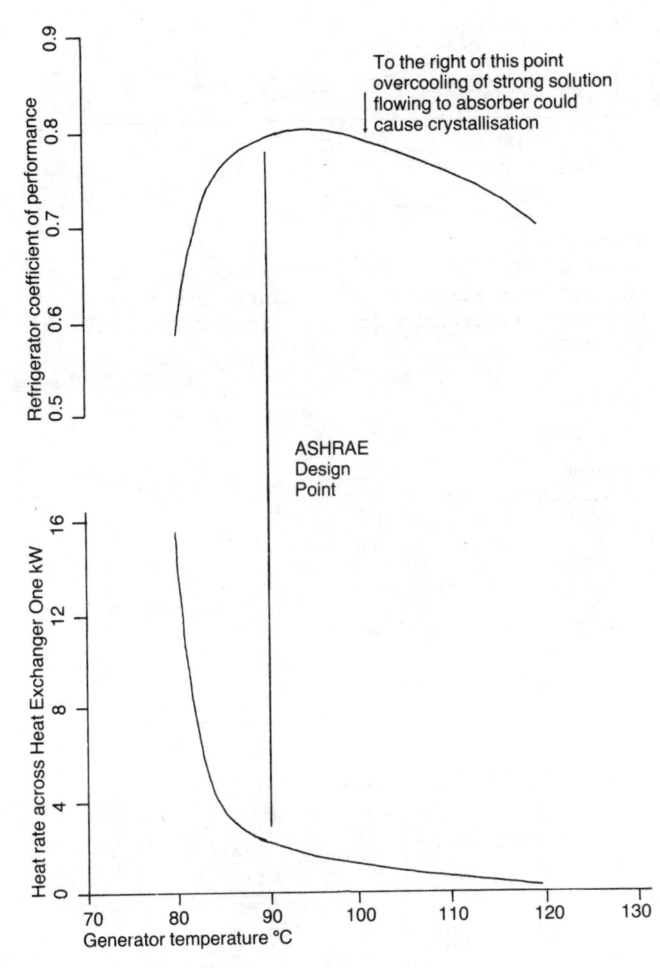

Figure 2
Graphs of coefficient of performance and heat transfer rate across heat exchanger one versus generator temperture, of the air conditioning example given by ASHRAE (6)

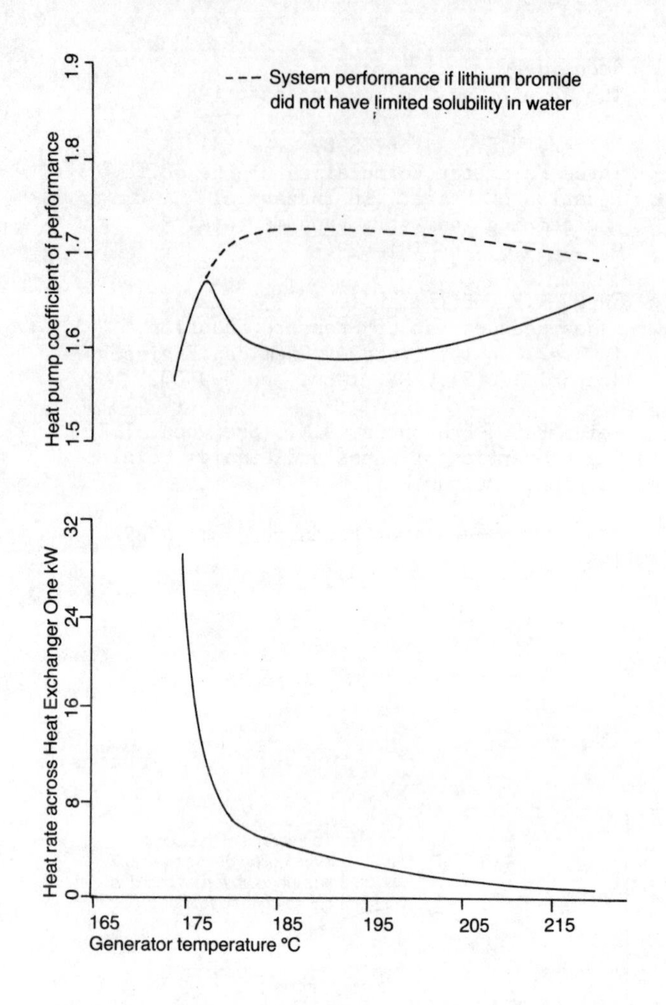

Figure 3
Graphs of coefficient of performance and heat transfer rate across heat exchanger one versus generator temperature for the first industrial application

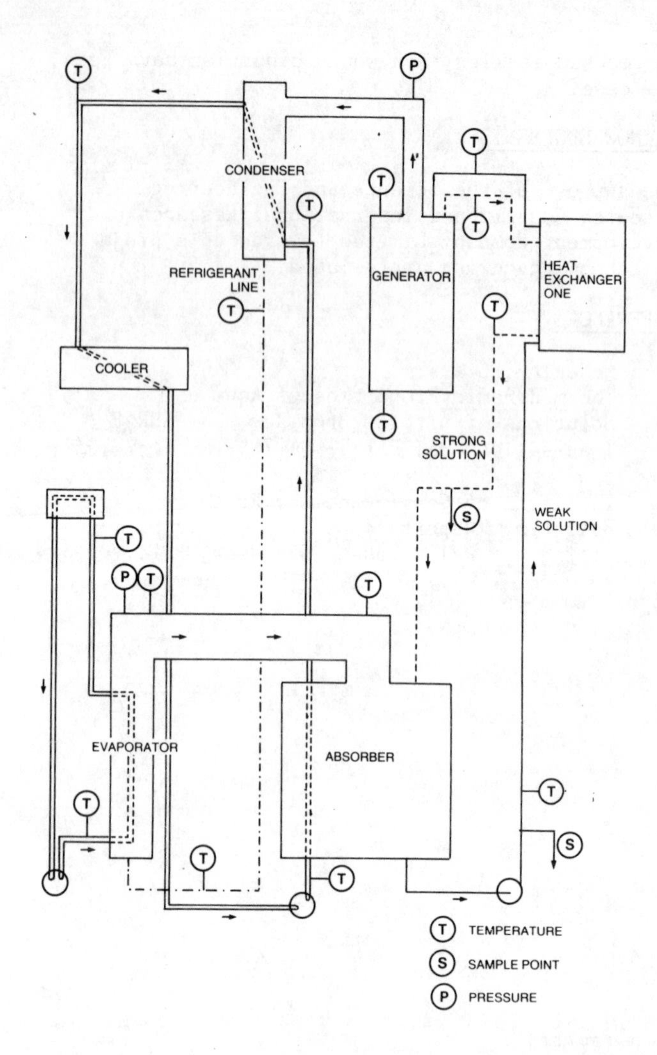

Figure 5
Location of measuring points

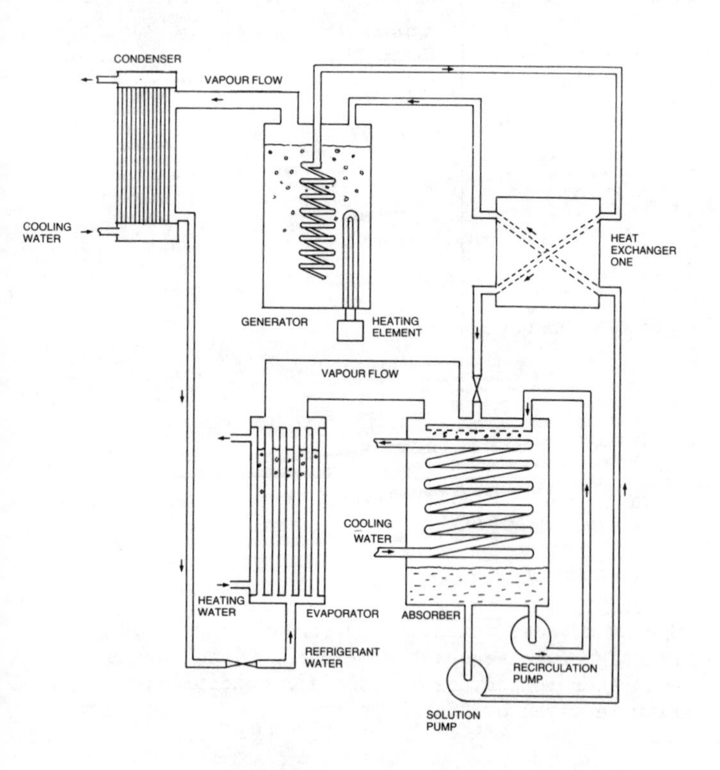

Figure 4
Schematic diagram of test rig

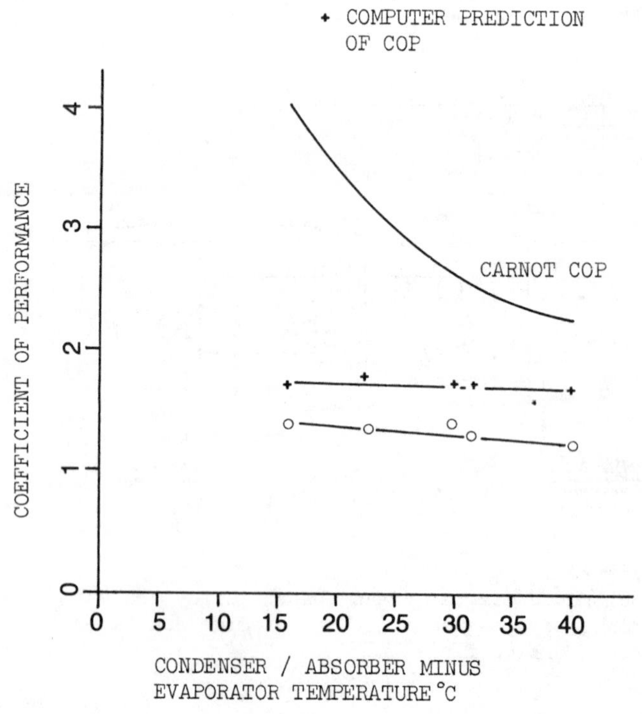

Figure 6
Coefficient of performance versus temperature lift

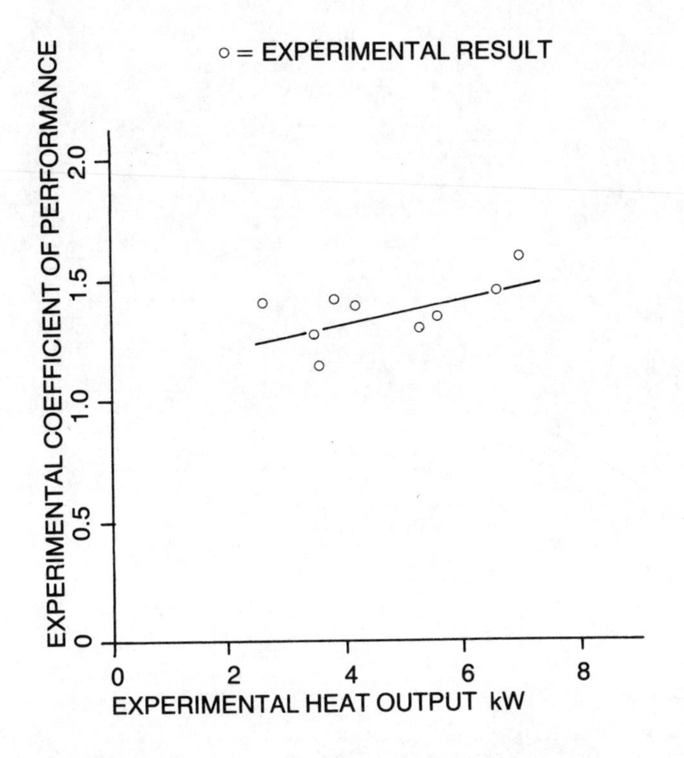

Figure 7
Experimental coefficient of performance for 60/30 set of results versus heat output

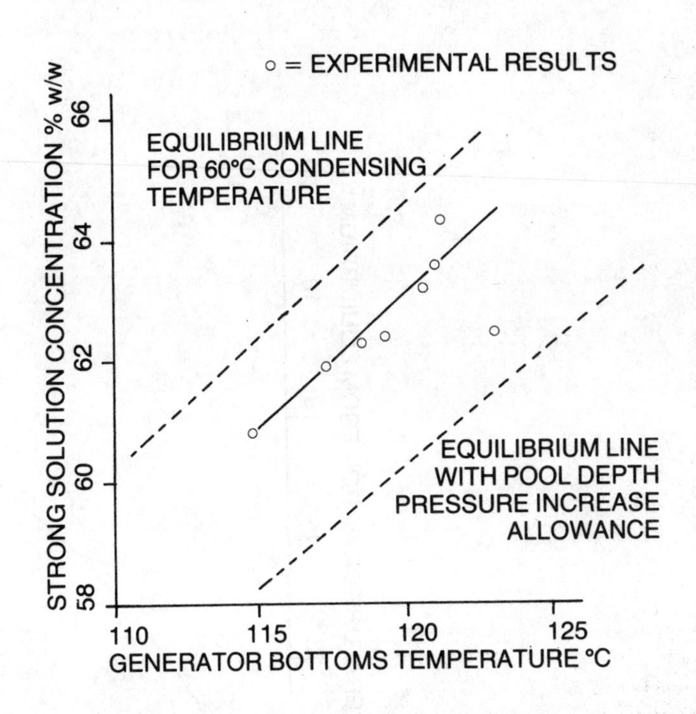

Figure 9
Generator bottoms temperature for 60/30 set of results versus strong solution concentration

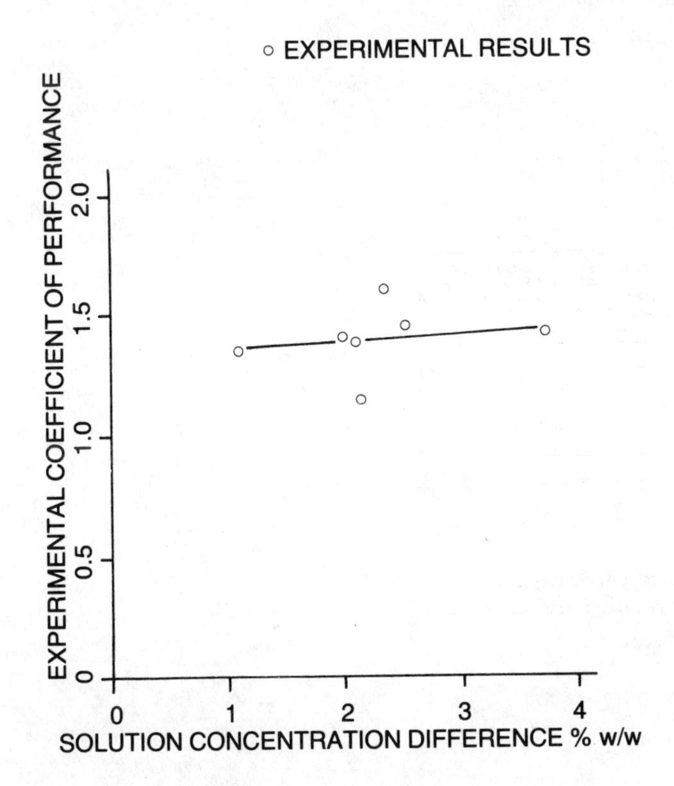

Figure 8
Cop versus solution concentration difference for 60/30 set of results

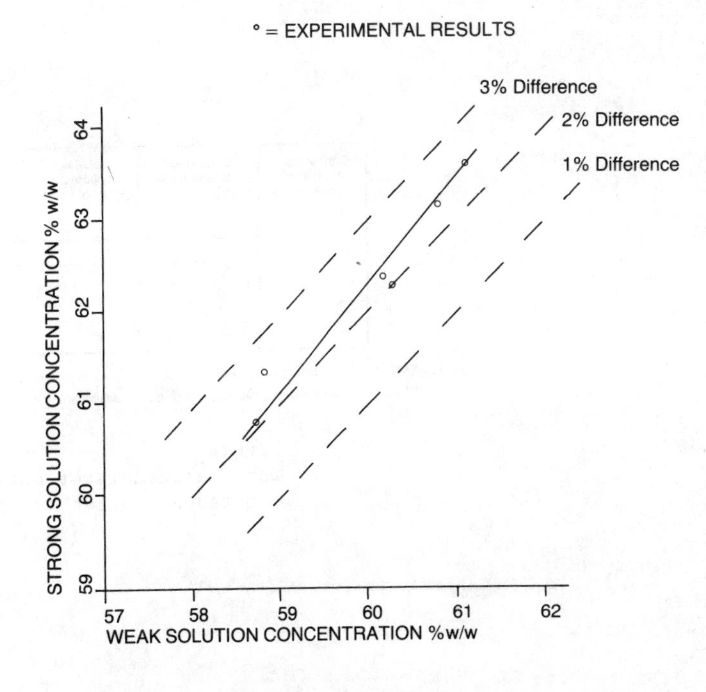

Figure 10
Strong versus weak solution concentration for absorbers at equivalent running conditions

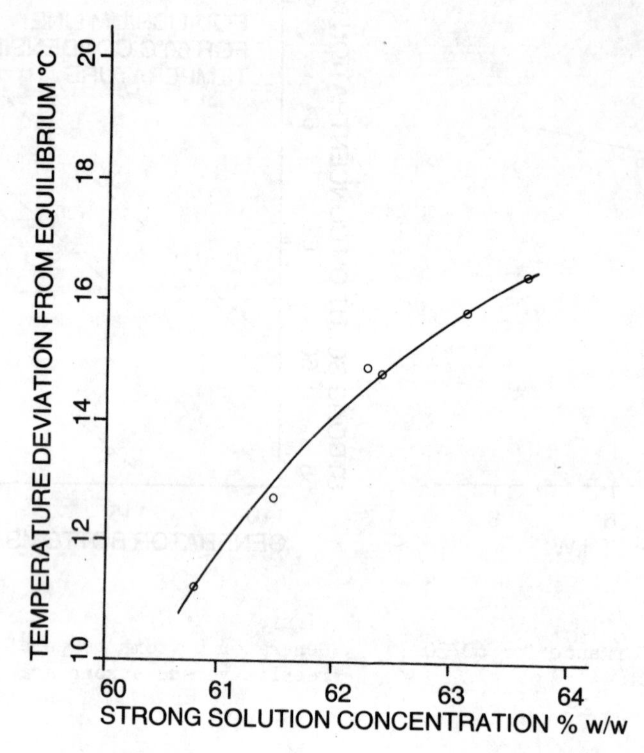

Figure 11
Temperature deviation from equilibrium of weak solution leaving the absorber versus entering strong solution conentration

Condenser/ absorber temperature °C	Evaporator temperature °C	Generator bottoms temperature °C	Temperature in strong solution line between Heat Exchanger One and absorber °C	Approximate strong solution when known % w/w
55	30	121	55	64.5
60	30	127	66	-
60	30	130	63	-
70	35	139	72	65.1
80	40	149	83	-

@ Note that these solutions have already lost crystals of solid lithium bromide and are consequently weaker that the average strong solution originally in the equipment.

Figure 12
Table of generator bottom temperature limited by outset of crystal formation

3rd International Symposium on the
Large Scale Applications of Heat Pumps
Oxford, England : 25-27 March 1987

PAPER J3

DYNAMIC BEHAVIOUR OF A DISTILLATION COLUMN WITH INTERNAL HEATING BY THE BOTTOMS PRODUCT

Dieter Butz and Karl Stephan
Institut für Technische Thermodynamik
und Thermische Verfahrenstechnik
Universität Stuttgart
Federal Republic of Germany

SYNOPSIS

The distillation column of an absorption heat pump operating with internal solution reflux has been studied. The bottoms product is not withdrawn from the bottom of the column, but guided from the lowest tray upward in a heat exchanger spiral through all trays of the stripping section, leaving the column on the feed tray. On its path it gives off heat and thereby furnishes a part of the heating duty. In addition, the stripping section is completely surrounded by a constant temperature medium (flue gas of a gas burner).
The column has been modelled by a stage-to-stage procedure. The selected working fluid is a binary mixture of ammonia and water.
The steady state of the simulated column is compared to experimental results obtained with a generator of an absorption heat pump.
The dynamic analysis of the column involves step changes of the input variables.

INTRODUCTION

Absorption heat pumps show a wide variety of construction details. These often follow the specific expertise a certain manufacturer has gained in his field of interest. However, a single component can not always be replaced by an alternate design without affecting the whole process.
This brings up the interest to model the different components and to simulate the process of an absorption heat pump.
In this paper a model for the generator of an absorption heat pump is presented.
If a heat pump were always to operate at its design conditions, a steady-state evaluation would suffice. However, in order to maintain a certain but arbitrarily changeable heat delivery it is desirable to control the process in an "optimal" way so that the long-term COP reaches a maximum. Especially for smaller machines with no large thermal capacity the on/off control is a less advantageous strategy. To better understand the real behaviour there is a need for a dynamic model which includes differental equations in time.

THE COLUMN DESIGN

The basic construction of the generator under consideration represents a distillation column. The specific design is sketched in Fig. 1. The apparatus consists of a cylinder which can be heated by a gas burner along a given length of its total height. In this heating section fins are affixed to enhance the heat transfer. The trays inside the column are made of sheet metal.
The present column acts mainly as a stripper because the strong solution is fed on the top tray. The weak solution leaving the bottom tray is - unlike in most designs - guided upward within a spiral inside the column while heating the hold-up of each stage. It leaves the generator on the top tray as a subcooled liquid.
The enriching section of the generator merely consists of a partial condenser. The heat of condensation is removed by the comparably cold strong solution coming from the solution pump. This results in a reduction of the heat which has to be supplied by the gas burner. On the other hand the heat delivered by the heat pump (that could be utilized for heating) decreases, but not to the same extend as the energy savings of the burner. So this design is advantageous over processes without this feature.
After leaving the partial condenser the strong solution is further heated up to the feed temperature in another apparatus before entering the feed tray.

THE DYNAMIC MODEL

The model is based on an equilibrium stage as shown in Fig. 2. In general there is a liquid phase "L" in contact with a vapour phase "V" by an interface "*" across which heat and mass are transferred. In addition, heat and mass can be added to or withdrawn from either of the two phases "L" and "V". In our special case there is only one feed added to the column and no side stream leaving any tray.
The peculiarity of the present column design is taken into account by a third phase "C" which transfers heat to the liquid "L". Heat exchange with the vapour phase is neglected because of the much lower hold-up and the smaller heat transfer coefficient on the vapour side.
The transient mass and energy balances for tray i (i = 0,...,I; counted from top to bottom) with a K component mixture read as follows since there is no mass and energy accumulation in the interface:

Phase "L":

$$\frac{d(x_{ki}M_i^L)}{dt} = x_{Fki}L_{Fi} + x_{ki-1}L_{i-1} - x_{ki}L_i - j_{ki} \qquad k=1,\ldots,K-1 \qquad (1)$$

Held at St. Catherine's College Oxford, England. Organised and sponsored by BHRA, The Fluid Engineering Centre, Cranfield, Bedfordshire, MK43 0AJ England.

$$\frac{dM_i^L}{dt} = L_{Fi} + L_{i-1} - L_i - \sum_{k=1}^{K} j_{ki} \quad (2)$$

$$\frac{d(h_i^L M_i^L)}{dt} = h_{Fi}^L L_{Fi} + h_{i-1}^L L_{i-1} - h_i^L L_i - \sum_{k=1}^{K} \bar{h}_{ki}^L j_{ki} + \dot{q}_i^C - \dot{q}_i^L + \dot{Q}_i^{Br} \quad (3)$$

Phase "V":

$$\frac{d(y_{ki} M_i^V)}{dt} = y_{Fi} V_{Fi} + y_{ki+1} V_{i+1} - y_{ki} V_i + j_{ki} \qquad k=1,\dots,K-1 \quad (4)$$

$$\frac{dM_i^V}{dt} = V_{Fi} + V_{i+1} - V_i + \sum_{k=1}^{K} j_{ki} \quad (5)$$

$$\frac{d(h_i^V M_i^V)}{dt} = h_{Fi}^V V_{Fi} + h_{i+1}^V V_{i+1} - h_i^V V_i + \sum_{k=1}^{K} \bar{h}_{ki}^V j_{ki} + \dot{q}_i^V + \dot{Q}_i^V \quad (6)$$

Phase "C":

$$\frac{d(h_i^C M_i^C)}{dt} = h_{i+1}^C L_I - h_i^C L_I - \dot{q}_i^C \quad (7)$$

interface "*":

$$0 = \dot{q}_i^L - \dot{q}_i^V + \sum_{k=1}^{K} (\bar{h}_{ki}^L - \bar{h}_{ki}^V)\, j_{ki} \quad (8)$$

with the liquid and vapour mass fractions x and y, the tray hold-up M, the liquid and vapour mass flow rates L and V, the mass flow rate j, the heat flow rates $\dot{q}$ and $\dot{Q}$, the specific enthalpy h, and the specific partial enthalpy $\bar{h}_k$ of component k. The superscripts L and V represent the liquid and vapour phase, and C the weak solution or bottoms product with the composition of the bottom tray. The subscript F indicates a feed to either the liquid or vapour phase and the superscript Br the heat input by the gas burner. Tray number 0 represents the partial condenser and tray number I the bottoms.
The absorption heat pump under consideration is filled with a two component mixture of ammonia and water, K=2. This reduces the mass balances for each of the phases L and V to one component and the overall balance. Phase C has already been assumed incompressible in the above equations. Thus its mass balances can be omitted entirely.
To have the temperature explicitly in our relations we make use of the total differential of the specific enthalpy as a function of temperature T, pressure p, and mass fraction x of ammonia (component number 1)

$$dh = c_p dT + \left(\frac{\partial h}{\partial p}\right)_{T,x} dp + (\bar{h}_1 - \bar{h}_2)\, dx \quad (9)$$

where $\bar{h}_1$ and $\bar{h}_2$ denote the specific partial enthalpies of ammonia and water in the mixture, respectively. In the following derivations we will neglect the influence of pressure by setting dp=0. Differentiation of the products on the left hand side of the above set of equations and insertion of the appropriate overall and/or component mass balances and eq. (9) yields for tray i

Phase "L":

$$M_i^L \frac{dx_i}{dt} = L_{Fi}(x_{Fi} - x_i) + L_{i-1}(x_{i-1} - x_i) - j_{1i}(1-x_i) + j_{2i} x_i \quad (10)$$

$$\frac{dM_i^L}{dt} = L_{Fi} + L_{i-1} - L_i - j_{1i} - j_{2i} \quad (11)$$

$$M_i^L c_{pi}^L \frac{dT_i^L}{dt} = L_{Fi}\left[h_{Fi}^L - \bar{h}_{2i}^L - (\bar{h}_{1i}^L - \bar{h}_{2i}^L) x_{Fi}\right] + L_{i-1}\left[h_{i-1}^L - \bar{h}_{2i}^L - (\bar{h}_{1i}^L - \bar{h}_{2i}^L) x_{i-1}\right] + \dot{q}_i^C - \dot{q}_i^L + \dot{Q}_i^{Br} \quad (12)$$

Phase "V":

$$M_i^V \frac{dy_i}{dt} = V_{Fi}(y_{Fi} - y_i) + V_{i+1}(y_{i+1} - y_i) + j_{1i}(1-y_i) - j_{2i} y_i \quad (13)$$

$$\frac{dM_i^V}{dt} = V_{Fi} + V_{i+1} - V_i + j_{1i} + j_{2i} \quad (14)$$

$$M_i^V c_{pi}^V \frac{dT_i^L}{dt} = V_{Fi}\left[h_{Fi}^V - \bar{h}_{2i}^V - (\bar{h}_{1i}^V - \bar{h}_{2i}^V) y_{Fi}\right] + V_{i+1}\left[h_{i+1}^V - \bar{h}_{2i}^V - (\bar{h}_{1i}^V - \bar{h}_{2i}^V) y_{i+1}\right] + \dot{q}_i^V + \dot{Q}_i^V \quad (15)$$

Phase "C":

$$M_i^C c_{pi}^C \frac{dT_i^C}{dt} = L_I (h_{i+1}^C - h_i^C) - \dot{q}_i^C \quad (16)$$

interface "*":

$$0 = \dot{q}_i^L - \dot{q}_i^V + \sum_{k=1}^{2} (\bar{h}_{ki}^L - \bar{h}_{ki}^V)\, j_{ki} \quad (17)$$

These equations can be simplified by assuming equilibrium stages, i.e. $T_i^L = T_i^V = T_i$. This implies that temperature is no longer an independent state variable but a function of pressure and composition (either of the liquid or vapour phase; we will choose the liquid). Since pressure has been set constant composition remains the only independent intensive variable of state. Therefore we can write

$$\frac{dT_i}{dt} = \left(\frac{\partial T_i}{\partial x_i}\right)_{p_i} \frac{dx_i}{dt} \quad (18)$$

The partial derivative on the right-hand side of this equation has to be evaluated from the equation of state analytically or numerically, whichever is more convenient.
Another consequence of the equilibrium stage concept is that the vapour phase composition is fixed by the equilibrium relation $y_i^*=y_i^*(T_i,x_i)$. In order to include nonequilibrium behaviour of the gas phase, a Murphree tray efficiency η_i is introduced via

$$\eta_i = \frac{y_i - y_i^+}{y_i^* - y_i^+} \tag{19}$$

with the vapour composition $y_i^+=y_{i+1}$ on the operating line.
Under most operating conditions vapour phase dynamics can be neglected since the liquid hold-up exceeds the vapour hold-up by at least two orders of magnitude. This makes the vapour balances quasi-stationary and the time derivatives vanish. Moreover, the heat exchange between vapour phase and flue gases can be omitted. Only in the partial condenser this has to be considered.
In the generator under investigation the hold-up of phase "C" is by an order of magnitude less than that of the liquid phase. This means that also this equation reduces to an algebraic one yielding an implicit expression for T_i^C since h_i^C and $\dot{q}_i^C$ are functions of T_i^C. Therefore this temperature will be determined by a separate iteration like the caloric properties.
For calculating hold-ups in the column some knowledge of the hydrodynamics is required (e.g. Francis weir formula). The liquid hold-ups, however, are not very different on the different stages and do not change much during operation of the column. Therefore we will assume constant liquid hold-up from now on, causing the overall liquid mass balance to reduce to an algebraic equation as well.
With this set of assumptions the only dynamic equation remaining is the component mass balance of phase "L". All other equations degenerate to a set of two major underlying algebraic equations which is highly non-linear and therefore needs iteration.
By appropriate combination of the remaining equations the flow rates across the interface can be eliminated resulting in

$$\begin{aligned} M_i^L \frac{dx_i}{dt} = {} & L_{Fi}(x_{Fi} - x_i) + V_{Fi}(y_{Fi} - x_i) \\ & + L_{i-1}(x_{i-1}-x_i) + V_{i+1}(y_{i+1}-x_i) \\ & - V_i(y_i - x_i) \end{aligned} \tag{20}$$

$$0 = L_{Fi}+V_{Fi}+L_{i-1}+V_{i+1}-L_i-V_i \tag{21}$$

$$\begin{aligned} 0 = {} & L_{Fi}\left[h_{Fi}^L-\bar{h}_{2i}^L-(\bar{h}_{1i}^L-\bar{h}_{2i}^L)x_{Fi}\right] \\ & +V_{Fi}\left[h_{Fi}^V-\bar{h}_{2i}^L-(\bar{h}_{1i}^L-\bar{h}_{2i}^L)y_{Fi}\right] \\ & +L_{i-1}\left[h_{i-1}^L-\bar{h}_{2i}^L-(\bar{h}_{1i}^L-\bar{h}_{2i}^L)x_{i-1}\right] \\ & +V_{i+1}\left[h_{i+1}^V-\bar{h}_{2i}^L-(\bar{h}_{1i}^L-\bar{h}_{2i}^L)y_{i+1}\right] \\ & -V_i\left[h_i^V-\bar{h}_{2i}^L-(\bar{h}_{1i}^L-\bar{h}_{2i}^L)y_i\right] - c_{pi}^L\left(\frac{\partial T_i}{\partial x_i}\right)_{p_i} M_i^L\frac{dx_i}{dt} \\ & + \dot{q}_i^C + \dot{Q}_i^{Br} + \dot{Q}_i^V \end{aligned} \tag{22}$$

These are three equations for the three variables x_i, L_i, V_i.
So far no specific working fluid has been selected. This means that any two-component mixture can be used provided the properties needed in the above equations are known.

THE SIMULATION ALGORITHM

For simulation purposes it is most convenient to have all equations in a standard form:

$$\frac{d\vec{x}}{dt} = \vec{f}(\vec{x},\vec{z},\vec{u},\vec{p}) \tag{23}$$

$$0 = \vec{g}(\vec{x},\vec{z},\vec{u},\vec{p}) \tag{24}$$

Here the vector $\vec{x}$ represents all state space variables characterised by differential equations, $\vec{z}$ those characterised by algebraic expressions, $\vec{u}$ the inputs, and $\vec{p}$ all constant parameters. $\vec{f}$ and $\vec{g}$ stand for the right-hand sides of the respective equations for $\vec{x}$ and $\vec{z}$.
In our example $\vec{x}$ contains the liquid concentrations on all trays, $\vec{z}$ the liquid and vapour mass flow rates inside the column, $\vec{u}$ the feed mass flow rates, compositions, and temperatures, and $\vec{p}$ all "constant" parameters such as heat transfer coefficients, caloric and thermal quantities, hold-ups, and others determined by their respective equations.
If all information is supplied in this manner, standard numerical techniques can be applied to solve the problem. For integration of the above set of equations we used a modified Euler algorithm /1/ with Newton-Raphson iteration.

THE STEADY STATE OF THE COLUMN

For a sample calculation we chose the normal operating point of a 25 kW residential water/water-absorption heat pump. The useful heat is supplied at 45°C whereas the evaporator is fed with water of 10°C. The working fluid is a binary mixture of ammonia and water.
The generator is fired by a 18.5 kW gas burner with 4.2 kW losses so that 14.3 kW are transferred to the column. The flue gas temperature around the column was measured to be about 350°C. The heated length extends over the 14 lowest trays and the bottoms. Each tray contains a liquid hold-up of 0.28 kg, the bottom hold-up is 0.81 kg, the liquid hold-up in the partial condenser 1.9 kg. The total height of the generator is 0.8 m with an internal diameter of 0.144 m. The upper 7 trays are not immersed in the combustion chamber so that here $\dot{Q}_i^{Br}=0$.
A partial condenser tops the column. Rich solution coming from the solution pump at an estimated mass fraction of ammonia of 0.44 and a mass flow rate of 22 g/s is guided through a coil while it is heated up from 44.5°C to 62.4°C. After further elevation of the temperature to 104°C this stream is fed on the top tray of the generator. The bottoms product is withdrawn from the lowest tray and cooled down on its way up inside a spiral within the column. The operating pressure is 20 bar, the vapour outlet temperature 86.3°C, its mass flow rate 7.5 g/s.
The reflux ratio was set to 1.2, the tray efficiency to 0.8.
All caloric and thermal properties are calculated using the fundamental equation for ammonia-water mixtures proposed by Ziegler /3/. The heat

exchange without mass transfer between the different phases is determined according to the relations given in the VDI-Wärmeatlas /2/, sections Ga through Ge.
With this information at hand the following simulation results were obtained. At steady state the total heat supplied by the gas burner is 14.35 kW while the weak solution gives off 2.08 kW. Thus the total heat input amounts to 16.43 kW.
The outlet conditions of the vapour and the weak solution are (mass flow rate; temperature; mass fraction of ammonia):

vapour 7.77 g/s; 75.9°C; 0.9883
weak solution 14.23 g/s; 140.0°C; 0.1405

From these numbers it follows that 1 kg vapour is generated per 2.83 kg feed.
In the partial condenser the strong solution is warmed up from 44.5°C to 58.6°C, taking up 1.43 kW. The bottoms temperature is 172°C.
The steady state temperature and concentration profiles are shown in Fig. 3. The most significant feature is that there is little change in the column temperature and concentrations along the section without external heating. Only the temperature of the weak solution does not exhibit this shape.
In accompanying experiments inlet and outlet temperatures have been measured. The mass flow rates were determined by metering valves. The calculated values are in reasonably good agreement with the measurements. Obviously the calculated heat transfer coefficients deviate from the real ones since the calculated outlet temperatures of vapour and strong solution are both too low.
However, the vapour outlet is very sensitive to composition changes. At the calculated concentration its equilibrium temperature (at a tray efficiency of 100%) is 84.9°C which is in good accordance with the measured value. Changing this composition by just 1 % results in an equilibrium temperature that differs by 10 K from the above value. This directly influences the driving temperature difference and thus the heat transfer to the strong solution. Thus already minor inaccuracies have a large impact.

DYNAMIC BEHAVIOUR OF THE COLUMN

Starting from steady state, step changes of the input variables were applied to depict the dynamic behaviour of the generator.
In a first example a feed temperature increase of 5 K was chosen. The resulting profiles along the column are presented in Fig. 4. The time-interval between the curves is three minutes. Thus the four lines cover a period of 9 minutes, starting from steady state. The direction of the changes is indicated by an arrow.
A new steady state temperature profile builds up from top to bottom. It is almost parallel to the original curve. The temperature of the weak solution increases all over the column, somewhat quicker in the upper part than in the lower one. The reason is that the temperature difference for heat transfer is first diminished in the top section where the feed enters. So this stream cannot give off as much heat any more. Secondly, the bottoms gradually heats up until the new steady state is attained. Thus we have a higher inlet temperature and a slightly lower temperature difference for heat transfer. The heat exchange is reduced by 0.23 kW.
Due to the assumption of equilibrium stages the concentrations on the trays decrease by an amount corresponding to the increase in the boiling temperature.
To better visualize the changes at the generator outlets Fig. 5 shows how the temperatures and mass fractions deviate from their steady state values. In this figure time is counted in steps with one time step equalling 90 s (1.5 min).
This representation shows that the temperature at the weak solution outlet T_1^C reacts in a similar way as the bottoms temperature T_I and approaches a lower final deviation, thus giving off less heat as before. During that process of around 10 time steps or 900 s (15 min) the vapour outlet temperature T_0 slightly increases by 1 K with a small time lag which corresponds to the slightly lower concentration. The liquid concentration in the bottoms is reduced by about 0.0074.
As a second example the response of the generator to a step change of +10 % in the feed mass flow rate is analysed. This stream acts as the cooling medium in the partial condenser before it is heated up to the feed temperature. Therefore we also have to increase its mass flow rate. This causes the heat transfer coefficient inside the condenser coil to rise.
The results are presented in Figs. 6 and 7 in an analogous manner as before.
The spatial profiles show that the temperature decreases rapidly along the heated section. Since we assumed constant liquid hold-up a change in the feed flow rate is felt instantaneously throughout the column. The internal liquid flow is increased and therefore the temperatures on the trays are reduced.
The temperature in the combustion chamber remains constant during the whole simulation. This means that at the higher flow rate the contents of the column cannot be heated to the former temperature. Consequently all temperatures fall and the concentrations on the trays rise. In addition, the weak solution can give off 0,34 kW more heat because due to the higher flow rates both in the column and in the spiral the heat transfer coefficients increase, causing a lower temperature difference $T_I - T_1^C$.
The vapour outlet does not change noticeably.
The new steady state is reached after about 10 time steps or 900 s.
These results show that the column considered here responds fairly slowly to input step changes. Critical outlet conditions can therefore easily be anticipated and control action can be taken in time.

CONCLUSIONS

A model for the dynamic behaviour of the generator of a residential absorption heat pump has been established. It consists of a distillation column with internal heating by the bottoms product.
Starting from the transient balances for mass and energy we derived a set of equations for simulation with a standard integration algorithm.
The steady state of the programme was compared with experimental results showing good agreement.
Dynamic simulations followed to visualize the influence of step changes in the input variables. Two examples were picked to demonstrate the behaviour of the generator.
With the model at hand all relevant variables in this system can be predicted as functions of time. This reveals the opportunity to judge different operating conditions and to find out the most desired one for the specific application.

ACKNOWLEDGEMENTS

The authors express their gratitude to the German

Research Foundation (DFG) for financial support of the work. Thanks are due to Martin Arnold for carrying out some sample calculations.

REFERENCES

/1/ Seinfeld, J.H.; L. Lapidus; M. Hwang: Review of Numerical Integration Techniques for Stiff Ordinary Differential Equations. Ind.Eng.Chem. Fundam. 9 (1970) 2, pp. 266-275

/2/ VDI-Wärmeatlas, 4th ed., VDI-Verlag, 1984

/3/ Ziegler, B.: Wärmetransformation durch einstufige Sorptionsprozesse mit dem Stoffpaar Ammoniak-Wasser. PhD Thesis ETH Zürich No. 7070, 1982

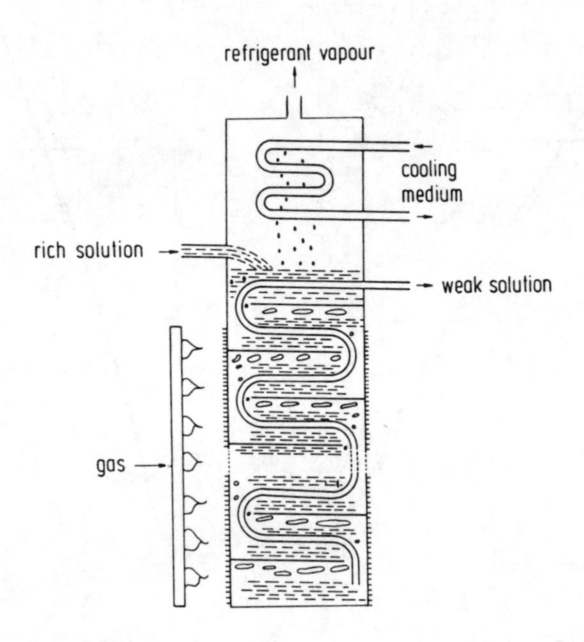

1 Principle design of the generator

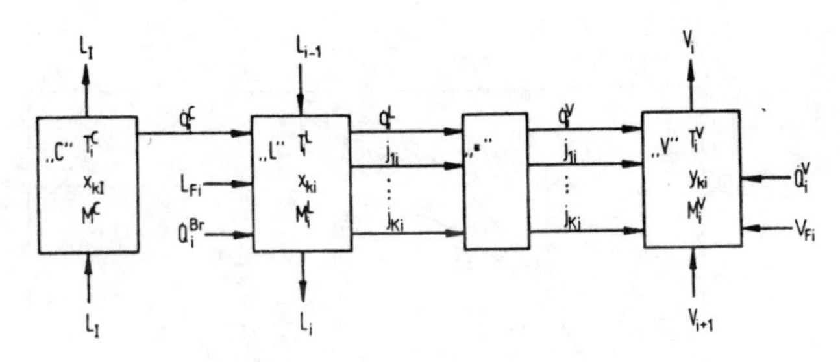

2 Equilibrium stage of the model

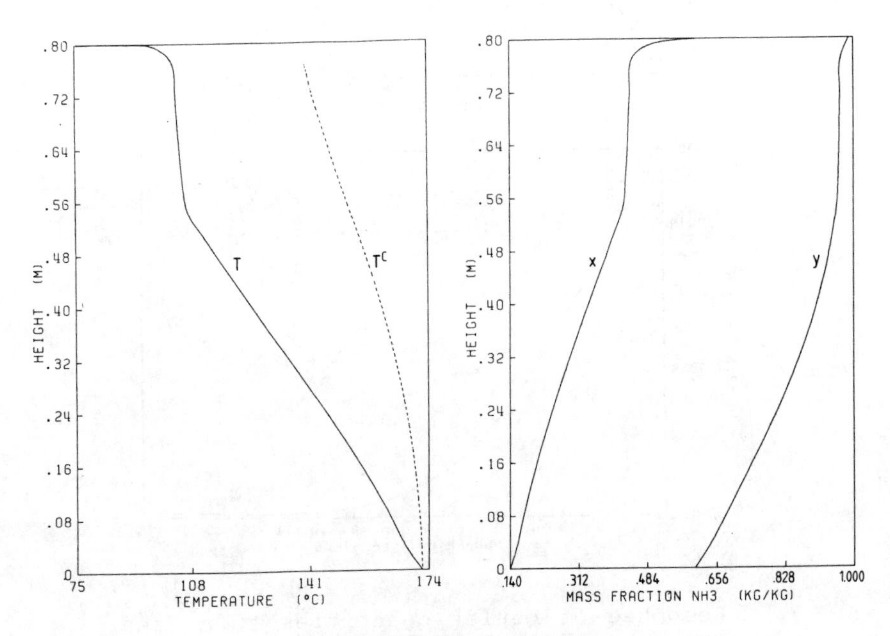

3 Steady state profiles

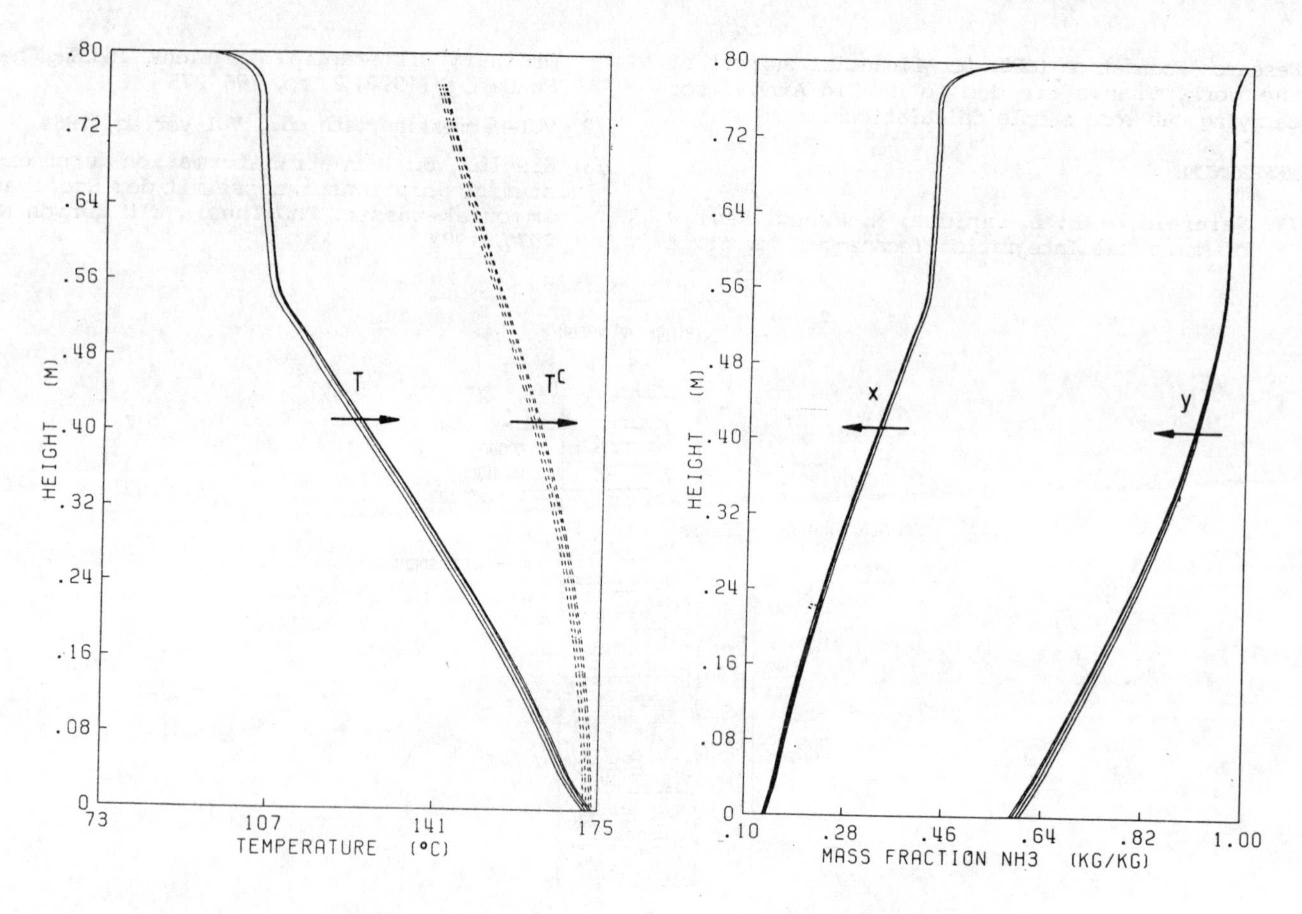

4 Spatial profiles for a feed temperature increase of 5 K

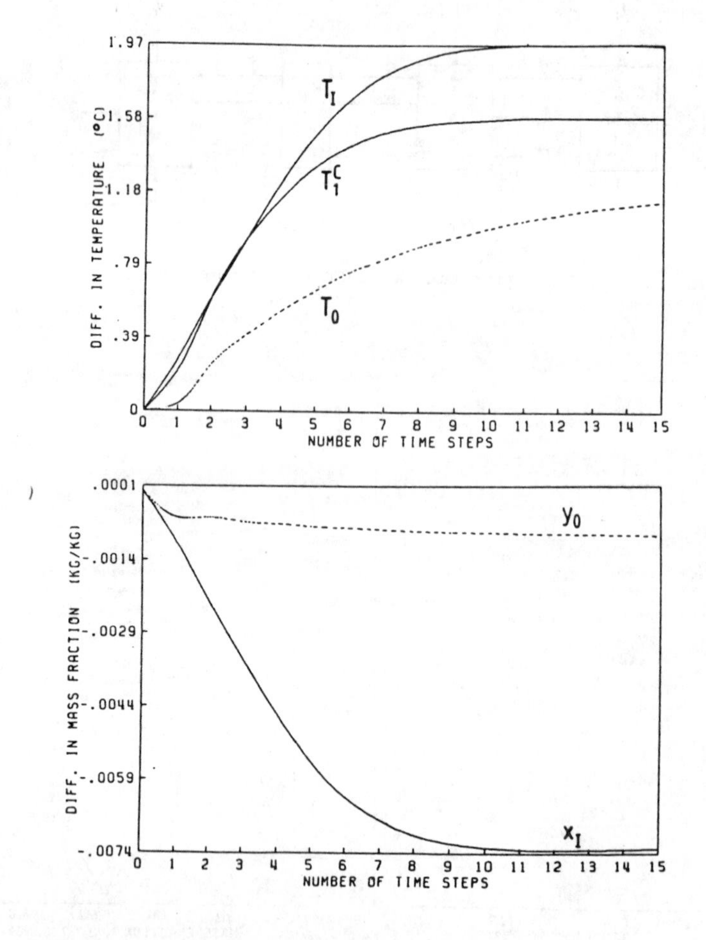

5 Response of outlet quantities to a feed temperature increase of 5 K

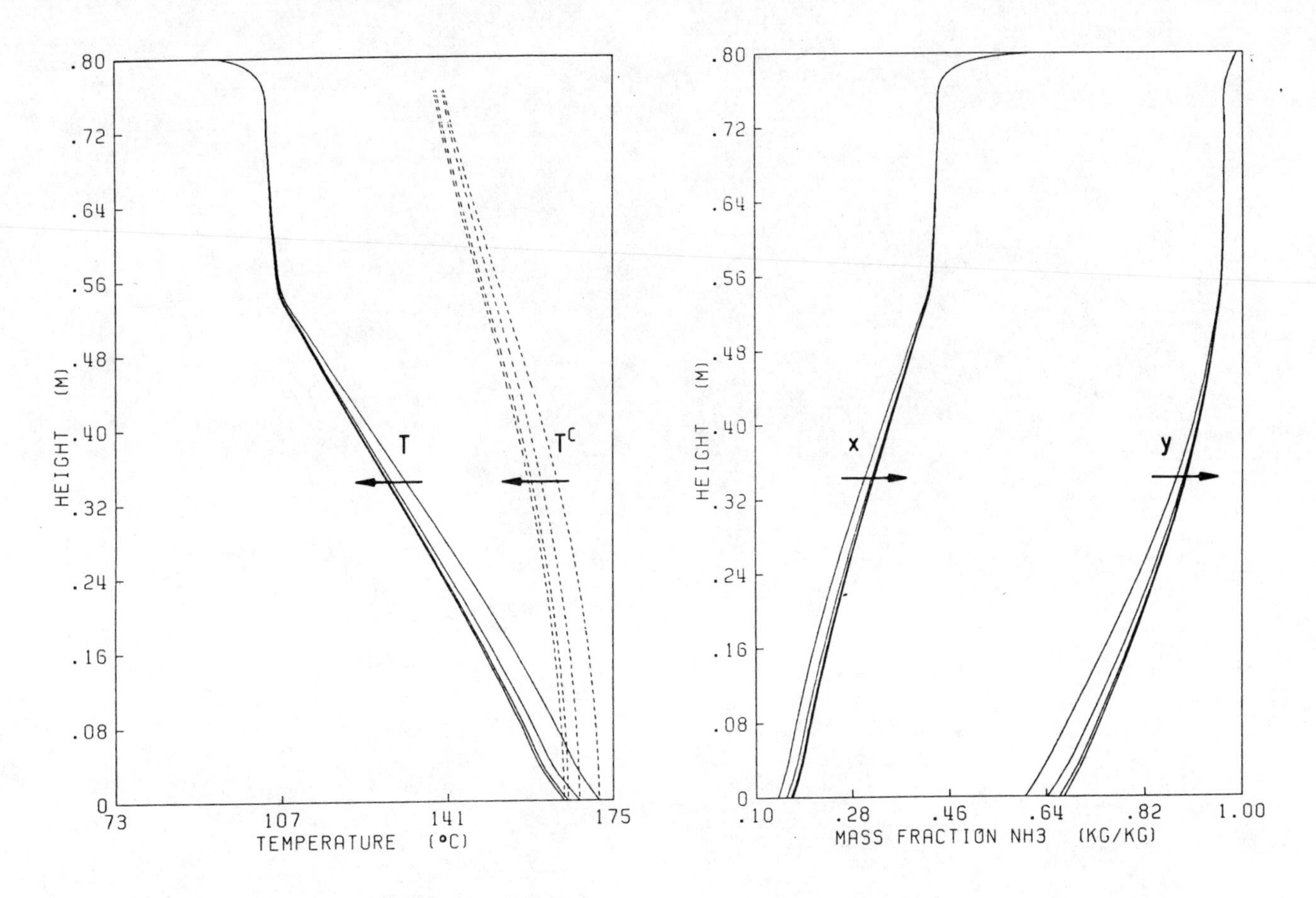

6 Spatial profiles for a feed mass flow rate increase of 10 %

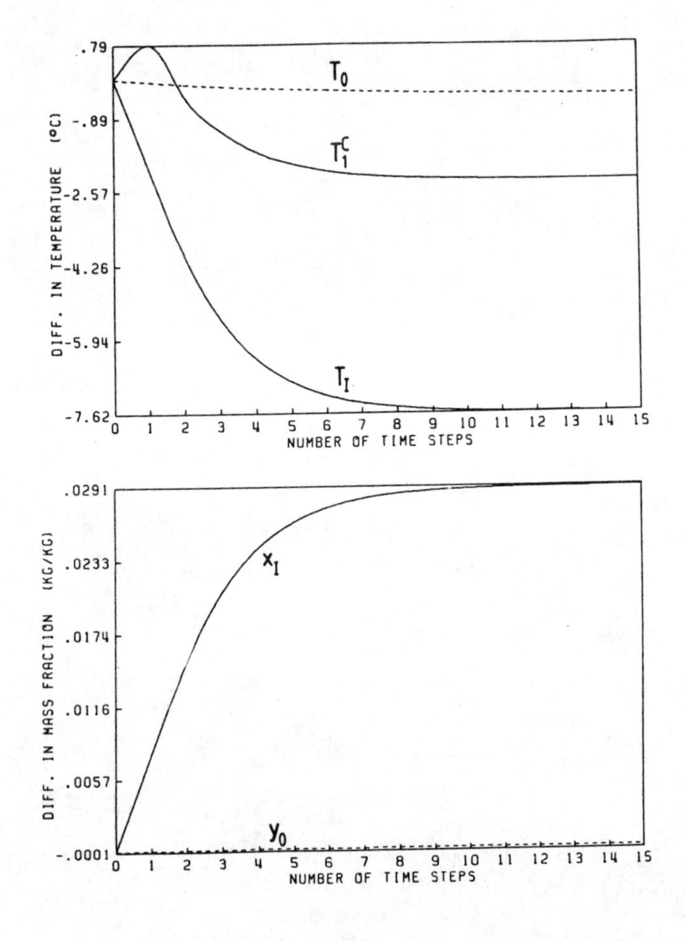

7 Response of outlet quantities to a feed mass flow rate increase of 10 %

3rd International Symposium on the

Large Scale Applications of Heat Pumps

Oxford, England : 25-27 March 1987

PAPER P1

DESIGN AND OPTIMIZATION OF A HIGH TEMPERATURE CASCADED HEAT PUMP AT 155°C

Q. S. YUAN, J.C. BLAISE and C. MISSIRIAN
Electricité de France
77250 Moret sur Loing, France

SUMMARY

This article describes the design and the optimization of a cascaded high temperature heat pump. The heat pump is divised in two stages and uses refrigerant R114 and water as fluids. It can produce saturated steam at 155°C by recovering the heat from waste heat with an interesting coefficient of performance.
The heat pump recovers the waste heat by evaporation of R114 and increases its temperature. The evaporator of the water heat pump draws this heat from the R114 condenser and raises it to a condensing temperature of approximately 155°C. The steam produced can be used either directly or indirectly.
The discussion focusses initially on the effect of cold source(air temperature, moisture and flow rate), the optimization of the intermediate temperature(condensing temperature of R114) and then on the efficiency of the system. It also examines the influence of the temperature of the hot source on efficiency. It is confirmed, for a given temperature of hot source, that:
-from an energy point of view, there is an optimum intermediate temperature
-the coefficient of performance is not very sensitive to the intermediate temperature.
-there is no optimum value for setting the size of compressors.

NOMENCLATURE

Cpa	: specific heat capacity of air, (kJ/kg°C).
Cp1, Cp2	: specific heat capacity of steam at corresponding temperature, (kJ/kg).
H	: enthalpy, (kJ/kg).
Lv	: latent heat, (kJ/kg).
Mf	: refrigerant mass flow rate, (kg/s).
Me	: water flow rate, (kg/s).
P	: power; Pf1 refrigerant evaporating power; Pf2 refrigerant condensing power; Pe1 water evaporating power; Pe2 steam condensing power, (kW).
Ps	: saturated steam pressure, (bar).
te, t	: air temperature at the inlet and after mixture, (°C).
Te	: refrigerant evaporating temperature, (°C).
Vva	: air volumetric flow, (m3/s).
Wf	: compression power of refrigerant cycle, (kW).
We	: compression power of water cycle, (kW).
Y1	: air moisture, (kg/kg).
ρa	: air density, (kg/m3).
$\rho v1$, $\rho v2$	: steam density at corresponding temperature, (kg/m3).
ΔM	: liquid injection flow, (kg/s).
ηis	: isentropic efficiency of compressor.

SUBSCRIPTS

1	evaporation.
2	condensation.

The other numerical subscripts correspond to points of Figure 1.

1. INTRODUCTION

During drying operation, a large quantity of the energy supplied is in the form of evaporation mist. At present, heat recovery is only practised on sensible heat and a small percentage of latent heat.
Electricite de France has launched a project to meet industry's need to reduce drying cost. Its aim is to construct an installation which can produce steam at low and mean pressure of 2 to 5 bar on the basis of the energy recovered from an evaporative system.
Heat pumps and mechanical vapour recompression are among the energy recovery systems currently used in industry. The advantages of heat pumps are as follows:
-flexibility of installation.
-availability of cold sources.
The mechanical vapour recompression leads to a high condensing temperature and an enhanced coefficient of performance through the small temperature difference that occurs. However, in the specific case of drying, it may be observed that:
-heat pumps utilising conventional refrigerants, especially fluorocarbons, can only reach a maximum condensing temperature of 120 C.
120°C.
-when mechanical vapour recompression is directly integrated in a drying unit, there are some disadvantages: large suction volume (low evaporating temperature), presence of air. In addition, the coefficient of performance degrades as compression ratio rises.
The design of conventional and high temperature heat pumps has been quite thoroughly studied. Their operation does not raise any special difficulties. The temperature thresholds associated with the fluids can only be exceeded by using other fluids which can operate with satisfactory characteristics at high temperature. Water is one of the best fluids at high temperatures. Its critical temperature is 373.97°C. In addition, it is particularly advantageous thanks to its high latent heat, its very satisfactory heat transfer coefficient and its low cost. The two technologies can thus be combined to recover the heat and produce steam as stated above. Then the heat pump operates in its

Held at St. Catherine's College Oxford, England. Organised and sponsored by BHRA, The Fluid Engineering Centre, Cranfield, Bedfordshire, MK43 0AJ England.

normal operating range and the mechanical vapour recompression becomes a closed cycle and is known as water heat pump(Figure 1a).
We have therefore focussed on the cascaded combination of a heat pump operating with a fluorcarbon refrigerant and a heat pump operating with water.
The possibility of heat recovery from the cold source has been examined together with the influence of the cold source on the operation of the system; the effect of intermadiate temperature on the overall performance of the system was also studied.
The findings of the studies are that the system is viable and worthwhile. The operations involved in setting the size and technological considerations are also brought to light. Future work will consist of conducting experimental studies and validating the theoretical model.

2. STUDY OF THE COLD SOURCE

During the evaporation phase, a hood is used to collect the mist. The presence of air during the operation is inevitable. The quantity of air depends on the design of the evaporation system and its environment. On mixing with the mist, the air plays a major role in recovery. Most of the energy contained in the mist is in the form of latent heat. The only condition for its use is to check that the temperature at which the energy is drawn remains below the dew temperature. This temperature is determined by the quantity of air. It is therefore indispensable to examine the relationship between the recoverable energy and the properties of the mist. For a flow rate of steam originating from the evaporation system, the mixing characteristics are dependent on the air flow rate, the temperature and the moisture of air. The phenomenon which occurs in the hood is that wet air is heated by cooling and partly condensing of the steam. The energy balance is written:

$$P1 + P2 = P3 + P4$$

in which

$$P1 = Vva \cdot Cpa \cdot \rho a \cdot (t - te)$$

$$P2 = Vva \cdot Cp1 \cdot \rho v1 \cdot (t - te)$$

$$P3 = Vvv \cdot \rho v2 \cdot Cp2 \cdot (100 - t)$$

$$P4 = Vvv \cdot \rho v2 \cdot Lv \cdot (1 - Ps/1.013)$$

P1: power needed to heat the air from te to t, (kW).
P2: power needed to heat the vapour from te to t, (kW).
P3: cooling power of the steam produced by the evaporation system from 100°C to t, (kW).
P4: condensation power of this steam, (kW).

Recovery is dependent on the one hand, on the mixing characteristics which determine the quantity of energy, and on the other hand, on the operation of heat exchanger.
Computational results are given in figures 2,3,4,5 and 6.
A thermal output corresponds to a quantity of energy withdrawn from the cold source. Figure 2 indicates the zone in which the cold source contains enough energy. It is best for the system to operate in that zone. In other case, the cold source can only supply enough energy in the hatching zone(figure 3, limited by the wet air and minimum moisture curves).
According to the computational results, latent heat supplies larger part of energy. Table 1 lists some numerical values for indicative purposes.

The conclusion can be drawn from figures 4,5 and 6, that:
-The lower the temperature at heat exchanger outlet is, the greater the energy recovered is. Besides, the efficiency of the system in terms of energy must be taken into account. It is favourable when the evaporating temperature, which depends on the temperature of the cold source, is high. There is a compromise between recoverable energy and the coefficient of performance.
-Among other parameters, the air flow rate is one of the most important. The air moisture and temperature have a limited effect on the quantity of recoverable energy.

3. MODELLING OF THE CASCADED HEAT PUMP

The assumptions below have been made:
-non isentropic compression with constant isentropic efficiency.
-pressure drops in the heat exchanger, pipings and valve compressors negligible.
-heat losses in the heat exchanger, pipings and compressors negligible.
-expansion isenthalpic.
Figure 1b presents the thermodynamic cycle of the refrigerant and water. The R114 is superheated before suction to avoid liquid slugging because its isentropic curve crosses saturated curve during an isentropic compression.
In the case of water, the thermodynamic cycle differs completely from the refrigerant cycle. Compression begins with saturated steam and ends with high superheating of steam. A liquid injection is necessary during compression to reduce superheat. The two cycles are combined by equality of the refrigerant condensation power and water evaporation power.
The characteristics of the fluids are calculated by correlations developed for real fluids.

Refrigerant cycle

Power withdrawn from cold source is:

$$Pf1 = (H3 - H1) \cdot Mf$$

Compression power is:

$$Wf = (H4' - H3) \cdot Mf/\eta is$$

in which H4' is enthalpy at the end of isentropic compression.

The power of the refrigerant at the condenser is:

$$Pf2 = (H4 - H5) \cdot Mf$$

Water cycle

As above, we have evaporation power:

$$Pe1 = Me \cdot (H7 - H6)$$

Compression power is:

$$We = Me \cdot (H8' - H7)/\eta is$$

Power at the condenser is:

$$Pe2 = Me \cdot (H8 - H10)$$

The quantity of water to be injected in order to desuperheating the steam is:

$$\Delta M = Me \cdot (H8 - H9)/(H9 - H10)$$

Coupling of these two cycles provides knowledge of the Pf1 needed to obtain a power Pe2. The overall coefficient of performance is defined as:

$$COP = Pe2/(Wf + We)$$

Figure 7 presents the general diagram of the computation.

4. MODELLING RESULTS

The different values of flow rate in the water loop were computed. The present article only discusses the results for a flow rate of 1 ton per hour. Condensing temperature of the refrigerant were varied between 90°C and 125°C, taking account of the present technology of heat pumps and compressors sizes.
Modelling results have shown that, for a fixed production temperature, the overall compression power is the lower, the higher the evaporation temperature of the refrigerant.
For a given evaporating temperature of refrigerant and a production temperature, the compression power of refrigerant cycle increases as condensing temperature rises; in parallel, the compression power of the water cycle decreases. The overall compression power depends on the relative role played by each parameter. There is therefore an optimum value for each temperature couple between evaporation of the refrigerant and steam condensation. This is shown in figure 8.
The coefficient of performance does not vary greatly in relation to the condensing temperature of the refrigerant(figure 9). From the standpoint of the energy efficiency of the system, the loop can thus be operated with any value of intermediate temperature. However, there are technical limits to the choice of this temperature.
For a given production temperature, the coefficient of performance increases as the temperature of the refrigerant rises. For instance, the coefficient of performance increases by 13% for a rise in temperature from 70°C to 80°C. Because of the characteristics of the refrigerant, this change is not linear with respect to its evaporating temperature.
From a thermodynamic point of view, it is preferable to have a high evaporating temperature of the refrigerant. The cold source sets this temperature.
Figure 10 shows variations in the coefficient of performance as a function of the condensing temperature of steam. The coefficient of performance degrades as the temperature rises. This phenomenon is due to the increase in the compression ratio of water cycle.
It can be observed that if a cold source which guarantees an evaporating temperature of 70°C is available, the system can produce steam at 150°C with a coefficient of performance which is always above 3 (figure 10).
When evaporating temperature is 60°C, with steam production at 150°C, the coefficient of performance is of the order of 3 if the condensing temperature of the refrigerant is below 100°C. The coefficient of performance can reach 3.5 when the cold source guarantees an evaporating temperature of 70°C. The coefficient of performance also increases when the steam production temperature is lower.
Figure 11 presents curves with identical coefficient of performances. If the operating point is below the curve, the coefficient of performance is higher than that indicated in the corresponding curve. The dotted straight line is the limit between the two heat pumps. This figure gives a general idea of the coefficient of performancy of the system.
Figure 12 shows the suction volume as a function of the condensing temperature of the refrigerant, It can be observed that the values obtained for steam are always higher than those obtained for the refrigerant. This stems from that, at low pressure, steam has substantial specific volume. For instance, the specific volume of saturated steam at 100°C is 1.695 m3/kg and only 0.00934 for the R114.
From the point of view of size, it is advantageous for the system to operate at high condensing temperature of the refrigerant to reduce the size of the steam compressor. In contrast to thermodynamic efficiency, there are no optimum values for setting the size of compressors.

5. CONCLUSION

The study shows that the cascaded high temperature heat pump can be used to produce steam from waste heat with a high coefficient of performance (between 2.5 and 3.5).
For each case, there is an optimum intermediate temperature. However, the temperature of the cold source should be as high as possible.
There is no optimum value for setting the size of compressors. A compromise with the coefficient of performance is required.
The modeling results for this heat pump are shortly to be checked by an experimental installation.

Air-steam mixture temperature at the outlet of heat exchanger (°C)	50	60	70	80
Recuperable energy in form of latent heat / Energy recuperable (%)	83.1	86	88.9	90.8
Recuperable energy in kWh per ton of steam produced	694	629	525	354

Characteristics: Air 35°C moisture 0.02kg/kg
Ratio between air volumetric flow and steam volumetric flow 46.4%

Table 1. Percentage of latent heat as function of air-steam mixture at the outlet of heat exchanger.

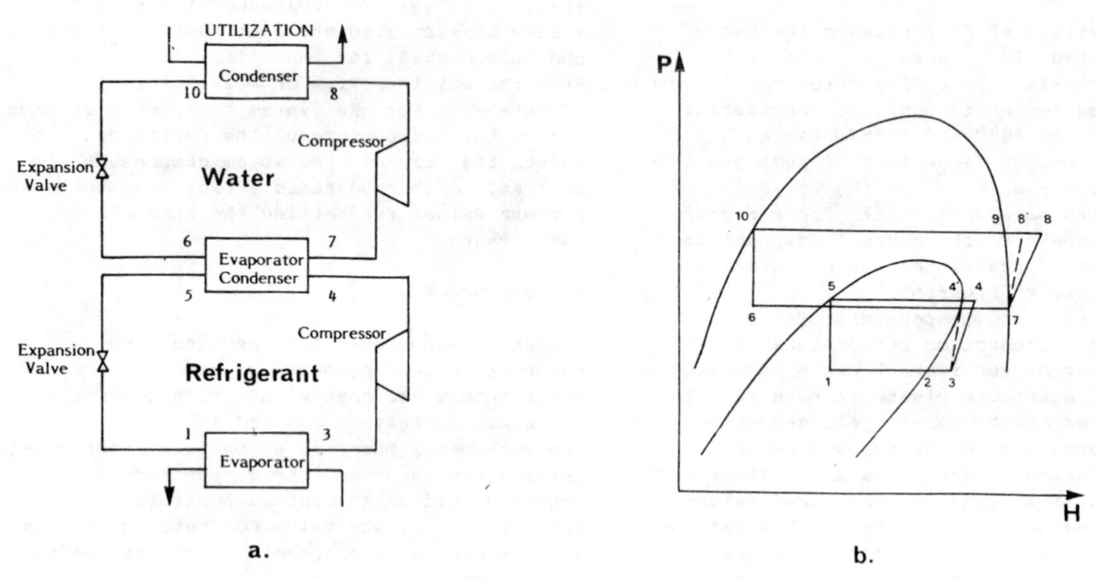

Figure 1. a. Cascaded heat pump.
b. Thermodynamic cycles.

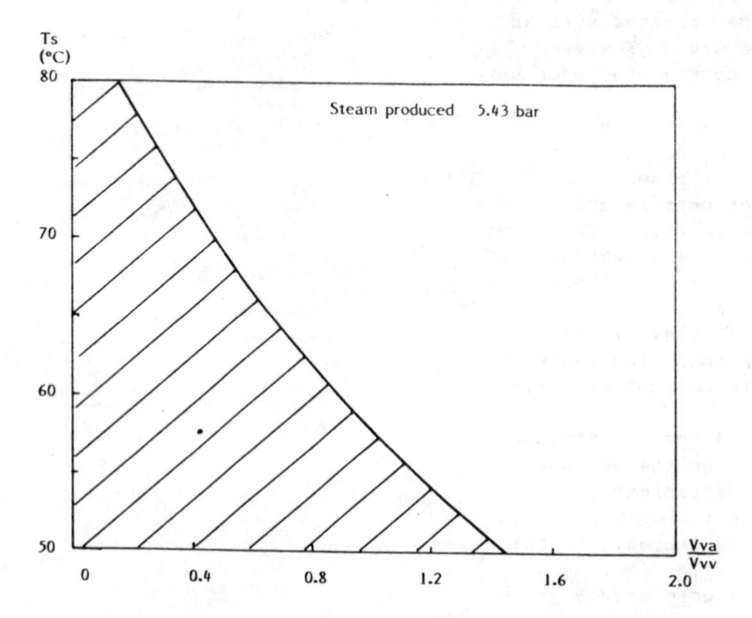

Figure 2. Working area (hatching zone).

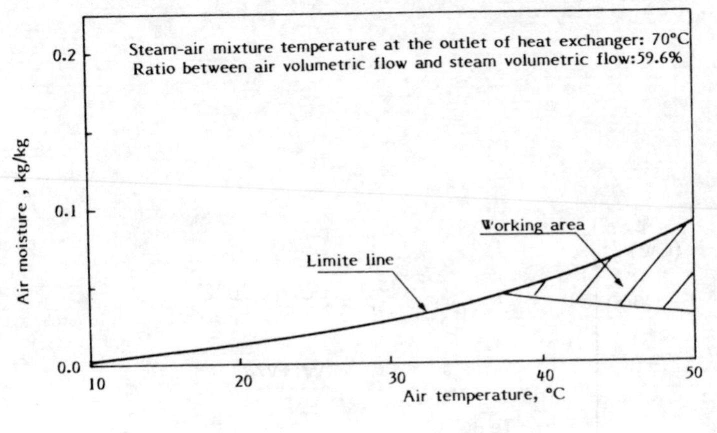

Figure 3. Limitation of working area.

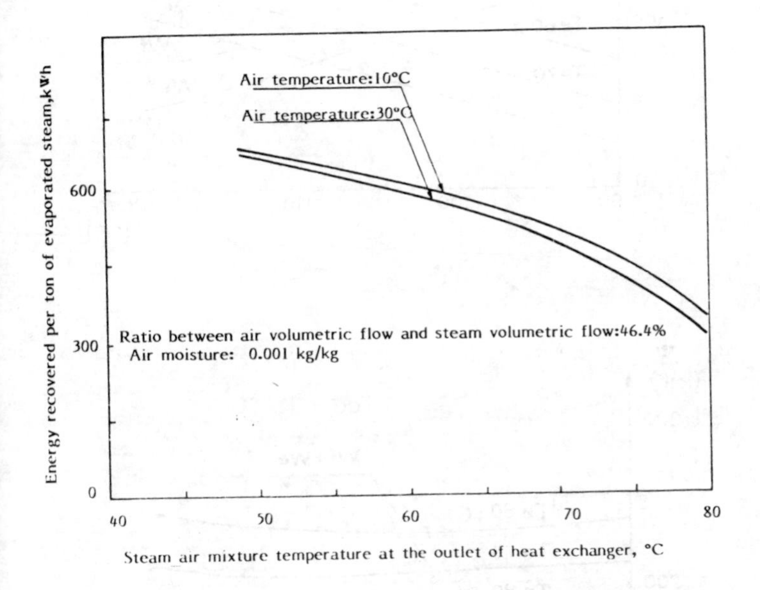

Figure 4. Influence of air temperature on energy recovered.

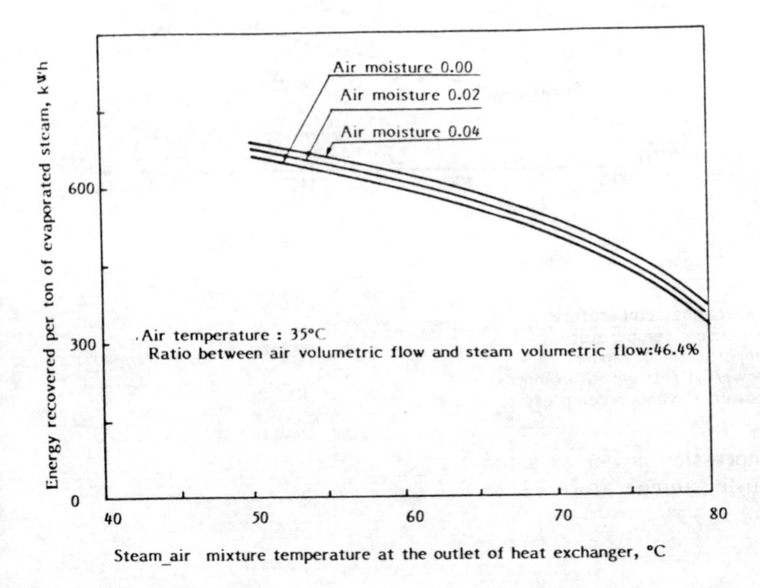

Figure 5. Influence of air moisture on energy recovered.

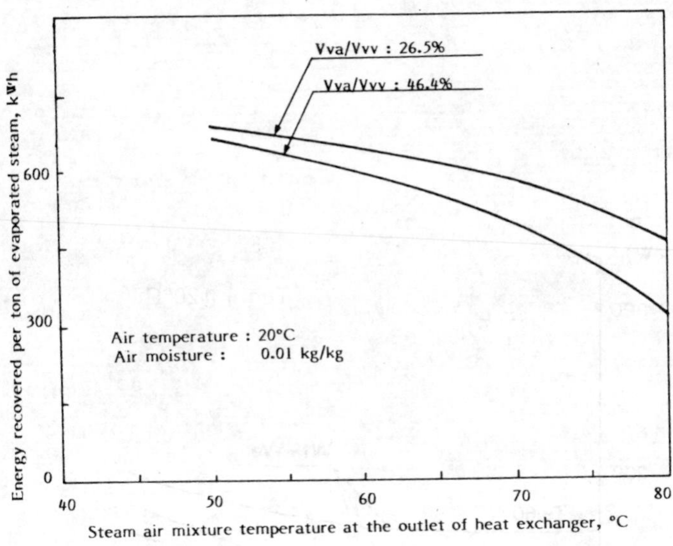

Vva/Vvv : Ratio between air volumetric flow and steam volumetric flow

Figure 6. Influence of air flow on energy recovered.

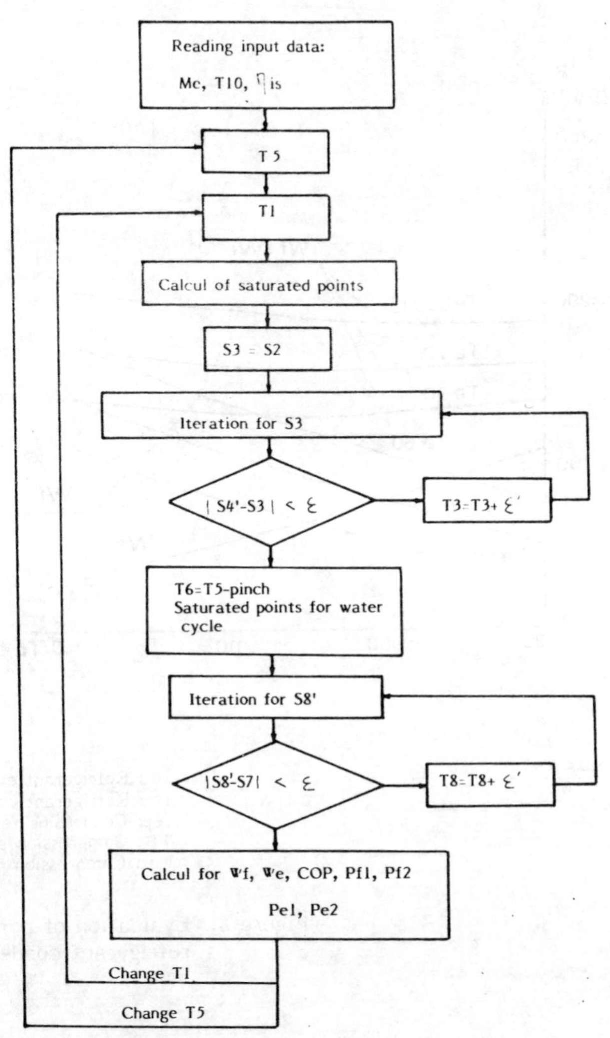

Figure 7. General bloc diagram of calcul.

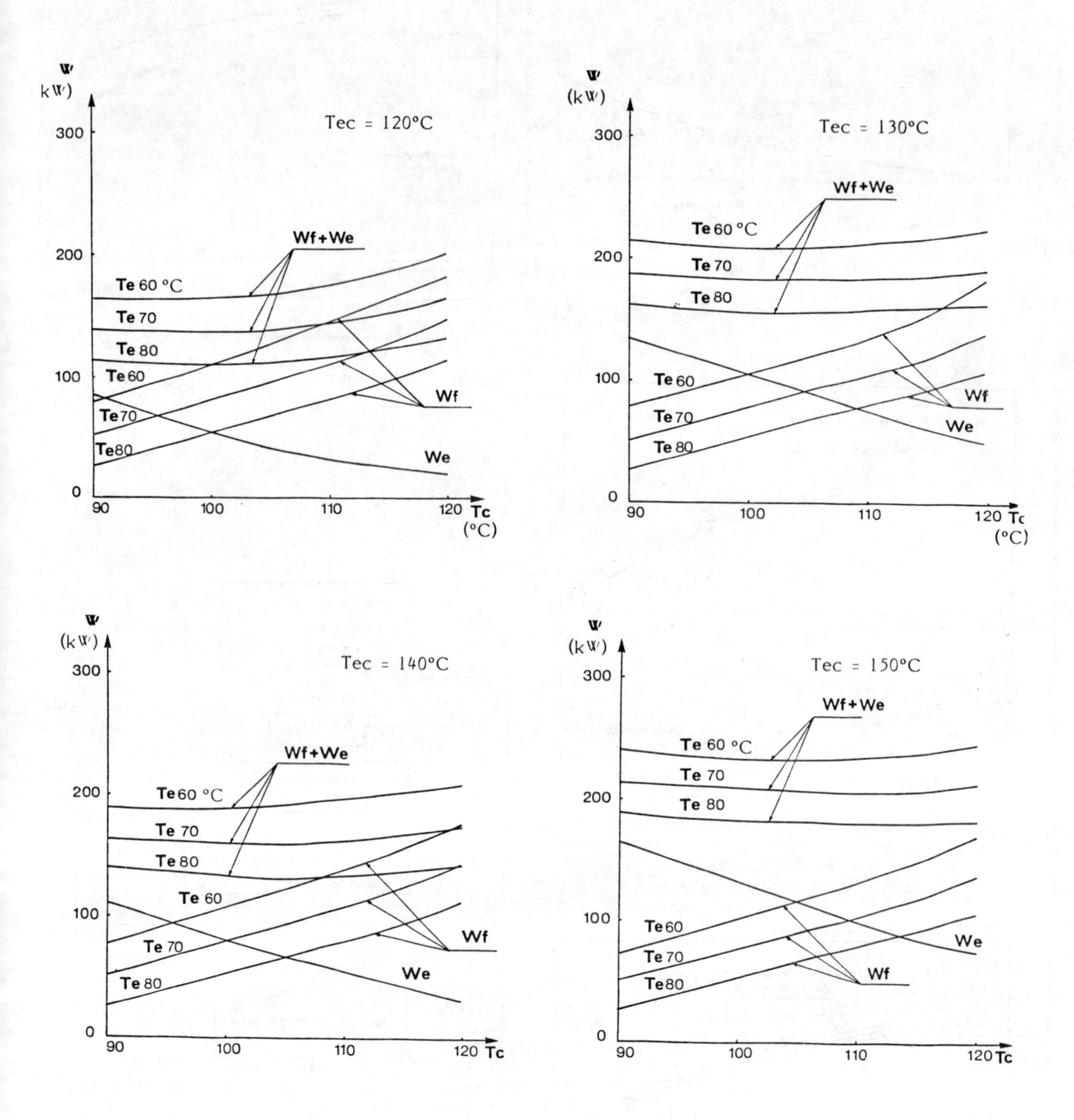

Te : Refrigerant evaporating temperature.
Tc : Refrigerant condensing temperature.
Tec: Condensing temperature of vapour.
Wf : Compression power of refrigerant compressor.
We : Compression power of vapour compressor.

Figure 8. Evaluation of compression power as a function of refrigerant condensing temperature.

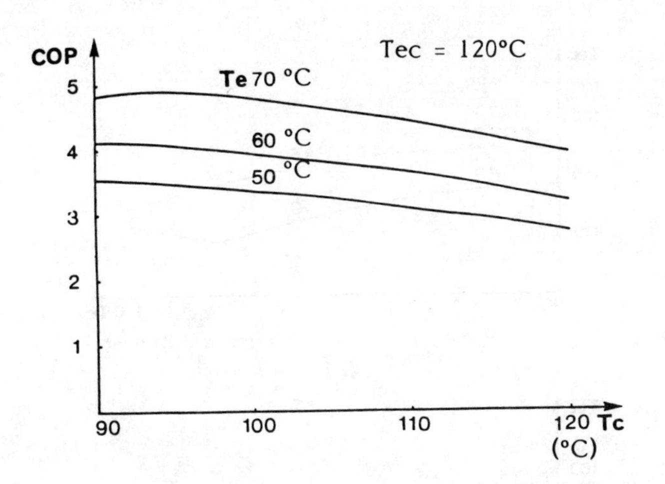

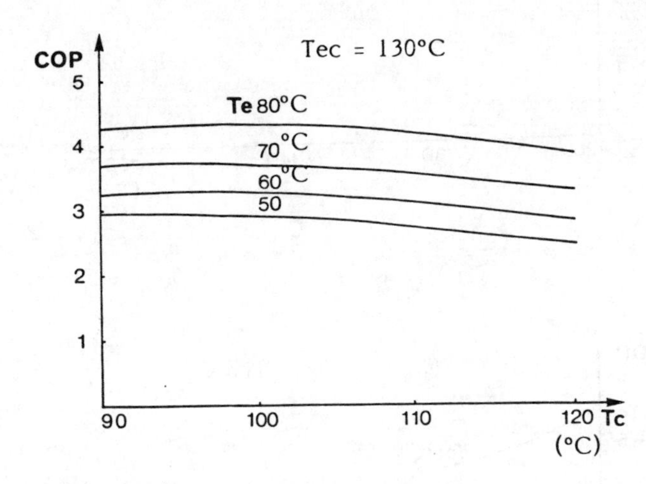

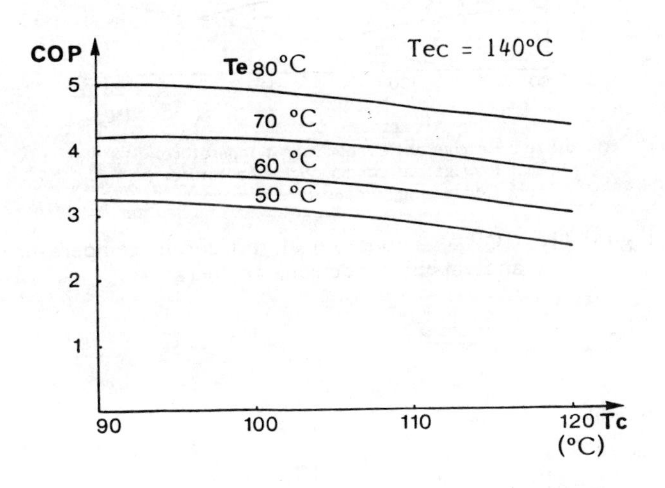

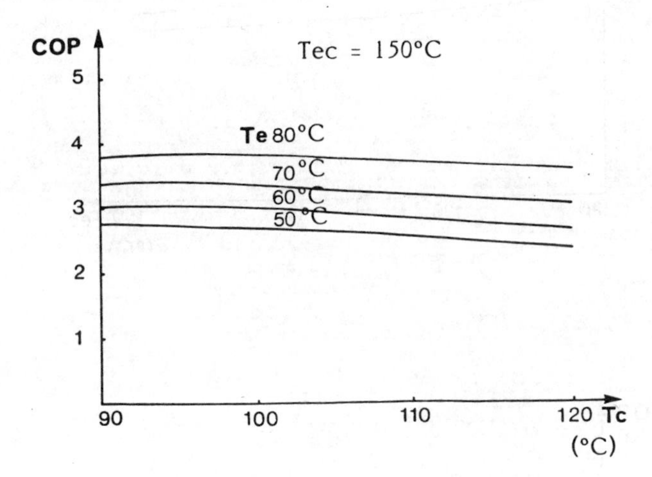

Te : Refrigerant evaporating temperature
Tc : Refrigerant condensing temperature
Tec : Condensing temperature of vapour.

Figure 9. Evaluation of COP as a function of refrigerant condensing temperature.

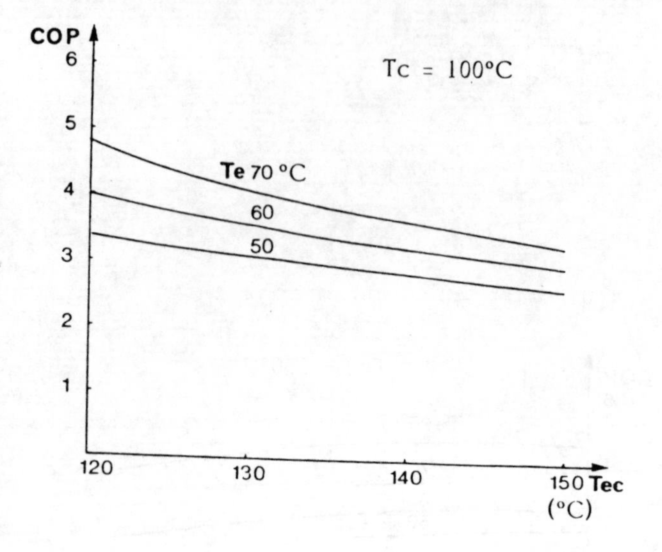

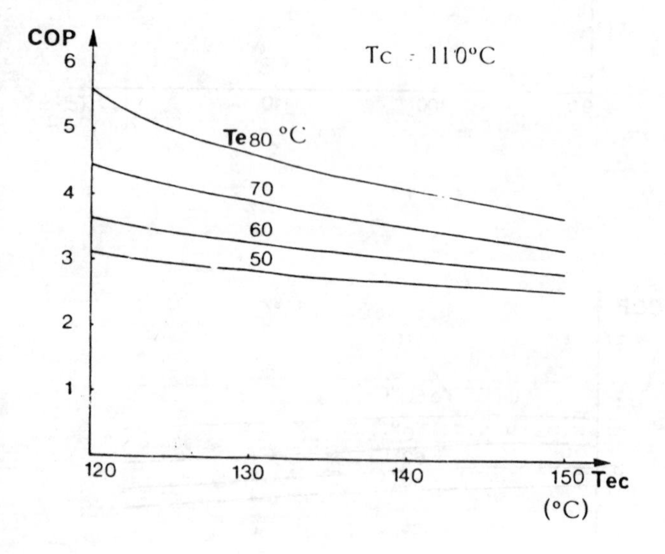

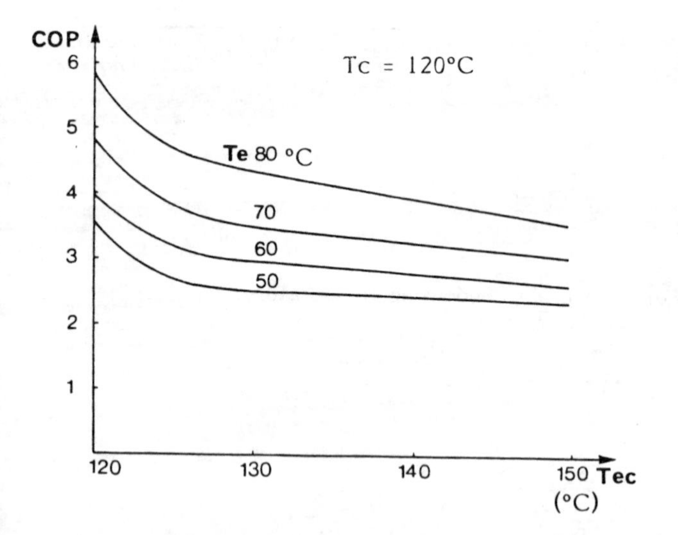

Te : Refrigerant evaporating temperature
Tc : Refrigerant condensing temperature
Tec : Condensing temperature of vapour

Figure 10. Evaluation of COP as a function of vapour condensing temperature.

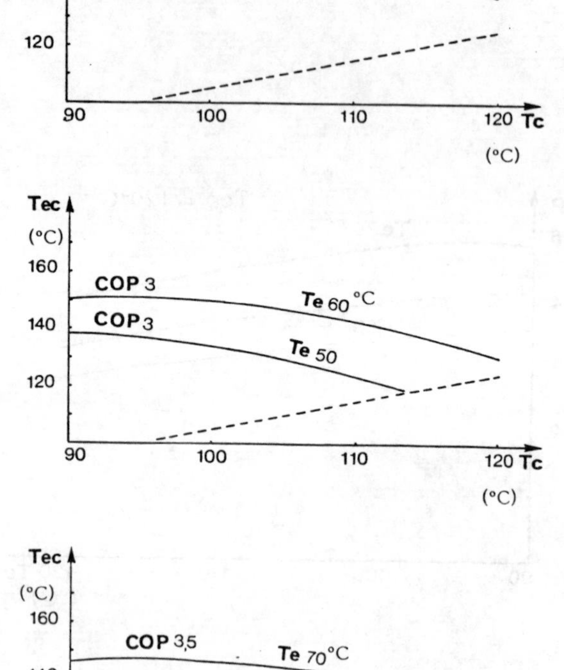

Te : Refrigerant evaporating temperature.
Tc : Refrigerant condensing temperature.
Tec: Condensing temperature of vapour.

Figure 11. COP as a function of refrigerant temperature and vapour condensing temperature.

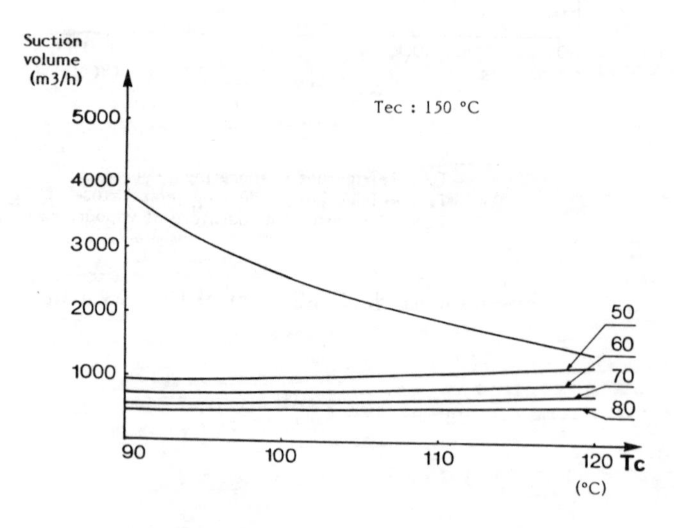

Te : Refrigerant evaporating temperature
Tc : Refrigerant condensing temperature
Tec : Condensing temperature of vapour.

Figure 12. Evaluation of suction volume as a function of refrigerant condensing temperature.

3rd International Symposium on the

Large Scale Applications of Heat Pumps

Oxford, England: 25-27 March 1987

PAPER P2

THE POTENTIAL OF ELECTRICALLY ENHANCED EVAPORATORS

P. H. G. Allen Ph.D., C.Eng
Thermo-fluids Engineering Research Group,
The City University, London ECIV OHB

P. Cooper, Ph.D., MSc., DIC
Built Environment Research Group,
Polytechnic of Central London, London NW1 5LS

Summary

This paper describes the extension of recent work on the Electrohydrodynamic (EHD) enhancement of the shell/tube condenser using Freons R12 and R114 to the case of nucleate pool boiling in evaporators of the same type using R114. Although apparently considered unpromising, this area of work gave remarkably useful results on two counts:

1) The virtual elimination of "boiling hysteresis" by using relatively modest transient voltages in an optimised electrode system.

2) An increase of an order of magnitude in nucleate boiling heat transfer coefficient with higher, continuous voltages and a 'lo-fin' profile tube.

The effect of EHD in overcoming the effect of oil contamination of the Freon was also examined and the mechanism of enhancement described.

NOMENCLATURE

A	=	heat transfer surface area	(m^2)
h	=	heat transfer coefficient	(W/m^2K)
$\dot{q}$	=	heat flux	(W)
$\dot{q}''$	=	heat flux density	(W/m^2)
T	=	temperature	(K)
ΔT	=	$T_w - T_s$	(K)

Subscripts

0	=	zero-field condition
m	=	arithmetic mean value
s	=	saturation value
w	=	wall value

1. INTRODUCTION

The notion of using an electric field to enhance boiling heat transfer was first reported in a UK patent as long ago as 1916 [1]. More than four decades were to pass before this Electrohydrodynamic (EHD) enhancement technique was investigated quantitatively by Bochirol, Bonjour and Weil [2]. In their study substantial increases in heat transfer were observed upon the application of an intense electric field to film boiling at the surface of a fine, horizontal, heated wire in a variety of insulating liquids. This particular EHD phenomenon was due to the electrical destabilization of the vapour film around the wire and was effected by electrical forces acting on the vapour-liquid interface. These forces arise from the difference between the dielectric permittivity of the liquid and vapour phases and are of the type called "dielectrophoretic". The instability was later found to be of a wavelike nature leading to a "rewetting" of the heat transfer surface and an increase in heat flux.

Since 1960 many further studies of EHD enhanced film boiling have been carried out and these have been reviewed by Jones [3] along with work on EHD enhanced condensation and single-phase heat transfer. However, most of this research has dealt with situations inappropriate to the needs of heat exchanger manufacturers, not only because film boiling is not a mode of heat transfer used under most normal operating conditions but also since electrode and heat transfer geometries have not been representative of those in practical evaporators. The present study appears to be the first to deal with the details of EHD enhancement of nucleate pool boiling.

EHD heat transfer enhancement is thought to be best applied to electrically insulating fluids. Thus, vapour recompression plant, Organic Rankine Cycle (ORC) engines, etc. using fluorinated hydrocarbons would seem a potential engineering application for this technique since the working fluids have exceptionally good dielectric properties. Furthermore, such plant can often be of considerable size, incorporating large shell/tube evaporators and condensers. Since practical implementation of EHD techniques requires the addition of electrodes and a high voltage source to the normal heat exchanger, it is thought that this would be most feasible/cost-effective in large units.

The work reported here represents the latest stage in a continuing research programme aiming at the development of full-scale practical shell/tube EHD enhanced condensers and evaporators. An earlier stage included optimization of a shell-side electrode system consisting of plates and/or rods (covered by patent in the UK [4] and internationally). Three possible electrode/tube arrangements in a horizontal tube bundle are shown in Figure 1. There followed an experimental and theoretical investigation of EHD enhanced condensation heat transfer which has been described elsewhere [5][6] and in which refrigerant-side heat transfer enhancement by up to a factor of three was observed. Although the tenor of previous work by

Held at St. Catherine's College Oxford, England. Organised and sponsored by BHRA, The Fluid Engineering Centre, Cranfield, Bedfordshire, MK43 0AJ England.

others indicated that condensation heat transfer would be more amenable to EHD enhancement, it has been found, in fact, that EHD techniques have even greater potential when applied to nucleate boiling.

2. APPARATUS

A schematic diagram of the EHD assisted boiling rig is shown in Figure 2. The evaporator comprised a single, horizontal, integrally finned tube within a concentric brass shell and was designed to model the operation of a tube in the bottom row of a shell/tube evaporator. The dimensions and fin profile of the "lo-fin" tube are given in Figure 3. The tube was supplied by a manufacturer of large shell/tube condensers and evaporators. The shell was of 63.5mm internal diameter and had two 50mm diameter sight glasses mounted at its mid-section to facilitate visual observation of the EHD phenomena. A permanent record of the latter was made using a video camera and recorder. The refrigerant circuit of the rig was designed for a maximum working pressure of 15bar (gauge).

The electrode system used for quantitative investigations was made from a copper wire mesh cylinder of 38mm diameter with brass stress relieving rings attached to each end. This was supported concentrically with respect to the experimental tube axis by "Tufnol" insulating rings. During qualitative experiments, unobstructed visual observation of the experimental tube was facilitated by a plate-and-rod electrode system corresponding to a single tube section of the arrangement shown in Figure 1A. Two conductive glass inserts were held in the centre of the two electrode plates through which the experimental tube could be seen. Full details of electrode systems, apparatus and experimental procedures are given in reference [6].

The high voltage supply (of up to 30kV) to the electrodes was provided by either one of two "SAMES" electrostatic generators for direct current (d.c.) potentials or from a 100kVA high voltage transformer and regulator for alternating current (a.c.) potentials. The electrical supply was fed through a high resistance potential divider and then into the EHD evaporator through a specially modified sparking plug in the shell wall which made a spring loaded contact with the cylindrical electrode. The entire apparatus (excluding some instrumentation) was sited in a Faraday cage. Current drawn by the EHD process from the high voltage sources was measured using a battery-powered "Levell" picoammeter suspended from the roof of the Faraday cage and connected electrically between sparking plug and potential divider. Electrode potentials were monitored by means of electrostatic voltmeters.

All temperature measurements were made using copper/constantan thermocouples. Surface temperature of the experimental tube was monitored at eight locations. To avoid the intense electric fields at the surface the thermocouple leads were routed through the inside of the experimental tube.

The liquid Freon saturation temperature was taken as that of the vapour exiting the EHD boiler and was measured by means of a thermopile immediately adjacent to (but outside) the evaporator shell. Measurement of the Freon liquid temperature per se was precluded by the intense electric fields within the fluid. Fortuitously, other researchers (e.g. Hahne and Muller [7]) have found that, for correlating pool boiling data, the most effective way to characterise wall-to-liquid superheat is to use the system saturation temperature, T_s, measured above the liquid surface. This was the situation assumed in the present study. Local wall-to-saturation superheats, ΔT_w, were calculated from the measured differential e.m.f.s generated between the thermopile and the individual tube surface thermojunction concerned.

The tube was heated by hot water pumped from a tank containing a 9kW immersion heater, its output varied by means of a three-phase 'variac'. Total heat flux, $\dot{q}$, through the tube wall was calculated from the heating water bulk temperature drop between inlet and outlet, the water flow rate (measured by a variable-gap flowmeter) and published data for the density and specific heat of water. Bulk inlet and outlet temperatures were measured by means of copper/constantan thermojunctions. Thorough mixing of the heating water was ensured by the use of vortex inducing spirals and copper gauze baffles positioned upstream of each thermojunction.

The use of water rather than electricity to heat the experimental tube resulted in the wall-to-liquid Freon superheat varying along the tube as the heating water cooled. Thus, all experimental results are quoted with respect to an arithmetic mean superheat, ΔT_m. Although not ideal for investigating the fundamental characteristics of EHD enhanced boiling phenomena this situation was certainly more representative of conditions in a commercial shell/tube evaporator than one using electrical heating. It is known that the boiling heat transfer coefficient at a given wall-to-liquid superheat varies considerably with the orientation of the heat transfer surface to the vertical [8]. To investigate this effect, tube surface temperature was measured at four points around its circumference at the tube mid-section. ΔT_m was then calculated assuming this relative temperature profile to be the same at any cross-section along the tube.

The liquid chosen for the experimental investigation was R114 (dichlorotetrafluoroethylene) because of its excellent dielectric properties, low toxicity and widespread commercial use in heat pump and Rankine cycle plant. Several experiments were also conducted to investigate the effect of electric stress on heat transfer to oil contaminated R114. The oil used was Shell Clavus 68. This was simply mixed on a weight-to-weight (w.w.) basis with the total charge of R114 in the rig.

To facilitate degassing and to ensure fully mixed, isothermal, initial conditions, the evaporator was heated before each experiment by means of self-limiting ("Freezguard") heating tape strapped to its shell. The condenser unit (see Figure 2) was cooled by water from a thermostatically regulated refrigeration unit.

3. RESULTS

3.1 Zero-field heat transfer

Initially, heat transfer under zero-field conditions was investigated. Results for a system saturation temperature, T_s, of 21.5°C, graphed as mean heat transfer coefficient, $h_{m,0}$, against mean heat flux density, q_m, are shown in Figure 4. All results presented in this paper are calculated with respect to the total (developed) external heat transfer surface area, A, of the experimental tube ($A=0.152m^2$). As the study of boiling hysteresis was of particular interest, experiments were conducted with both increasing and decreasing wall-to-liquid superheat, ΔT_m. Boiling (or nucleation) hysteresis manifests itself as a given threshold ΔT required to activate vapour generation at nucleation sites. Below this critical value of ΔT, heat transfer is by natural convection alone. The actual value of ΔT required to activate nucleation depends on a number of factors, particularly those relating to roughness and contamination of the heat transfer surface. In the present study, the lo-fin tube was installed in the as received condition, with no special treatment other than cleaning with a wire brush and degreasing with acetone.

The results in Figure 4 illustrate the considerable difference between heat transfer rates obtained for increasing and decreasing $\dot{q}''_m$. At T_s=21.5°C a local superheat of approximately 7.5°C was required to activate ebullition on the tube surface; this threshold decreased with increasing system saturation temperature. In some engineering plant with limited temperature drops across evaporators this may indeed cause problems.

Also shown in Figure 4 for comparison are results from two other similar studies [7][9].

3.2 Effect of electric stress on boiling in R114

Two main EHD phenomena were observed, viz.:

a) EHD elimination of boiling hysteresis
b) EHD enhancement of nucleate boiling

EHD elimination of hysteresis resulted from the electrical activation of nucleation sites. Visual observations showed that following zero-field superheating of the experimental tube by a few degrees Centigrade with heat transfer operating in the natural convective mode application of a modest electrode potential (5-8kV, say) resulted in immediate activation of ebullition. This electrical stimulation of nucleation sites appeared to be similar to thermal activation since sites remained active after removal of the electric field. Moreover, by applying a sufficiently intense field (V>10kV) for a short duration (less than one second) it was possible to jump from a point on the increasing $\dot{q}''_m$ heat transfer hysteresis loop to one for the same superheat on the decreasing $\dot{q}''_m$ curve (i.e. with a dramatic increase in heat transfer coefficient).

Such a technique for EHD elimination of boiling hysteresis may be relatively simple to implement in many engineering heat transfer situations. The required duration of applied electric stress is short and the energy required extremely small. In practice, this has the following implications:

a) a very simple high voltage generator may be used (e.g. similar to spark ignition on a domestic gas cooker)
b) the need for a sophisticated electrode/insulation system capable of withstanding continuous electric stress is reduced
c) relatively conducting working fluids (including water) may be amenable to this technique.

Continuous application of electric stress to the superheated lo-fin tube produced substantial increases in nucleate boiling heat transfer. Quantitative results are presented in Figure 5 where curves A and B represent heat transfer rates under zero-field conditions and curves C and D are for applied electrode potentials of 10kV and 27kV, respectively. As with all EHD phenomena observed in the present study, a.c. electric fields and d.c. fields of either polarity were equally effective. It can be seen from Figure 5 that EHD enhancement of refrigerant-side nucleate boiling heat transfer of up to an order of magnitude was achieved with the complete elimination of hysteresis. This is the first study to observe such a dramatic enhancement and this is thought largely due to the type of heat transfer surface geometry utilized.

With electric field strength increasing from zero, visual observations indicated that EHD forces modified the bubble dynamics of the boiling process, so that: a) bubble departure diameter decreased; b) bubbles leaving the tube surface were dispersed quickly by the combined effects of the gravitational and electric fields; c) strong electric field inhomogeneities at the surface of the lo-fin tube distorted and trapped bubbles between tube fins at high field strengths. The latter phenomenon is probably the major mechanism responsible for the largest enhancement of nucleate boiling to have been reported to date for geometries other than the simple fine heated wire.

Figures 6 to 9 illustrate the action of field inhomogeneities around the lo-fin tube on bubble growth and motion (for V>20kV) as observed by the naked eye and using video recordings. The areas of high and low electric field strength are shown schematically in Figure 6. Dielectrophoretic forces tend to move matter of low permittivity (i.e. vapour) to areas of low field strength. Since the vast majority of active nucleation sites on the lo-fin tube were at the fin roots, where the degree of wall-to-liquid superheat was greatest, bubbles generated there were trapped in the inter-fin spaces (Figure 7). Forced to rise by buoyancy forces these bubbles were compelled to follow the curvature of the tube (Figure 8). At low rates of heat transfer a column of vapour could be seen to be trapped on the top half of the tube in each inter-fin space (Figure 9). These columns would grow until buoyancy forces were sufficient to overcome the restraining EHD forces and a bubble would be released from the top of the tube. The lower liquid-vapour interfaces would oscillate as a result (point X in Figure 9).

It is suggested that the EHD forces, by keeping the bubbles close to the hottest part of the heat transfer surface, greatly increased turbulence/mixing and led to the substantial EHD induced rise in heat transfer. The electrical and heat transfer geometries of the EHD evaporator appear to have achieved an electrical enhancement of boiling heat transfer similar to that produced mechanically by certain extended surface configurations (e.g. as in the case of the GEWA-T-Tube which has fins with a "T" cross-section [10]).

Qualitative studies on a smooth EHD evaporator tube showed considerably less EHD boiling heat transfer enhancement than for the lo-fin tube. This was probably due to the fact that EHD forces tend to move bubbles radially away from the surface of a smooth tube minimizing any disruptive or "scouring" influence on the thermal boundary layer around the tube.

The electrical power consumed by the EHD process was extremely small. In fact, no detectable change between the electrode potential vs supply current characteristic for boiling and isothermal conditions was observed. For a positive d.c. electrode potential of 25kV at T_s=21.5°C the current supplied to the EHD evaporator was approximately 8.8μA, representing a power input of 0.22W. Such an electrode potential could induce an increase in total heat flux through the lo-fin tube of about 200W for ΔT_m=2°C, giving an "EHD amplification factor" of the order of 1000.

3.3 EHD enhancement of boiling in refrigerant-oil mixtures

Most compressors in heat pumps require a lubricant oil which, by necessity, is mixed with the heat transfer fluid. The presence of this less volatile component leads, in general, to a degradation in the thermal performance of the condenser and (more significantly) of the evaporator. At the evaporator heat transfer surfaces oil concentration increases as the more volatile refrigerant is vaporized. This leads to the formation of a barrier to diffusion of refrigerant, and therefore heat energy, from the bulk of the fluid to the evaporator tube wall. The behaviour of refrigerant/oil mixtures under these conditions is complex (e.g. foaming may occur and, at low oil concentrations, heat transfer may be increased marginally). However, it is clear that in practical systems performance is substantially degraded (see [11], for example).

The experimental run equivalent to that shown in Figure 5 was repeated with 5% w.w. Shell Clavus 68 oil added to the R114 refrigerant charge. Almost identical results were obtained as for pure R114. The addition of 10% w.w. oil resulted in very significant heat transfer degradation. Comparison of Figures 5 and 10 shows that under zero-field conditions:

a) the presence of 10% oil suppressed nucleation and further increased the degree of boiling hysteresis observed in pure R114.
b) for decreasing heat flux density 10% w.w. oil contamination reduced heat transfer coefficients by at least 50% compared with the case of pure refrigerant at any wall-to-liquid superheat.

Application of electric stress resulted in the same EHD effects observed in pure R114. Boiling hysteresis could be completely eliminated with the brief application of an electrode potential of approximately 10kV. The effect of continuous application of the maximum electrode potential used in the present apparatus is shown in Figure 10. The substantial EHD induced increases in heat transfer observed were accompanied by a dramatic reduction in foaming of the R114/oil mixture.

4. PRACTICAL APPLICATIONS

The nature of the two types of EHD phenomena observed in the present study suggest that EHD enhancement could be profitably incorporated in large shell/tube evaporators and condensers using fluorinated hydrocarbon working fluids. Of particular interest are systems utilizing small temperature differences across their evaporators. These may be characterized by poor heat transfer coefficients or start-up problems due to incomplete activation of the available nucleation sites. ORC (Organic Rankine Cycle) engines, now becoming more cost-effective and therefore more widely used for low-grade heat recovery, are one example of this situation. Another example is an OTEC (Ocean Thermal Energy Conversion) power station which uses the temperature difference between the warm surface waters of tropical oceans and cold water pumped from about 1000m below the surface. Although this technology is in its infancy, several successful demonstration projects have been realised. These units rely on a gross temperature difference across the ORC engine of no more than 30°C, giving refrigerant-side evaporating and condensing superheats of less than 5°C. Thus, EHD elimination of nucleation hysteresis could be a very desirable feature. Moreover, the size of the heat exchangers in a full-scale plant would be enormous (with an effective heat exchange surface area of 2.5km^2 for a 400MWe plant [12]) and the financial benefits of using continuous EHD evaporator enhancement may then be very significant, particularly as it is thought that the heat exchangers will need to be made from Titanium!

Much remains to be done to assess the costs and benefits of incorporating a suitable electrode system in a practical heat pump, ORC or OTEC evaporator. To this end, the research programme is continuing at The City University, the next stage being the testing of a circuit containing two 9-tube EHD enhanced shell/tube evaporator/condensers.

5. ACKNOWLEDGEMENTS

The authors wish to thank the Science and Engineering Research Council for their support of the work reported, which was carried out in the laboratories of the Department of Electrical Engineering, Imperial College. The Council's Heat Pump Panel was particularly helpful in providing a forum for technical discussions with other researchers. Hall Thermotank Ltd. kindly supplied "lo-fin" tubing and their Mr. Ashmole, Mr. Little of Searle Heat Transfer Ltd. and Dr. Lopez-Cacicedo of the Electricity Council Research Centre have proved valuable sources of advice. The Central Research Fund of the University of London is thanked for providing one of the h.v.d.c. sources used in the experimentation. Dr K. E. Bett, Department of Chemical Engineering and Chemical Technology, Imperial College, gave invaluable advice on the design of pressure vessels.

6. REFERENCES

1) Chubb, L.W.:"Improvements relating to methods and apparatus for heating liquids", UK Patent Number 100796, 1916.

2) Bochirol, L., Bonjour, E. and Weil, L.:"Exchanges thermiques - Etude de l'action de champs electriques sur les transferts de chaleur dans les liquides bouillants", C. R. Hebd. Seances Acad. Sci. (Paris), 250, Jan. 1960, pp. 76-78 (in French).

3) Jones, T. B.:"Electrohydrodynamically enhanced heat transfer in liquids - a review", Advances in Heat Transfer, 14, Academic Press, 1978, pp. 107-148.

4) Allen, P. H. G. and Cooper, P.:"Improvements in or relating to heat exchangers", UK Patent Number 8522680, 1985.

5) Cooper, P. and Allen, P. H. G.:"The potential of electrically enhanced condensers", Proc. 2nd Int. Symp. on the Large Scale Applications of Heat Pumps, York, 1984, pp. 295-309.

6) Cooper, P.:"Electrically enhanced heat transfer in the shell/tube heat exchanger", PhD thesis, University of London, 1986.

7) Hahne, E. and Muller, J.:"Boiling on a finned tube and a finned tube bundle", Int. J. Heat Mass Transfer, 26, 1983, pp. 849-859.

8) Nishikawa, K., Fujita, Y., Uchida, S. and Ohta, H.:"Effect of surface configuration on nucleate boiling heat transfer", Int. J. Heat Mass Transfer, 27, 1984, pp. 1559-1571.

9) Henrici, H. and Hesse, G.:"Untersuchungen uber den Warmeubergang beim Verdampfen von R114 und R114-Ol-Gemischen an einem horizontalen Glattrohr", Kaltetechnik-Klimatisierung, 23, 1971, pp. 54-58.

10) Stephan, K. and Mitrovic, J.:"Heat transfer in natural convective boiling of refrigerant-oil mixtures", Proc. 7th Int. Heat Transfer Conf., Munich, 1982, paper PB12, 4, pp. 73-87.

11) McMullan, J. T., Hughes, D. W. and Morgan, R.:"Influence of lubrication oil on heat pump performance", Final Report to CEC, Contract no. EEA-4-028-GB, 1983.

12) Ford, G., Niblett, C. and Walker, L.:"Ocean thermal energy: prospects and opportunities", Occasional Paper 9, PREST, University of Manchester, 1981.

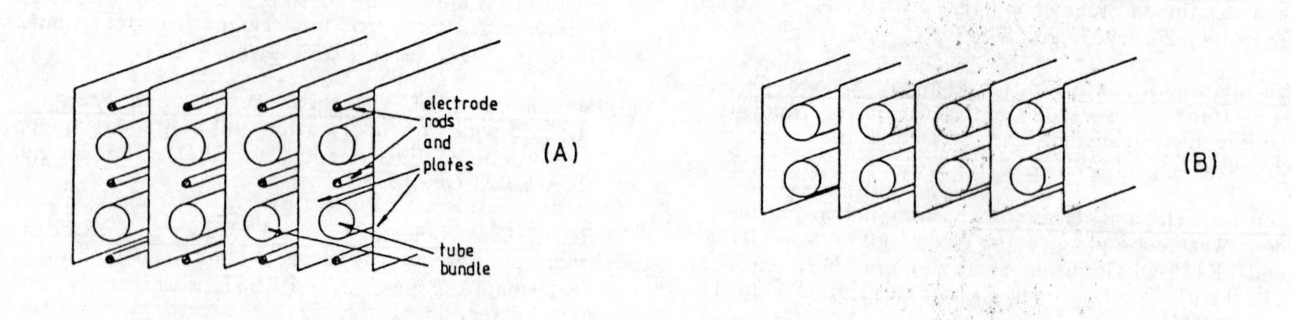

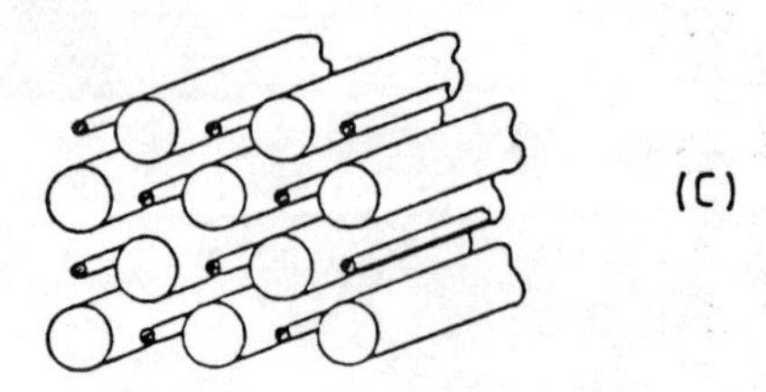

Figure 1 Three types of electrode system in a tube bundle

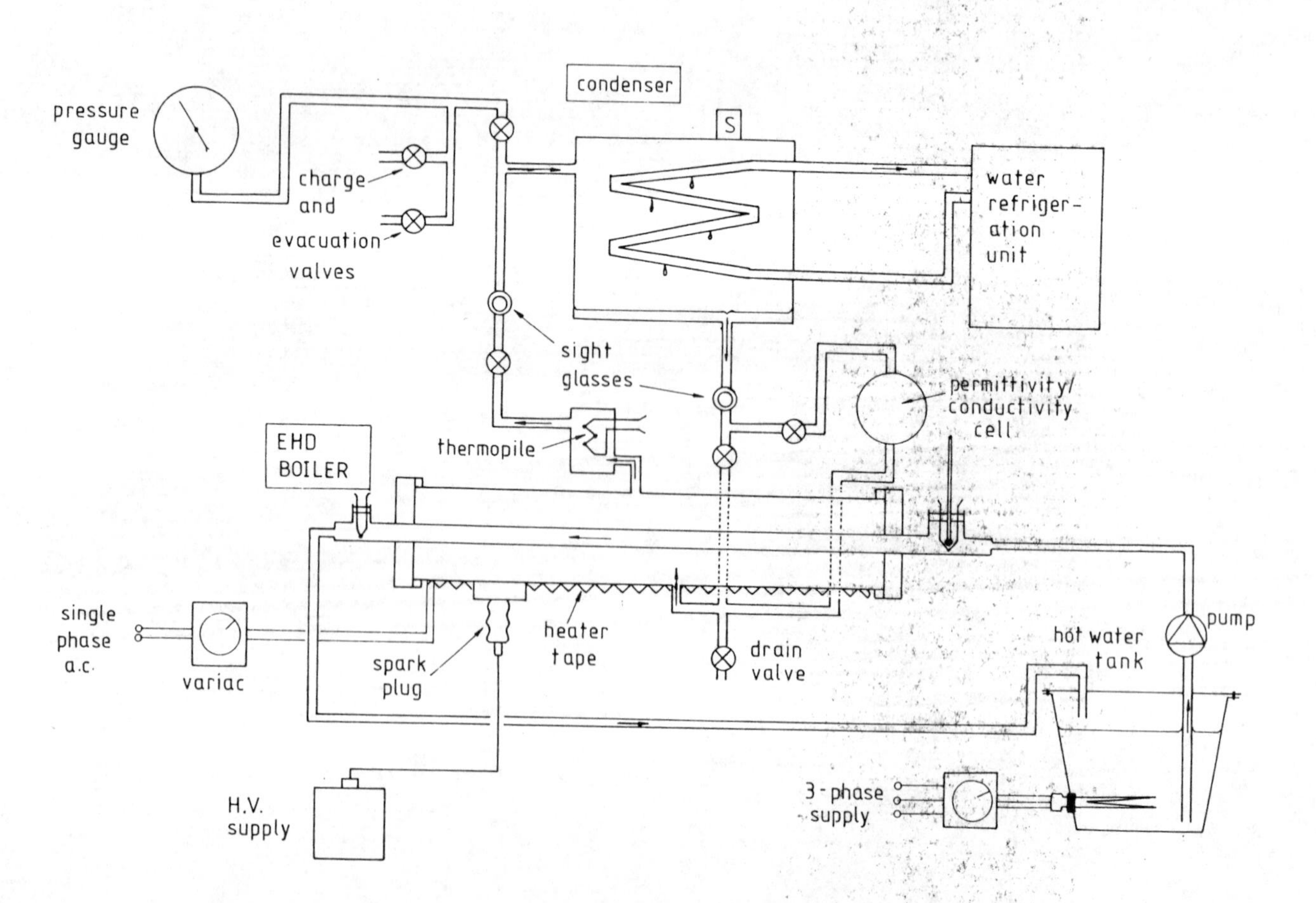

Figure 2 Single-tube EHD boiler test rig

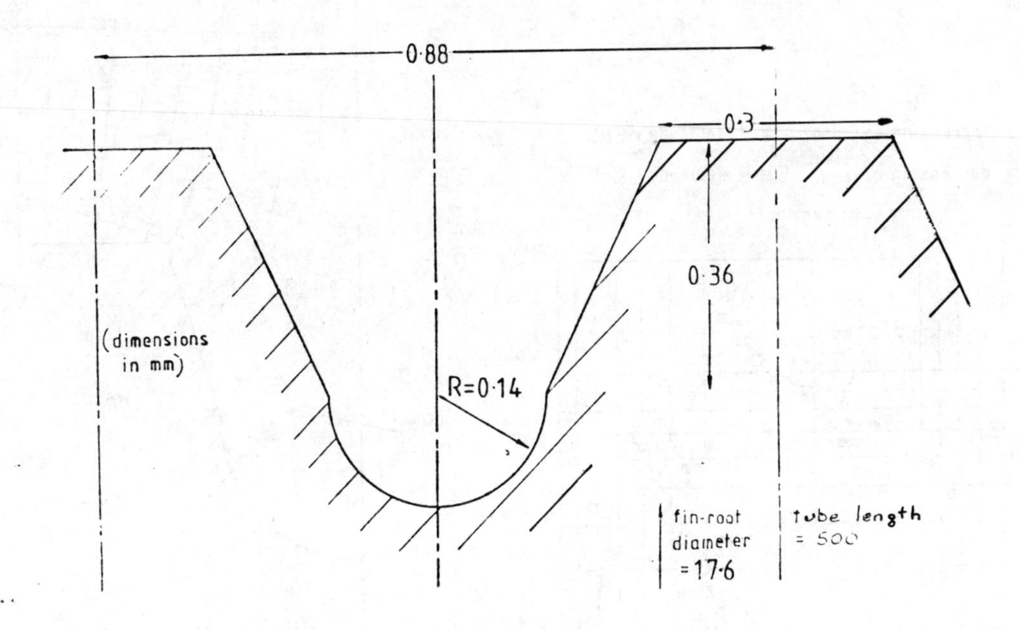

Figure 3 Cross-section of lo-fin tube profile

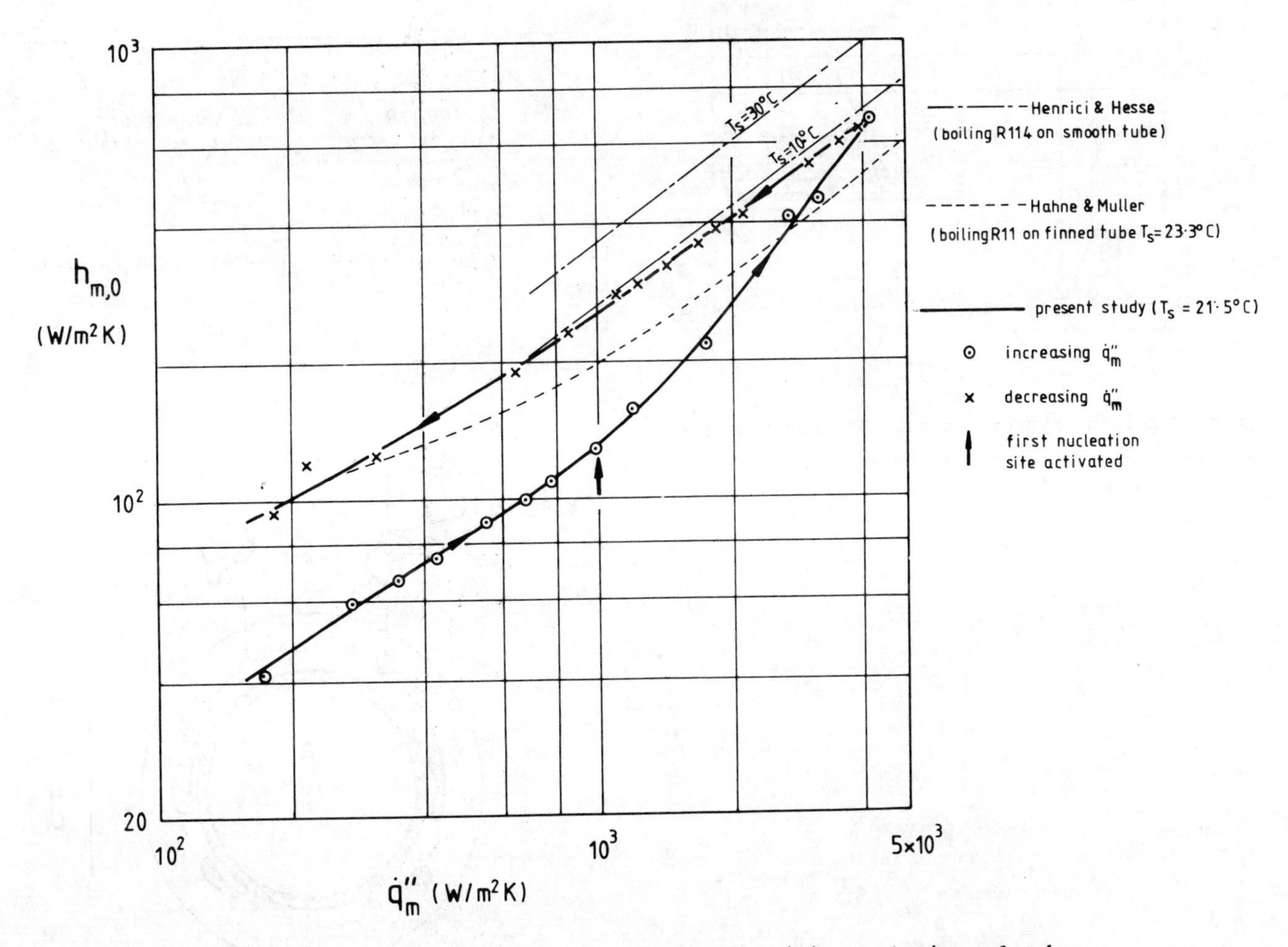

Figure 4 Zero-field heat transfer of Freons boiling on horizontal tubes

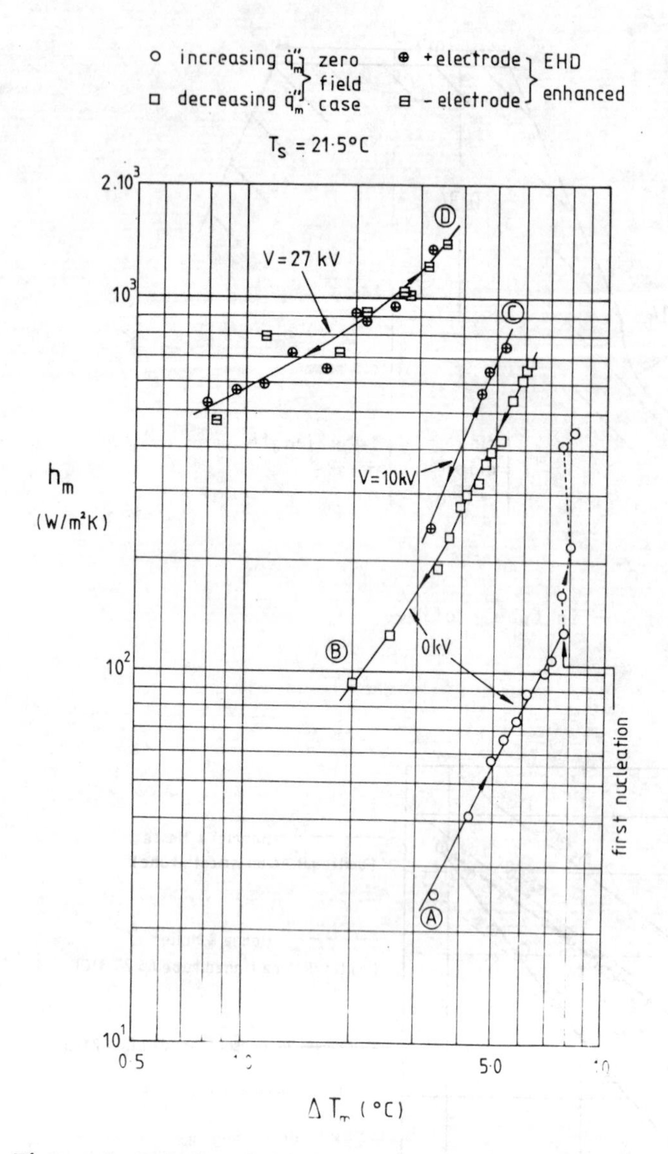

Figure 5 Boiling hysteresis and EHD heat transfer enhancement on horizontal lo-fin with R114 and cylindrical electrode

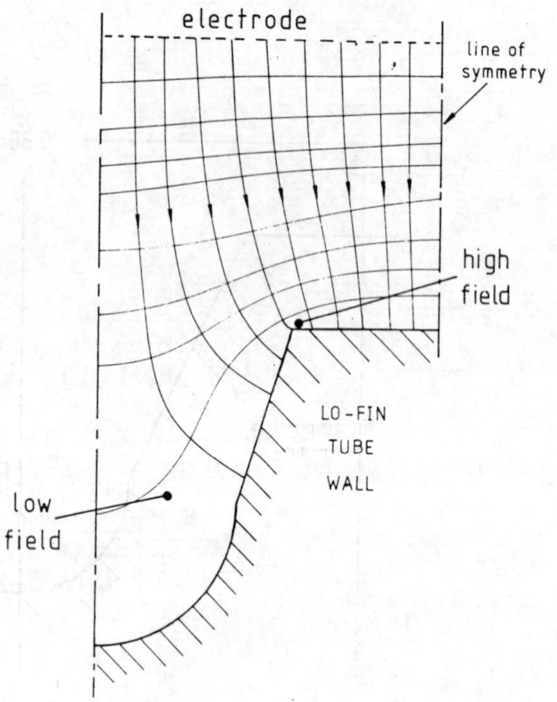

Figure 6 Electric field on a lo-fin tube

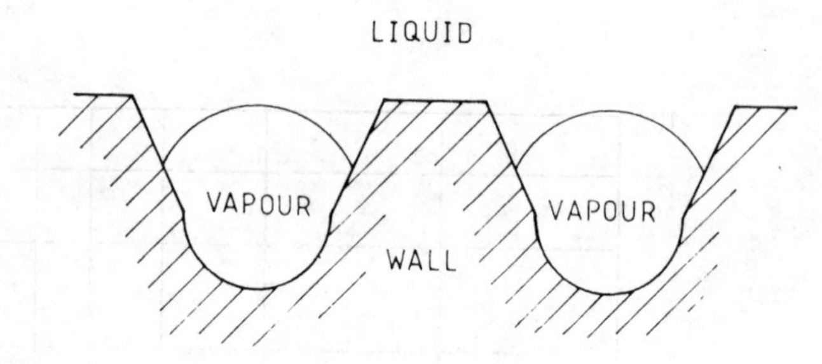

Figure 7 Vapour trapped in lo-fin tube interfin spaces

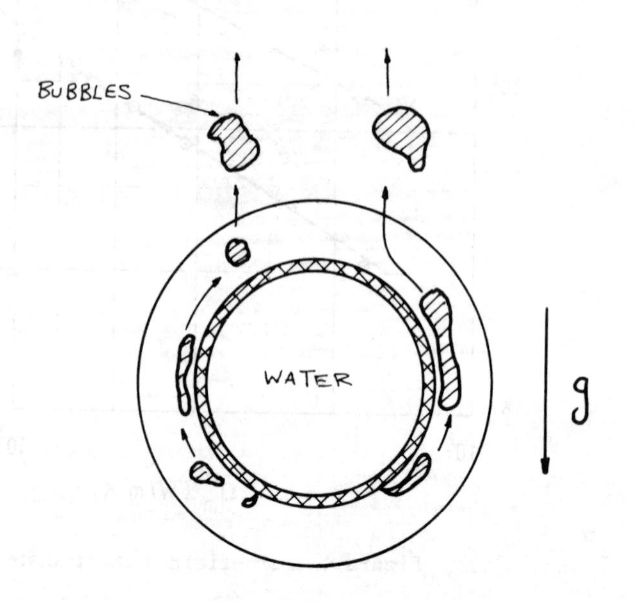

Figure 8 Bubble flow paths

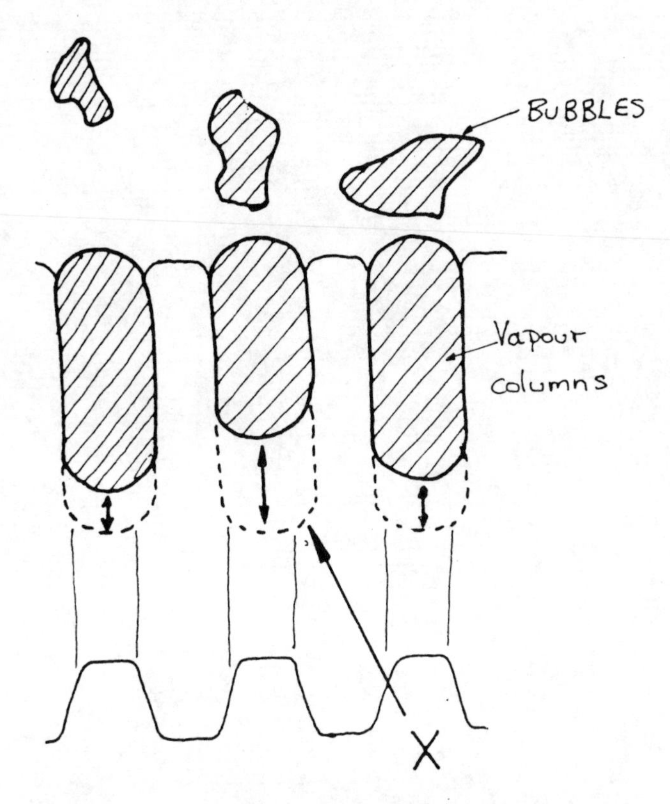

Figure 9 Vapour column oscillation

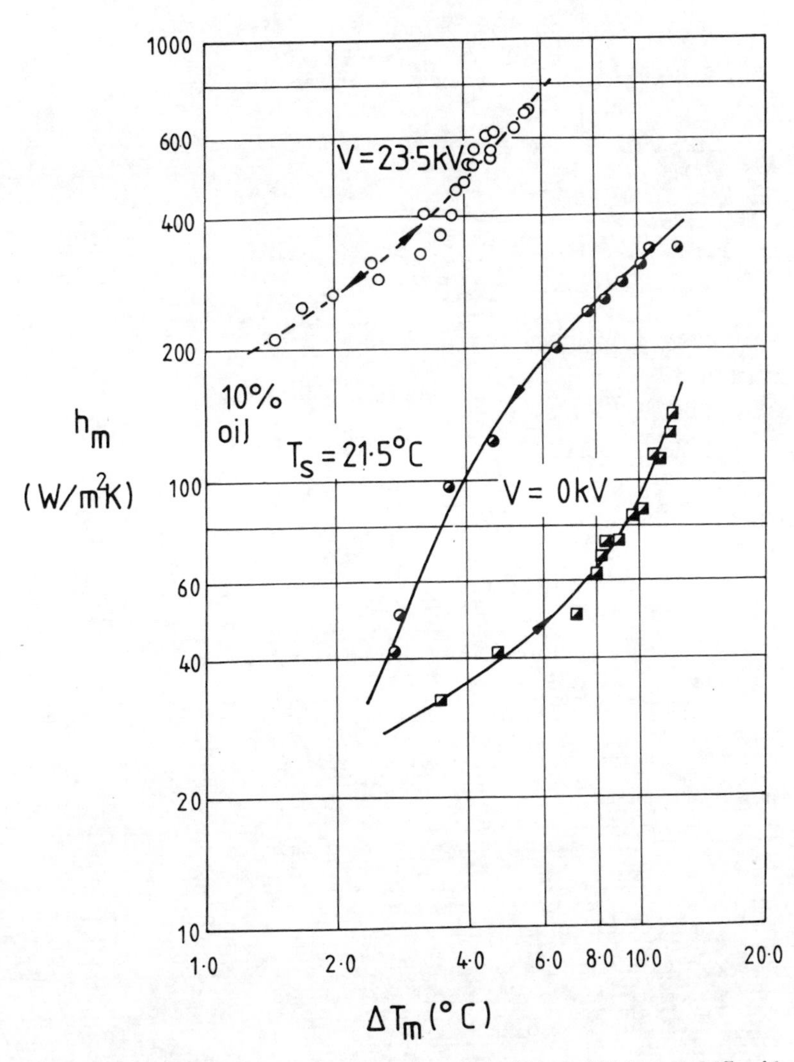

Figure 10 EHD heat transfer enhancement of boiling R114 and 10% oil (w.w.) mixture on horizontal lo-fin tube with cylinder electrode

3rd International Symposium on the

Large Scale Applications of Heat Pumps

Oxford, England : 25-27 March 1987

PAPER P3

AN ASSESSMENT OF HEAT PUMP TIMBER DRIERS

P.G.Baines, C.G.Carrington and I.R.Cox-Smith

Department of Physics, University of Otago, Dunedin, New Zealand.

SYNOPSIS

Closed cycle heat pump driers are widely used in the timber industry but their efficiency varies considerably depending on both the system design and the control strategy for the drying operation. This paper presents the results of a system study of a range of drier configurations in order to assess their inherent limitations. The sensitivity of drier performance to kiln conditions and to design parameters is examined and the principal loss mechanisms for different configurations are investigated. Some results from second law analysis are also presented. The validity of the analysis is examined by comparing model calculations with data obtained from an operational drier. The paper focuses on the opportunities for improving the operating efficiency of dehumidifier driers by design changes.

INTRODUCTION

The use of heat pump systems for timber drying is growing in many countries. However the practical performance of dehumidifier systems can be disappointingly low. In one particular drier, with a charge of 50 mm radiata pine dried from 70% to 20% moisture content, Miller[1] obtained a value of 1.15 kg/kWh for the specific moisture extraction rate. On the other hand, in a conventional heat-and-vent kiln, values equivalent to 0.71 - 0.99 kg/kWh were obtained by Miller, although these latter driers were not in fact electrically heated. Evidently the market for heat pump driers is driven to some extent by considerations other than the efficiency, such as the convenience aspect and the acceptability of the dried timber product.

Nevertheless the efficiency of heat pump driers remains an important issue. The purpose of this paper is to present an analysis of the loss mechanisms and limitations inherent in different drier configurations. Investigations of this type have been made already, notably by Geeraert[2] and by Zylla, Abbas, Tai, Devotta, Watson and Holland[3]. These studies are extended in this paper by the inclusion of different drier configurations and by the use of more complete modelling procedures and improved moist air property routines. The simulation allows a number of realistic details of the kiln and heat pump to be included and results from this aspect of the work have been compared with data from an operating commercial drier.

Our primary measure of efficiency for heat pump driers is the specific moisture extraction rate (SMER), expressed in units of kg of water extracted from the product per kWh of electrical energy supplied to the drier, including auxiliaries such as fans. However it should be remembered that heat pump driers are essentially "second law" devices. Thus, for every process within the system boundary, unnecessary irreversibilities contribute to the inefficiency of the drier. A quantitative investigation of these thermodynamic losses helps to give additional understanding of their causes because it provides a common standard against which they can be assessed. This analysis procedure acquires more significance as design changes improve the efficiency of driers because the remaining irreversibilities then assume greater relative importance.

PROCEDURES

This work is confined to chamber type timber kilns operating with product temperatures typically in the range 30-70°C, using "Freon" working fluids in closed Rankine cycle heat pump systems. For the simulation calculations presented here R12 has been used, but the general conclusions are not limited by this choice. The simulation work is, however, restricted to steady state operation.

We have adopted a "lumped component" approach to modelling, a minimal set of critical parameters only being specified for each component. This level of analysis is suitable for evaluating system choices. The compressor is characterised by the volumetric and isentropic efficiencies and the swept volume. Air-refrigerant heat exchangers are specified by the temperature difference between the saturated refrigerant and the leaving air stream while other heat exchangers are specified by their thermal effectiveness (Kays & London[4]). Refrigerant properties have been calculated using the routines of Riemer, Jacobs and Boehm[5] which are in good agreement with ASHRAE[8] compilations. For the thermodynamic properties of moist air, ASHRAE routines compiled by Wexler, Hyland and Stewart[6] have been used. These employ a third order virial equation of state for mixtures and incorporate other real-gas features. Additional moist air routines have been written as required using the ASHRAE programs as a base.

Moisture up-take in the product region is assumed to occur adiabatically. In order that the model should represent steady state operation in a consistent way, moisture evaporated from the product is balanced by providing an in-flow of liquid water to the product at the ambient surrounding temperature. Thus the energy required to preheat the moisture in the product to the kiln operating temperature is taken into account, but the energy required to preheat the dry product matter is not included.

For the purposes of second law analysis, as well as for the simulation of the drier, it is necessary to specify a definite state for the air of the surrounding environment. A representative ambient temperature of 10°C has been adopted and the relative humidity has been set at 100%. This latter condition is required for second law analysis because the model of the environment must represent a state of equilibrium between liquid phase water and the atmosphere.

Primary fixed parameters in the model calculations include the air dry-mass flow-rates for the evaporator and condenser, the compressor swept volume rate (as well as other compressor characteristics noted already) and the state of the air leaving the product region. To establish the steady state conditions for a given configuration an iterative procedure has been used, based on the requirements of mass and energy balance for the control volume of each specified component. The rate of ventilation to the atmosphere is determined by the need for energy balance. Generally ten or more iterative stages are required to consistently satisfy the fixed input data as well as the energy and mass balance criteria which define an acceptable solution.

BASIC DEHUMIDIFIER

Simple dehumidifier systems have been analysed

Held at St. Catherine's College Oxford, England. Organised and sponsored by BHRA, The Fluid Engineering Centre, Cranfield, Bedfordshire, MK43 0AJ England.

previously by Geeraert[2] and by Zylla *et al.*[3]. Here we present our analysis of this configuration for comparison with measured performance data and also to place other systems in context. Figure 1 illustrates the lay-out of a basic heat pump dehumidifier drier.

Detailed comparisons between a steady state simulation model and operating driers are difficult because the steady state is seldom realized in practice. However a limited comparison may still provide a useful indication of the validity of a model. The parameter most directly affected is the ventilation rate because during the drier heating phase the excess energy which would be released to the atmosphere by venting is reinvested in the product. However the condensate rate should not be substantially affected. In table 1 we give data for the condensate rate measured in an operating commercial timber drier to compare with the rate determined by the simulation. The model in this case included the fan power consumption and conduction losses from the chamber of the kiln. The isentropic and volumetric efficiency of the compressor have been determined as a function of the evaporating and condensing temperatures using the compressor manufacturer's performance data. The table indicates that under the conditions selected here the calculations are reasonably close to the practical performance of the drier. For this particular drier the evaporator air-flow was maintained at a fixed value.

OPTIMIZING OPERATING CONDITIONS

Early dehimidifiers often failed to make provision for independently setting the air-flow rates of the evaporator and condenser. However the importance of providing for this was emphasised strongly by Geeraert[2], because the criteria for optimum air-flow are different for the evaporator and condenser. We consider first the condenser and ignore the possible effect of changing the air-flow on the product itself. Quite simply, the principal effect of increasing the condenser air-flow is to reduce the evaporator-condenser temperature difference. Consequently, within limits determined by the compressor characteristics, the air-flow increase tends to increase the coefficient of performance (COP) of the refrigeration sub-system. This would improve the SMER of the drier except that higher condenser air-flows will, in general, increase the fan power input. In this situation the most appropriate criterion for selecting an optimum air-flow is an economic one. The selection of air-flow should take into account how changes in the SMER affect the operating cost of the drier in relation to the cost of the fan and the cost of alternative condenser configurations. These may also influence the SMER through the condensing temperature. This problem should be investigated within the context of a more detailed plant specification, as has been done, for instance, in the case of heat pumps for space heating[7], and it will not be pursued here further. We note however that relatively high condenser air-flow rates are practical in heat pump driers so that the temperature change of the air stream at the condenser may be quite small, less than 10°C for example. Thus it is common for dehumidifier timber driers to operate close to the isothermal limit. It is because timber products are generally well suited to this type of drying process that conventional Rankine cycle heat pump driers can be used here succesfully.

The choice of evaporator air-flow presents an optimisation problem different from that for the condenser because the evaporator fan power demand is usually not significant. While the refrigeration system COP generally increases as the evaporator air-flow rate is increased, it is only in a narrow range of flow rates that this has a positive impact on the SMER. Thus the conditions for optimising the SMER of the drier do not necessarily coincide with those for optimising the COP of the refrigeration sub-system. This is a consequence of the evaporator latent/total cooling ratio (LTCR) decreasing as the evaporator air-flow rate is increased. Figure 2a illustrates how the COP and LTCR change counteractively as the evaporator air-flow is varied. The effect of these changes on the SMER is shown in figure 2b. The fixed conditions relevant to these figures are listed in table 2. Here the compressor isentropic efficiency has been fixed in order to separate the influence of evaporator behaviour from compressor characteristics. A relatively low value has been used for the isentropic efficiency (50%) but no allowance has been made for fan power or kiln conduction losses.

Figure 2b shows that under many kiln conditions the SMER is very sensitive to the evaporator air-flow. The consequence of ignoring this at the design stage is that the efficiency of the drier may be no better than a vented drier, electrically heated. This condition occurs at low values of the humidity and higher values of the evaporator air-flow rate, where no moisture condenses on the evaporator surface. To further show the influence of changing the air-flow, table 3 lists the distribution of exergy losses within the drier for one operating condition at different values of the evaporator air-flow rate. At the optimum air-flow the relative exergy loss in the product is a maximum because of all the losses in a convective drier this is the least avoidable.

The optimum value for the evaporator air-flow rate decreases with increasing product temperature, but increases with increasing relative humidity. Figure 3 shows how the value of the SMER varies as a function of product exit air temperature and humidity, assuming that the evaporator air-flow is adjusted at all times to the optimum rate. Evaporator and condenser losses are principally those associated with heat transfer across a temperature interval, which can be reduced by sizing the heat exchangers to reduce the approach tempertaure differences, Δ_e and Δ_c for the evaporator and condenser respectively. Such changes would also reduce the losses in the compressor as well as the throttling losses at the expansion valve. Figure 4 illustrates how the SMER is influenced by changing Δ_c for three values of the product exit air relative humidity. Changes in Δ_e give rise to essentially the same variations in the SMER. These temperature gradients are clearly critical determinants of the system efficiency. Again fixed parameters in this case are listed in table 1.

USE OF AN ECONOMISER

The sensitivity of the dehumidifier performance to evaporator air-flow is primarily a consequence of the changing latent/total cooling ratio of the evaporator. A number of options are available to reduce this effect, for instance by the use of multiple heat pump stages. Alternatively the use of an economiser, a recuperative heat exchanger around the evaporator as shown in figure 5, to carry some of the sensible heat load also has the potential to improve the LTCR significantly. Results for this situation are given in figure 6, which shows the dependence of the SMER on evaporator air-flow rate. This figure illustrates how the selection of the optimum air-flow is much less critical than in figure 2b, the maximum for the SMER occurring at a larger value of the evaporator air-flow rate. The effect of this change on performance is greatest at low values for the relative humidity. The thermal effectiveness of the economiser used to generate figure 6 is 75%; other fixed parameters are given in table 2. We note here that a further improvement in the evaporator efficiency may be obtained by the use of a liquid refrigerant sub-cooler placed between the mixing region and the condenser.

COMPRESSOR LIMITATIONS

The results above for optimum evaporator air-flow are not directly applicable in practice partly because of limitations to the range of conditions for compressor operation. Consider, for example, low/medium temperature timber drying for which R12 is a common choice for refrigerant. Some heat pump compressors, particularly those which are adaptations of air-conditioning compressors, are restricted for use at saturated suction temperatures not greater than 25°C with R12. The isentropic efficiency of these compressors tends to deteriorate considerably at compression ratios much less than 2 or 3. Industrial quality compressors, on the other hand, tend to have higher efficiencies[9], and maintain them over a wider range of pressure ratios. Some are suitable for operation with R12 to saturated suction temperatures up to 35°C.

These evaporating temperature limitations have a direct effect on the maximum practical value for the evaporator air-flow rate. This is illustrated in figure 7 which shows the SMER as a function of evaporator air-flow for a dehumidifier with an evaporator economiser and sub-cooler. The simulation in this case takes account of the fan consumption and of conduction losses from the drier. A fixed value of 65% has been used for the compressor isentropic efficiency, the air temperature is 50°C and the relative humidity is 70%. Other fixed parameters are given in table 4. Lines of constant evaporation temperature have also been included in figure 7. These show that even at a moderate kiln temperature, 50°C, the effect of restricting the evaporator temperature to 25°C by limiting the air flow impairs the SMER significantly at high values of the relative humidity.

To illustrate the impact of this particular restriction in more detail figure 8 shows the SMER at the optimum evaporator air-flow as a function of kiln exit condition, the model here being based on the manufacturer's performance data for a light industrial quality heat pump compressor. In accordance with the application conditions for this compressor the evaporating temperature has been kept below 25°C by restricting the evaporator air flow as necessary. The impairment of the drier performance at higher kiln temperatures is evident. (Some dehumidifiers achieve this result by throttling the suction vapour.) However some improvement in this situation occurs when a good industrial quality compressor is used[9], one suitable for operation with R12 to a saturated suction temperature up to 35°C. This is shown in figure 9. We note that some of the difference between the results for these two compressors is attributable to the difference in the compressor efficiency. The mean isentropic efficiency of this compressor is around 80%, compared with 60% for the compressor which produced the results for figure 8. Table 5, which compares these two compressors further using exergy analysis, shows that the reduction in compressor losses is considerable. This table indicates that fan power is the major single contributor to the irreversibility in this drier.

DISCUSSION AND CONCLUSIONS

The model calculations have illustrated a number of significant factors which influence the specific moisture extraction rate of dehumidifier driers. These are relevant both in connection with the design of the drier and in the choice of a control strategy for the drier. Here we comment on some specific issues.

It is well known, as we have confirmed, that it is advantageous from an efficiency point of view for the drying process to be conducted at a high relative humidity. This need for a high humidity also implies that oversizing the heat pump for the amount of product to be dried should be avoided. On the other hand the penalty for undersizing is that drying cycles may become prolonged, although the average rate of product throughput may not actually suffer. With regard to control, the first stage of timber drying is often scheduled to have a high relative humidity so this requirement is consistent with normal practices. However for chamber type timber driers the control strategy must recognise that the kiln temperature and relative humidity usually cannot both be set independently early in the drying cycle. Consequently the authority of the parameter to be used to exercise control should be clearly defined, although the kiln condition at the conclusion of the drying cycle may target both temperature and humidity. As a practical matter the reliability of the control system is most important. The cost of installing back-up protection against failure will almost certainly be recovered many times over on the first occasion that the controler malfunctions.

With regard to drier design the importance of air-refrigerant temperature differences in determining the efficiency of the drier has been illustrated. In addition it has been shown how the problem of optimizing the evaporator air-flow rate can be managed by the use of a recuperative heat exchanger in association with the evaporator. An air-refrigerant sub-cooler also helps in this context; both this and the economiser yield improvements in the drier efficiency. However to have a significant impact, the effectiveness of these added heat exchangers should be high, 75% or more.

With increasing kiln temperature the SMER of the driers simulated tends to decrease once the saturated suction temperature limit of the compressor is reached. Thus at high humidities the suction temperature limitation, rather than that on the condensing temperature, may define the useful region of operation of the dehumidifier timber drier in practice. This also has implications for the control of the drier, for unless there are operational reasons for using higher temperatures, the kiln should be vented to prevent the saturated suction temperature from exceeding the compressor limit. If then the kiln humidity becomes too low as a result of venting, this indicates that the product load is too small. In this situation the vent should be closed and the compressor cycled on/off to maintain the required kiln condition. To attempt to control the drier condition under these circumstances using steam or water sprays while leaving the drier operating is clearly inappropriate.

Compared with lighter grades of compressor, the use of a good quality industrial compressor with a higher limiting saturated suction temperature both moves the maximum SMER point to higher kiln temperatures and also gives significant efficiency improvements. In this situation the fans may become an important source of irreversible losses in the drier and fan requirements should therefore be examined critically. However because of the need for air circulation in the timber stack the opportunity to reduce fan consumption is restricted.

Finally, we note that as driers become more efficient, the unavoidable loss due to evaporation irreversibilities in the product become a higher fraction of the drier consumption. In table 5 these losses are in the range 8-14%. This gives an indication of the opportunities remaining to improve the efficiency of such driers.

ACKNOWLEDGEMENTS

The authors gratefully acknowledge the following for financial support: N.Z. Energy Research and Development Committee (contract 3316), N.Z. Forest Service (contract 192), N.Z. Ministry of Energy, N.Z. University Grants Committee (grant 85/180), and the University of Otago Research Committee.

REFERENCES

[1] Miller, W.R., 1983, "Application of heat pumps to timber drying", University of Auckland, NZERDC Publication **P78**, ISSN 0110 5388, 29-40.

[2] Geeraert, B., 1976, "Air drying by heat pumps with special reference to timber drying", *Heat pumps and their contribution to energy conservation*, Camatini, E. and Kester, T. (eds.), NATO advanced study institute series, series E, Applied Sciences, No. 15, Leydon: Noordhoff, 219-246.

[3] Zylla, R., Abbas, S.P., Tai, K.W., Devotta, S., Watson, F.A. and Holland, F.A., 1982, "The potential for heat pumps in drying and dehumidification systems", *International Journal of Energy Research*, **6**, 305-340.

[4] Kays, W.M. and London, A.L., 1964, *Compact heat exchangers* (2nd ed.), New York: McGraw - Hill.

[5] Riemer, D.H., Jacobs, H.R. and Boehm, R.F., 1977, *A computer program for determining the thermodynamic properties of freon refrigerants*, NTIS No. UTEC-ME-76-212, University of Utah, Dept. of Mechanical Engineering, Salt Lake City, Utah, U.S.A.

[6] Wexler, A., Hyland, R.W. and Stewart, R.B., 1984, *Thermodynamic properties of dry air, moist air and water and S.I. psychrometric charts*, Reports from ASHRAE Research Projects 216-RP and 257-RP, American Society of Heating, Refrigerating and Air-Conditioning Engineers, Atlanta, GA.

[7] Carrington, C.G., 1978, "Optimizing a heat pump for heating purposes", *International Journal of Energy Research*, **2**, 153-170

[8] ASHRAE Handbook - 1985 Fundamentals Volume, American Society of Heating, Refrigerating and Air-Conditioning Engineers, Atlanta,GA.

[9] Villadsen, V. and Boldvig, F.V., 1980, "A balanced view of reciprocating and screw compressor efficiencies", *Proceedings of the 1980 Purdue compressor technology conference*, Purdue Research Foundation, 317-322.

	Condensate (kg/hr)	Temp. air exit product	R.H. air exit product	Moisture vent rate (kg/hr)	Compressor power (kW)	Temp. air into product	R.H. air into product
M	105	43.5	98		22.8	49.6	69
S	108	43.5	98	36	25.5	48.3	74
M	112	50.0	93		26.3	57.1	64
S	130	50.0	93	38	29.3	55.4	70
M	123	48.2	97		24.8	54.9	68
S	128	48.2	97	38	28.4	53.6	72
M	126	52.1	86		25.8	58.3	64
S	128	52.1	86	39	30.2	57.5	65

Table 1. Comparison of measured(M) and simulated(S) data for an operating dehumidifier timber drier.

Compressor swept volume	100 m^3/hr
Compressor volumetric efficiency	100%
Compressor isentropic efficiency..	50%
Condenser dry-air mass flow-rate	26 kg/s
Evaporator-leaving-air temp. diff.	5°C
Condenser-leaving-air temp. diff.	10°C
Ambient air temperature.........	10°C
Ambient air relative humidity...	100%

Table 2. Fixed data relevant to Figs. 2a,2b,3,4,6 and table 3.

Evap.flow-rate (kg/s)	0.79	3.4	26
Compressor..............	42.2	42.6	43.5
Condenser...............	10.4	15.7	22.2
Expansion valve......	11.8	10.3	8.8
Evaporator..............	12.0	9.3	13.0
Exhaust air to surroundings..	5.7	5.2	6.0
Product processes.......	7.6	9.4	1.8
Vent mixing............	4.4	4.0	4.6
Pre-condenser mixing	5.8	2.8	0.0
Condensate to surroundings..	0.1	0.7	0.0

Table 3. Exergy loss distribution (%) in a simple dehumidifier. Temp. 50°C, R.H. 70%.

Compressor swept volume	100 m^3/hr
Condenser dry-air mass flow-rate	16 kg/s
Evaporator-leaving-air temp. diff.	5°C
Condenser-leaving-air temp. diff.	10°C
Economiser effectiveness..............	75%
Sub-coolereffectiveness.................	75%
Ambient air temperature..........	10°C
Ambient air relative humidity....	100%

Table 4. Fixed data relevant to Figs 7,8,9.

Reference:	fig 8	fig 9
Compressor.....................	22.0	7.5
Condenser........................	7.6	13.1
Expansion valve...............	4.9	4.3
Evaporator.......................	8.0	7.8
Sub-cooler.........................	1.8	3.9
Condensed water to surroundings	0.2	0.9
Vent air to surroundings..........	4.8	3.6
Vent mixing.........................	2.8	2.2
Pre-condenser mixing..........	3.4	1.1
Condenser fans..................	23.5	26.2
Evaporator economiser..........	0.9	1.3
Kiln conduction heat losses....	3.2	4.0
Evaporator fan.................	0.7	0.8
Product fans....................	8.5	9.4
Product evaporation processes..	7.8	13.9
Optimum evap. air-flow (kg/s)...	1.0	5.6
Compressor isentropic efficiency.	58.3	84.2
Compressor volumetric effeciency	78.4	92.0

Table 5. Exergy loss distribution in two driers. Product leaving air condition: 50°C, 70% R.H.

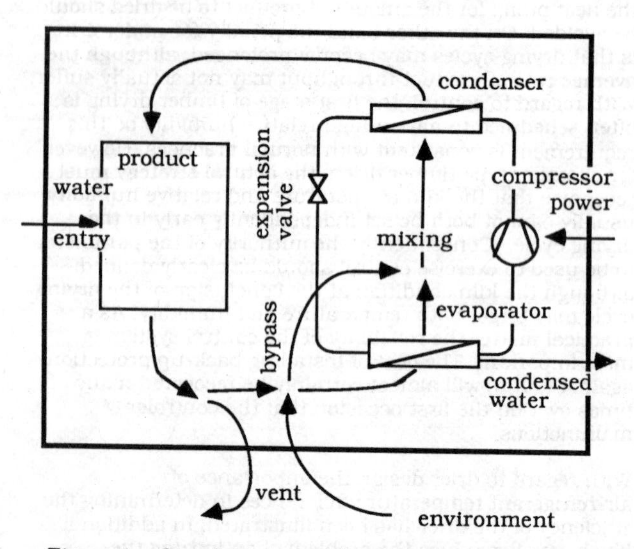

Figure 1 Simple Dehumidifier lay-out.

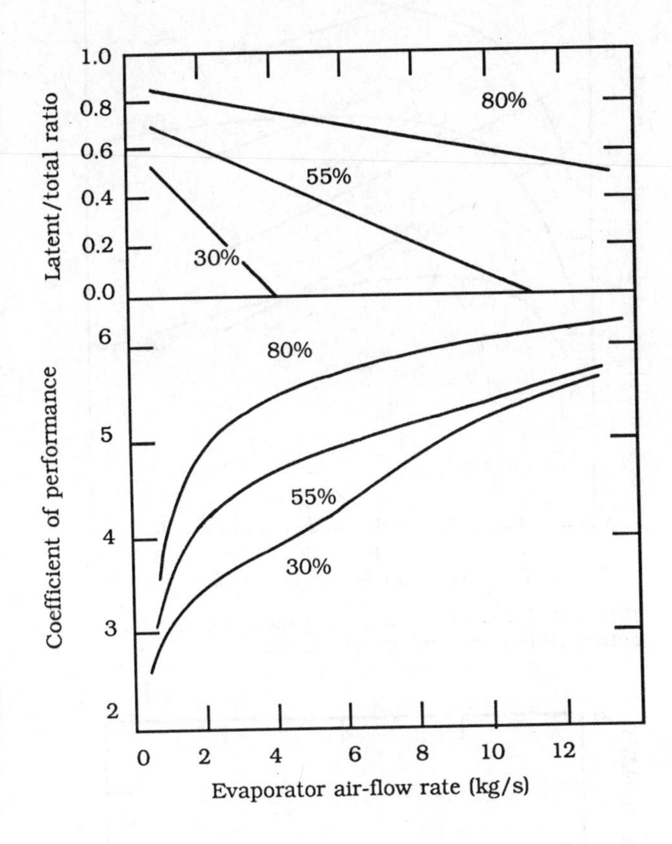

Figure 2a. Effect of evaporator air-flow on latent/total cooling ratio and COP. T=50°C. Relative humidity is indicated.

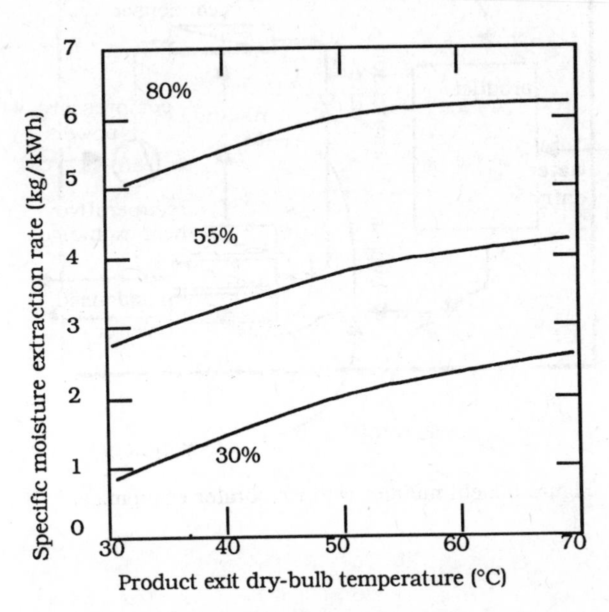

Figure 3. Optimum SMER dependence on temperature and relative humidity of air leaving product. Compressor efficiency fixed. (table 3.)

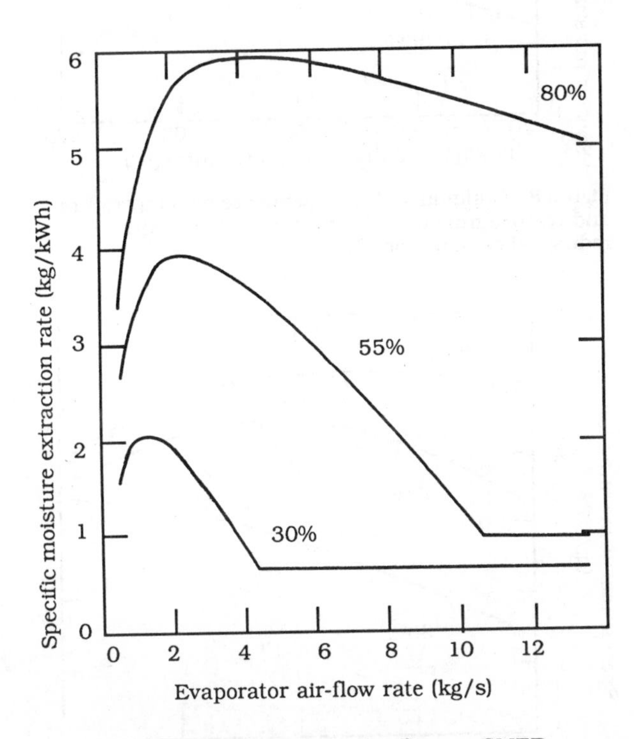

Figure 2b. Effect of evaporator air-flow on SMER.

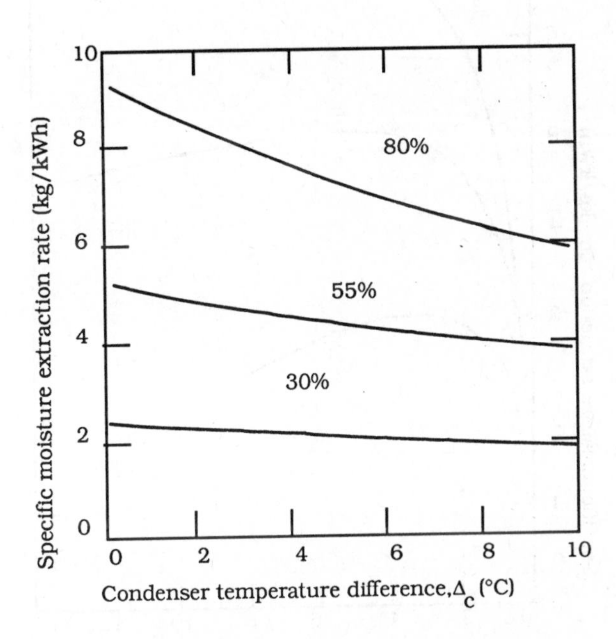

Figure 4. Influence of condensing temperature difference on SMER. Temperature of air leaving product 45°C. Evaporating temperature difference 5°C.

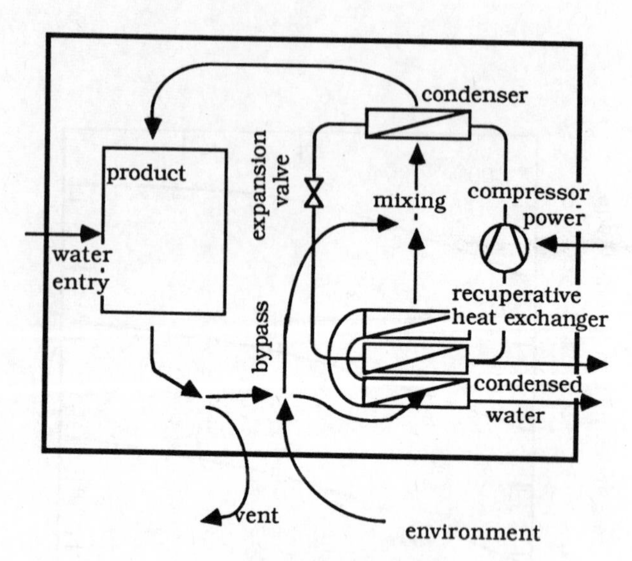

Figure 5 Dehumidifier with evaporator economiser

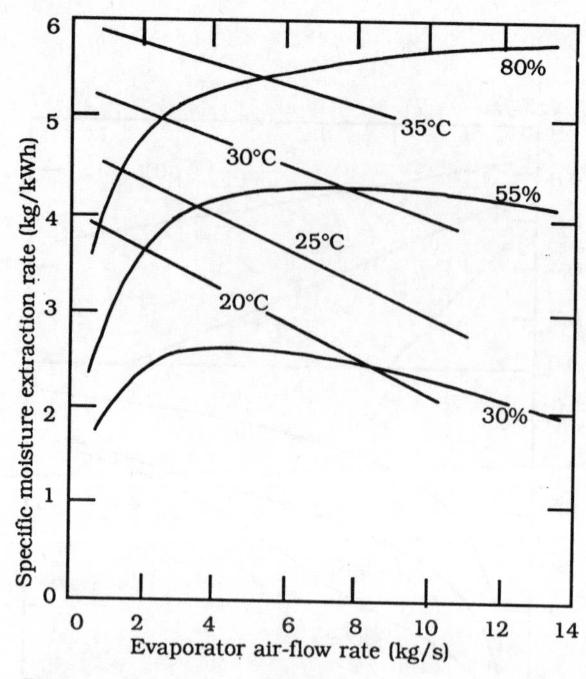

Figure 7. Effect of restricting the maximum suction saturated temperature on the SMER.

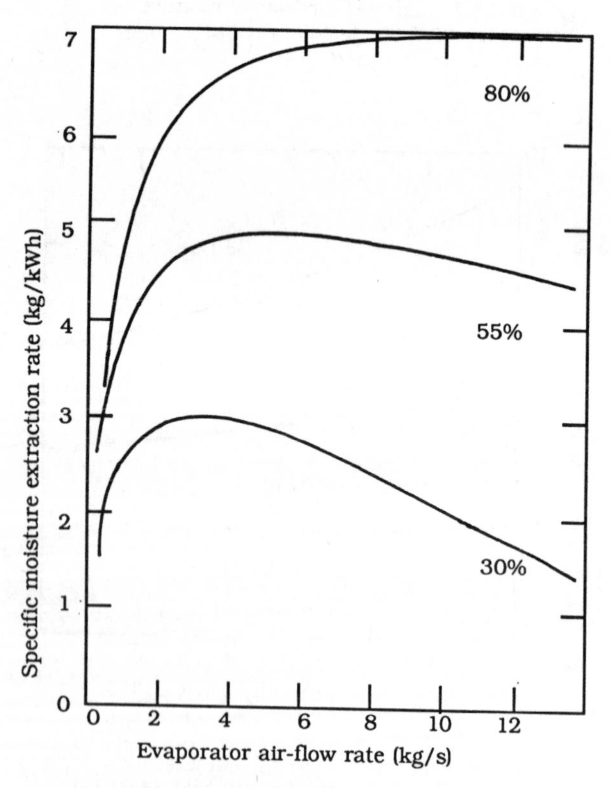

Figure 6. Dependence of SMER on evaporator air-flow. Evaporator with economiser.

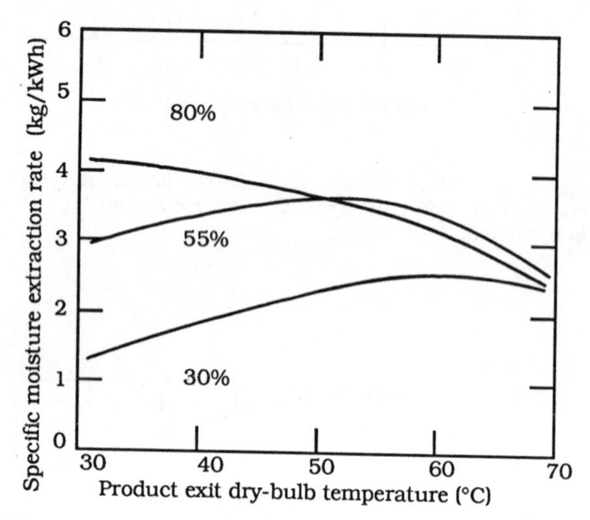

Figure 8. Optimum SMER dependence on temperature and relative humidity of air from product. Light industrial compressor. (Table 4)

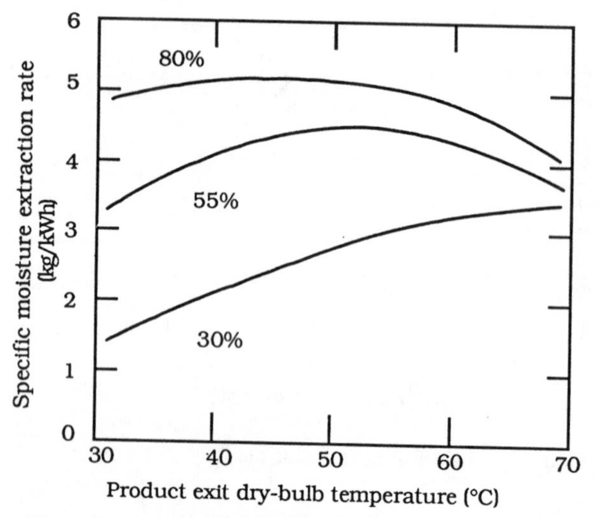

Figure 9. Optimum SMER dependence on temperature and relative humidity of air from product. Industrial quality compressor.(Table 4)

3rd International Symposium on the

Large Scale Applications of Heat Pumps

Oxford, England : 25-27 March 1987

PAPER H3

SUPER HEAT PUMP ENERGY ACCUMULATION SYSTEM JAPAN'S NEW R&D PROJECT ON HIGH PERFORMANCE HEAT STORABLE HEAT PUMP SYSTEM

T.Shimura

The author is a Director of the Energy Conversion & Storage Department of the New Energy Development Organization,Tokyo,Japan

SYNOPSIS

The Agency of Industrial Science and Technology of the Ministry of International Trade and Industry (MITI) of the Government of Japan initiated a new R&D project named "Super Heat Pump Energy Accumulation System" in October 1984. This project will continue for eight years until 1991 and a total budget for this project is estimated to be 10 billion Yen which is US$ 50 million at Yen200/US&.

This project aims to conduct research and development on heat-storable heat pump systems that utilize high performance compressor-driven heat pumps and chemical heat storage systems.

1. PROJECT OBJECTIVE

This project aims to develop a high performance heat pump and chemical heat strage system which will be used for conversion of excess electric power at night.

The R&D work will include the following areas:

a) high performance compressor-driven heat pumps
b) chemical heat storage systems
c) total system configuration

2. PROJECT TARGET

Targets to be achieved in this project are as follows:

2.1 Performance of Total System

Application		Output Temperature	Energy Efficiency
A.C. of Office Buildings	Heating	45°C	4.5
	Cooling	7	5.3
District A.C.	Heating	45	4.5
	Cooling	7	5.3
	Hot Water Supply	85	6
Industrial Processes	Heating from Waste heat	150	>2.3
		300	>2.3

	High Performance Type			High Temperature Type	
	For Heating Only Type	For Heating & Cooling Type		low Temp. Heat Source	High Temp. Heat Source
		Heating	Cooling		
Output Temp.	85°C	45°C	7°C	∿150°C	∿300°C
C.O.P.	8	6	7	3	3

2.2 Compressor-Driven Heat Pumps

Comparison of the targets of compressor-driven heat pumps in this project with the performance of conventional heat pumps is shown below.

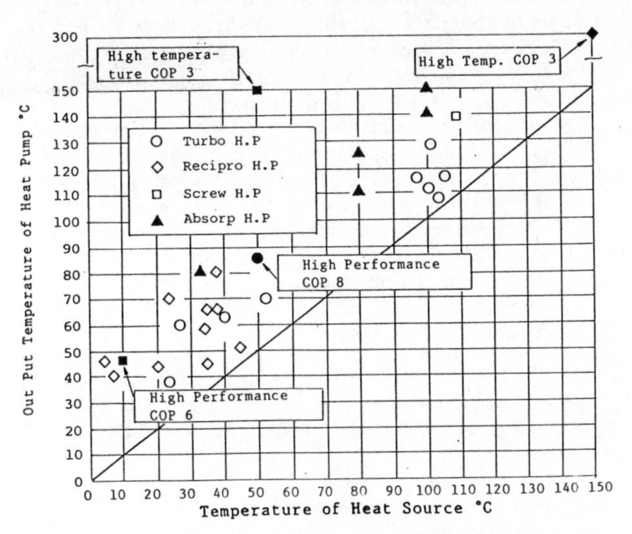

2.3 Chemical Heat Storage Systems

	High Temperature Storage	Low Temperature Storage
Output Temperature	∿200°C	10°C
Capacity of Storage	50 Kcal/kg (material)	30 Kcal/kg (material)
Heat Recovery Ratio	>75%	>75%

3. PROJECT PROGRAM

This project was started in 1984, and will be continued for eight fiscal years until 1991. The total costs related to this project are estimated by the Japanese Government at approximately 10 billion Yen, equivalent to US& 50 million at Yen200/US&.

As shown in Table 1, the project program will be conducted in three separate stages over the eight year time period.

Research from FY1984 to FY1986 has concentrated on compressors, heat exchangers, non-azeotropic refrigerants mixtures, for the compressor-driven heat pumps, and on reaction cycles, appropriate materials and systems for chemical heat storage systems.

During FY1987-1988, bench scales of 100kW thermal output class heat pumps and 10,000kcal class heat storage systems will be tested.

In FY1988-1991, pilot systems will be tested

Held at St. Catherine's College Oxford, England. Organised and sponsored by BHRA, The Fluid Engineering Centre, Cranfield, Bedfordshire, MK43 0AJ England.

combining systems of heat pumps and chemical heat storage systems. These systems will have a 1000kW class thermal output.

4. PROJECT ORGANIZATION

Starting in October 1984, four Japanese national research institutes began research and evaluation of working fluids, materials, etc. for basic research.

Since FY1985, development of the system and research on operation and research on the total system have been conducted by New Energy Development Organization (NEDO).

5. RESEARCH AND DEVELOPMENT OUTLINE

5.1 High Performance Compressor-Driven Heat Pumps

In order to achieve high preformance, essential technologies for the following three araes have been designated for study and development.

a) Adoption of improved thermodynamic cycles, that is, choice of appropriate non-azeotropic refrigerant mixtures, optimum usage of multi-staged compression and of counter-flow heat exchange

b) Use of improved compressor and recoverry of waste heat from driving motors by developing hermetic-sealed compressors and motors

c) Improvement of optimum control of system by inverter controlled motor

5.1.1 COP8 Heat Pump for Heating

High performance heat pump which can achieve COP 8 under the following conditions shall be developed.

i) output temperature 85°C

ii) heat source temperature 50°C

For this purpose, R&D undertakes to develop the following system.

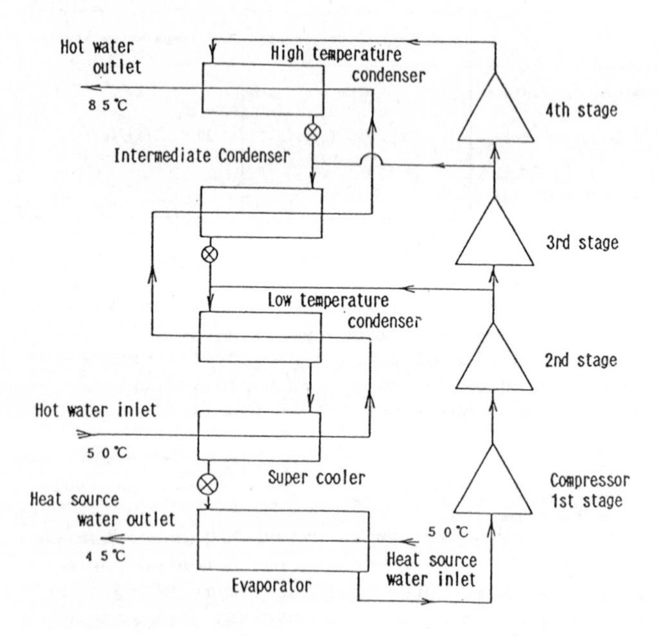

features:

This system consists of improved 4-stage turbo-compressors hermetically sealed with electric motor. We expect to attain minimum difference in temperatures between hot water heated by counter flow in 3-stage condensers and refrigerant in each condender by applying the most suitable mixture of non-azeotropic freons.

5.1.2 COP6 Heat Pump for Heating and COP7 for Cooling

High performance heat pump for heating which can achieve COP 6 and for cooling which can achieve COP 7 under the following conditions shall be developed.

i) heat pump for heating
   output temperature 45°C
   heat source temperature 10°C

ii) heat pump for cooling
   output temperature 7°C
   heat sink temperature 32°C

For this purpose, R&D undertakes to develop system shown in the following diagram.

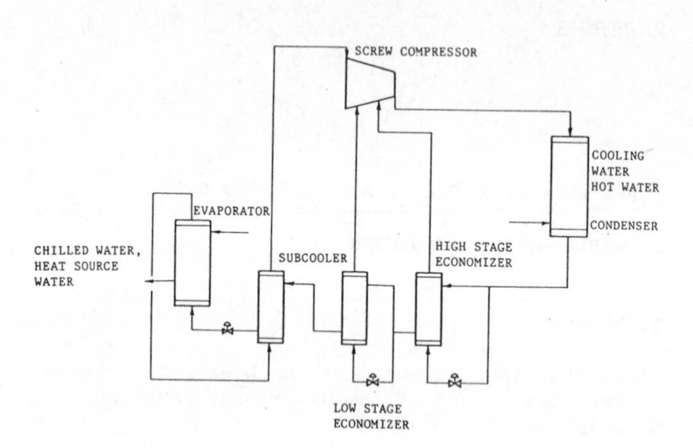

features

High performance will be achieved by employing two economizers for supercooling the refrigerants and by injecting a liquid-vapour mixture of refrigerants into the screw compressor at an appropriate intermediate position of the compressor.

5.2 High Temperature Output Heat Pumps

Two types of COP 3 heat pump systems for industrial application use will be developed. One is a cascaded two-staged screw heat pump producing 150°C from a 50°C heat source, and the other is a steam reciprocating heat pump producing 300°C from a 150°C heat source.

5.2.1 Heat Pump of 150°C Output

In order to attain of 100°C temperature increase to 150°C from 50°C, it is necessary to cascade two screw compressor-driven heat pumps with screw expanders for recovering power as shown in the following figure.

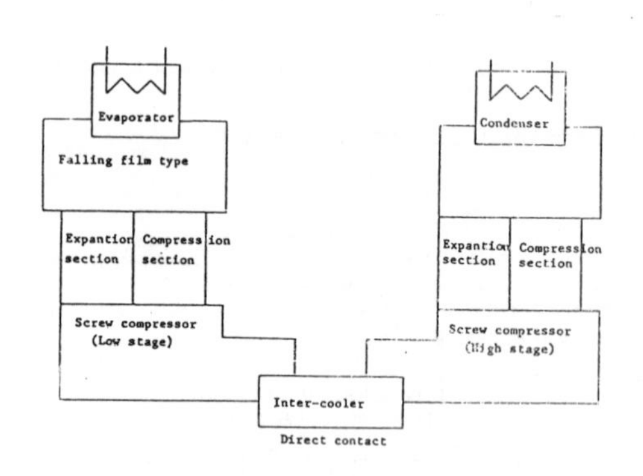

Most important for development of this type heat pump is the development of refrigerants which can be operated higher temperature than 150°C.

5.2.2 Heat Pump of 300°C Output

This system will employ water steam as refrigerant as shown in the following figure.

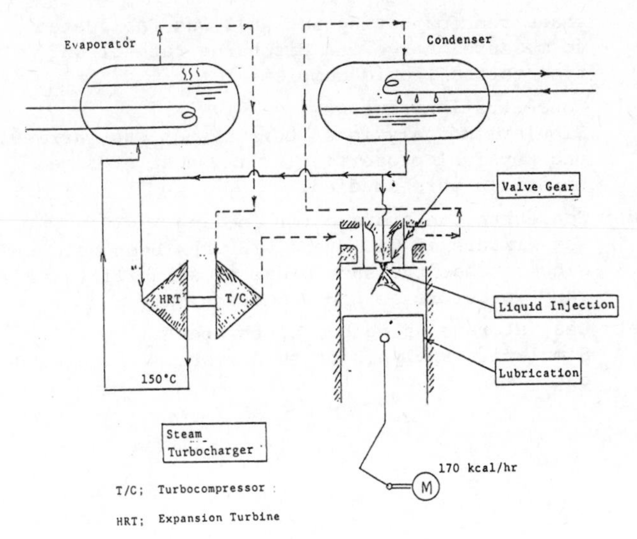

The system is composed of the following:

a) a high pressure condenser
b) a high speed reciprocating steam compressor with hot water injection
c) a low pressure sream turbine

5.3 Chemical Heat Storage System

Reactions for high temperature heat storage system (200°C) are as follows:

a) Hydration-dehydration reaction
Utilization of following reversible hydration reaction:
$CaBr_2 \cdot 2H_2O$(solid)----- $CaBr_2 \cdot H_2O$(solid) + H_2O

b) Absorption-resorption reaction of ammonia
$NiCl_2 \cdot 6NH_3$ (solid)----$NiCl_2 \cdot 2NH_3$ (solid) + $4NH_3$

$NaSCN \cdot 3.5NH_3$ (liq.)---$NaSCN \cdot 2.5NH_3$(liq.) + NH_3

c) Concentration storage
Heat is stored as the difference of concentration potential of the solution of triflouretanol (TFE) in E181 by the following reaction:
TFE·E181 (liq.)-----E181 (liq.) + TFE (vapor)

With regard to low temperature heat storage systems, the following two methods of R&D are under study:

a) Concentration storage
Heat storage media, which are mixtures of a couple of refrigerants, are separated through hydrophobic membranes using the thermal energy generated by a compressor-driven heat pump. The thermal energy is released by the hydration reaction when these media are diluted.

b) Heat storage using gas clathrates
A refrigerant of flourmethane, such as R11 or R12, is injected into water and small cristal of flourmethane and water, called gas clathrate, is crystalized at 3-10°C. Gas clathrate has large latent heat of crystalization equivalent to that of water. Gas clathrate can be crystalized by direct contact heat exchange between refrigerant and water.

6. PROGRESS

6.1 High Performance COP 8 Heat Pump for Heating

a) Research on refrigerant
We investigated the formula to estimate the performance of non-azeotropic refrigerant mixture.
A program was prepared to approximately the performance. Using this program, we computed to approximately the performance concerning a variety of non-azeotropic refrigerant mixture containing R-11 and R-113 as main components.

b) Research on counter flow heat exchanger
Test heat exchangers(for evapolation and condensation) to conduct basic experiments on heat transfer were designed and fabricated to investigate the details of non-azeotropic refrigerant mixture evaporation and condensation process. As the result of the testing, the condensing heat transfer coefficient scarcely showed a difference between non-azeotropic refrigerant mixture and single refrigerant over a wide range of refrigerant flow rates. Additionaly, regarding the evaporation heat transfer coefficient, the single refrigerant showed a higher rate than the non-azeotropic refrigerant mixture at a low refrigerant flow rate. At an increased flow rate, however, little difference was seen between the two cases.

6.2 High performance heat pump of COP 6 for heating and COP7 for cooling

a) Research on non-azeotropic refrigerant
Various non-azeotropic refrigerants of freon type were calculated for physical properties at different points of a single stage, using an amended BWR equation.
There have now been selected R-12/R-114, R-22/R-114 etc., as a mixed refrigerant which is found most appropriate for a good operation of this system.

b) Research on a high efficiency screw compressor
Effect of seal clearance and blowhole on an available compressor efficiency as found in case where liquid injection has not been applied was calculated, on the assumption of a constant displacement volume. Both efficiencies, one volumetric and the other adiabatic, tend to drop as the clearance between both rotors becomes larger, and only an adiabatic efficiency shows a drop depending on an increase in the biowhole, the former then being kept unchanged.

4.3 Heat pump of output 150°C

a) Research on high temperature refrigerant
Thermal stability of triflouretanol was carried out with oil-TFE mixed condition by means of a sealed tube test. During our testing for various lubricating oil, the hydrocarbon family synthetic oil was the most desirable.

b) Research on new screw compressor
We have produced two kinds of rotors in order to set off the axial direction load and to minimize shaftload as well as minimizing leakage of gas through the rotors in mesh, one for double shaft end discharge type rotors and the other for central discharge type rotors.
With double shaft end discharge type rotors, the shaft power was aparently decreased for its thrust force offset structure, but gas leakage from the rotor end was considerably increased under the operating condition with high pressure difference. With the central discharge type rotor with convential rotor profile, considerable defect could not be remove due to the structure for producing of gas confinement chamber with no connection with the discharge port. But, there is possibility to overcome the aforesaid difficulty by remodeling rotor profile.

4.4 Heat pump of output 300°C

Research on liquid injection system
In order to realize a very short time evaporation and to make clear mechanism of evaporation of droplet we have made element test.

4.5 Chemical heat storage system

a) Hydration-dehydration reaction
Studies for form of favorite hydrates and measurement of their vapour pressure are carried out. Screening of material for equipment are carried out and experimental apparatus for corrosion test is manufactured and now materials were testing.

b) Absorption-resorption reaction of ammonia
The properties of $NiCl_2 \cdot 6NH_3$, $NaSCN \cdot nNH_3$ and other complexes are investigated constantly, such as NH_3 vapor pressure, heat capacity, thermal conductivity and so on. Heat and mass transfer and also the operating condition were discussed for each component equipment. Liquid phase reactor having wet wall was fablicated to measure charge and discharge rate of NH_3 from/to the liquid complex.

c) Concentration storage
Simulation analysis of this system was carried, and physical properties of the heat storage materials were studied.

d) Concentration storage for cooling
The mixture of $CaCl_2$ and LiBr has been selected since it has the same order of solubility and vapor pressure as LiBr solution only.

e) Heat storage using gas clathrates
Simulation analysis of this system was carried-out.

Table 1 Super Heat Pump Energy Accumulation System R & D Schedule

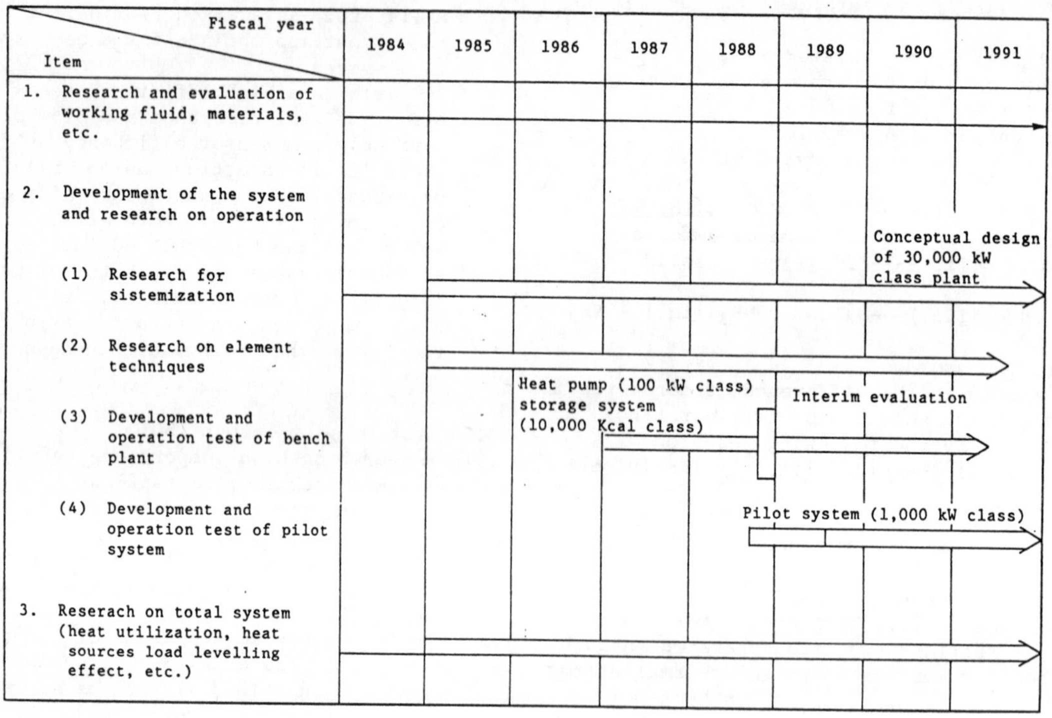

Item \ Fiscal year	1984	1985	1986	1987	1988	1989	1990	1991
1. Research and evaluation of working fluid, materials, etc.								
2. Development of the system and research on operation								
(1) Research for sistemization							Conceptual design of 30,000 kW class plant	
(2) Research on element techniques								
(3) Development and operation test of bench plant			Heat pump (100 kW class) storage system (10,000 Kcal class)			Interim evaluation		
(4) Development and operation test of pilot system					Pilot system (1,000 kW class)			
3. Reserach on total system (heat utilization, heat sources load levelling effect, etc.)								

3rd International Symposium on the
Large Scale Applications of Heat Pumps
Oxford, England : 25-27 March 1987
PAPER F1

District heating of commercial buildings by using waste heat sources of an industrial process.

H. Kling

Sulzer-Escher Wyss GmbH
Kemptener Strasse 11-15
D-8990 Lindau/B

The task faced

Reconciling the interests of ecology and economy, and finding answers that would reduce environmental offence besides making rational use of energy - this was the task faced by an enlightened municipality, a gas supply utility with imagination and an environment-conscious industrial enterprise.

At Goppingen the Deutsche Gelatine-Fabrik Stoess produces gelatine nearly the whole year round. Amoung other purposes this is employed in the food industry as binding agent. In the course of its production, great quantities of reject heat occur due to the evaporation process. Previously most of it was utilized in-plant, depending on the temperature level. To enable the cooling water to be recycled, its temperature had to be reduced in cooling towers to a level usable in the production process. That meant a large part of the residual heat remained unused, and in addition it imposed a heat and moisture burden on the environment.

The Stoess works and the Filstal gas utility therefore decided to exploit this waste heat. The heated coolant, available free of charge, is cooled by heat pumps instead of cooling towers. In the condensers of the heat pump installation and the heat exchangers of the engines driving it, the waste heat recovered in this way is brought to a temperature level suitable for heating purposes. Available as consumers of this heating via a district pipe network are a few communal buildings, including a high school in its first construction phase, the school swimming pool, a youth amenities building and the city engineers' building with the gas transfer station. All these buildings had to be heated by other means previously. A later extension will take in a business zone and existing apartment blocks.

Through this use of gas-engined heat pumps the environment is relieved of a heat load, the primary energy consumed is put to rational use, and at the same time electricity is saved in cooling.

- The youth building, sports centre with gymnasium, and the indoor swimming pool are heated with energy piped from the heating station.

Plant configuration

Two heat pumps with reciprocating compressors and direct-coupled gas engines were adopted. Each unit has a heat exchanger group consisting of evaporator, condenser, and heat exchanger for engine cooling water and exhaust gas. The two units together cover about 30% of the maximum heat demand, and they are able to supply the consumers from their heat source almost the whole year round. Over the year they cover about 76% of the demand. The rest is supplied by a peak-load boiler on extremely cold days. It also provides the supply when no waste heat is available, as for instance when the gelatine works are on holiday. On days when little heating is needed, the surplus energy is fed into two hot water accumulators. In this way undesirably frequent starting and stopping of the gas engines is avoided.

Plant diagram

1 Reciprocating compressor
2 Gas engines
3 Evaporator
4 Condenser
5 Hot water accumulator
6 Boiler
7 Heat exchanger
8 Consumers

Pipework:

blue	cooling tower water
yellow	refrigerant
red	hot water flow
orange	hot water return

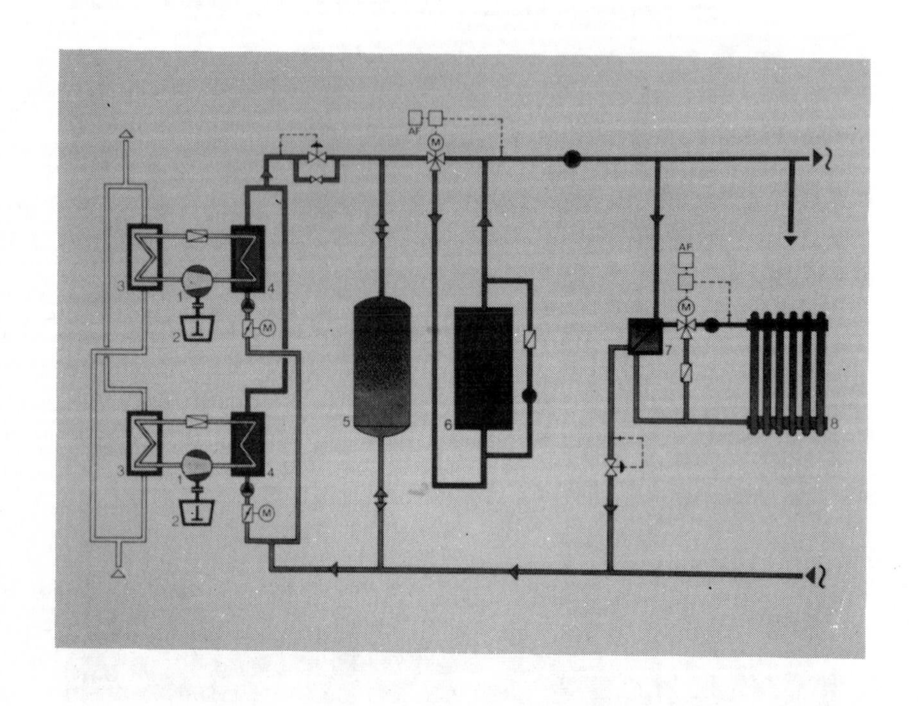

Function of the plant

The water from the cooling towers of the Stroess works is led through plastic pipelines from the factory premises to the evaporators of the heat pump units in a machine room about 100 m away. Inflow temperature lies between 35 and 40°C. In the evaporators this water is cooled to about 28°C and returned chemically unaltered. In the two reciprocating compressors, the heat energy transferred to the R12 refrigerant is "pumped" to a higher temperature level and surrendered to the district heating network in the condensers. The amount of heat is further increased with the waste heat from the two gas engines.

The heat radiated by the components located in the machine room is utilized as a further source of heat. A room air cooler transfers to the refrigeration circuit the heat absorbed from this source. The heat pump installation is controlled according to the accumulator temperature by a freely programmable system.

If more heat is demanded by the consumers than the two heat pumps can supply, the control system starts up the peak-load boiler as well. When consumption is low, output may be reduced to about 400 kW by shutting-down one unit and lowering the speed of the other.

The complete installation consists of two heat pumps and one boiler

Technical data of the complete installation

Heating system	80/50	°C
Maximum heat load	3 630	kW
Nominal thermal output of both heat pumps at 1 500 rpm	1 200	kW
Maximum thermal output of both heat pumps at 1 750 rpm	1 410	kW
Boiler output	3 000	kW

Each heat pump consists of:

1 Sulzer Escher Wyss reciprocating compressor direct-driven by gas engines

Type	6 W 110	
Refrigerant	R12	
Max. speed	1 750	rev/min
Min. speed	1 200	rev/min
Max. power input at shaft	110	kW
Evaporation temperature	+25	°C
Condensing temperature	+65	°C
Capacity regulation via speed	100-65	%

1 Gas engine

driving the compressor, 4-stroke Otto gas engine with 6 cylinders, unsupercharged, Type F 817 Gu

Max. speed	1 750	rev/min
Min. speed	1 200	rev/min
ISO standard rating at 1 750 rev/min	110	kW
Gas consumption under operating conditions	71	m^3/h
Lower calorific value of natural gas	10.13	kWh/m^3

1 Shell-and-tube evaporator

submerged type with externally finned tubes of SFCu

Refrigeration capacity	400	kW
Volume flow	34	m^3/h
Water inlet temperature	about 38	°C
Water outlet temperature	28	°C

1 Shell-and-tube condenser

with externally finned tubes of SFCu

Condensation capacity	505	kW
Volume flow	36	m^3/h
Water inlet temperature	50	°C
Water outlet temperature	62	°C

1 Heat exchanger

for removing the cylinder and lube oil heat of the gas engine

Heat transfer capacity	126	kW
Volume flow	36	m^3/h
Water inlet temperature	62	°C
Water outlet temperature	65	°C

1 Exhaust gas heat exchanger

Heat transfer capacity	74	kW
Volume flow	36	m^3/h
Water inlet temperature	65	°C
Water outlet temperature aprox.	66,7	°C

Mounted on a common base frame are the reciprocating compressor and gas engine. Maintenance is facilitated by the arrangement of the two heat pumps. The exhaust gas catalyzer is easily accessible too.

- The office building of the Göppingen municipal engineers' department is served by the district heating network.

- In the heating station, energy is recovered from the cooling tower water by heat exchanger and put into the heating water.

Environment protection

This installation is notable for its environment-friendliness in several respects. In the first place the neighbourhood of the factory suffers significantly less nuisance from cooling tower reject heat than before. Moreover the operator has installed a catalytic exhaust purifying system, which without using extraneous reducing agents lowers the emission of nitrogen oxides almost completely, that of carbon monoxide and residual hydrocarbons to a large extent. Prerequisite for optimal reduction of the noxious components NO_x, CO and HC is engine tuning with a an air:fuel ratio of $\lambda \leq 1$. This mixture ratio must be maintained within very close tolerances if the three-way catalyzer is to function effectively. On account of this, each gas engine is fitted with an automatic lamda control system.

Measurements by Ruhrgas AG, Dorsten yielded the following extremely good results:
Nitrogen oxides: reduced by 98-99% to 30-50 vol ppm.
Carbon monoxide: reduced by 93-97% to 90-360 vol ppm.
Residual hydrocarbons (99% CH_4): reduced by 50-96% to 500-900 vol ppm.
The energy balance of the gas-engined heat pump installation (coefficient of performance) showed only insignificant change (lying within the measuring accuracy) compared with the original state without catalyzer.

German Gas Industry's Prize for rational use of energy 1984

The gas-engined heat pump installation of the Göppingen district heating supply was awarded a distinction in the competition held every two years for rational energy utilization. A tribute was made to the conception here, whereby imaginative planning and the latest technology have combined to make a special contribution to energy saving by exploiting otherwise useless energy to cover public and private demands in the proximity of an industrial enterprise.

Further merit has been earned by this installation through the first use of a three-way catalyzer in the Federal Republic of Germany, achieving an exemplary and environment-friendly solution for the elimination of noxious gases.

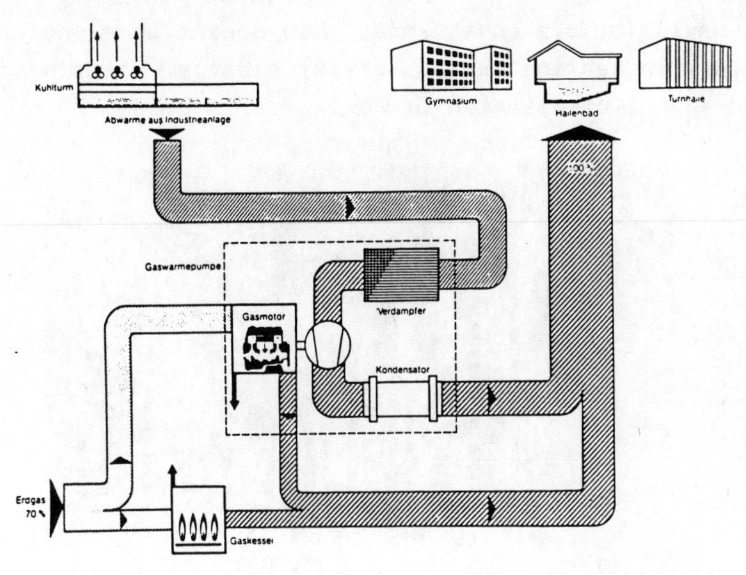

Energy flow

Cooling tower High school Swimming pool Gymnasium

Waste heat from industrial plant

Heat pump Evaporator

Gas engine Condenser

Natural gas 70%

Gasfired boiler

Operating results:

The following results were obtained during the first heating period from **November 1983 to October 1984:**

	Heat generated MWh	%	Gas consumption MWh	Operating hours h	Performance coefficient or utilization
Heat pump I	1 574	41.8	841	3 354	1.87
Heat pump II	1 314	34.9	692	2 736	1.90
Heat pumps I & II	2 888	76.7	1 533		1.88
Boiler	877	23.3	959	507	0.914
Totals	3 765	100.0	2 492		1.51

From these figures it will be seen that the two gas-engined heat pumps covered about 76.7% of the year's heat demand, while the boiler contributed 23.3%, although owing to the gelatine factory being on holiday between 22nd December 1983 and 8th January 1984 and between 25th July and 21st August 1984 the heat source was not available and the heat pumps were idle as agreed. This means a **fuel saving of 57%** on the heat pump operation alone and compared with heating exclusively by a boiler plant with an assumed annual utilization of 80%.

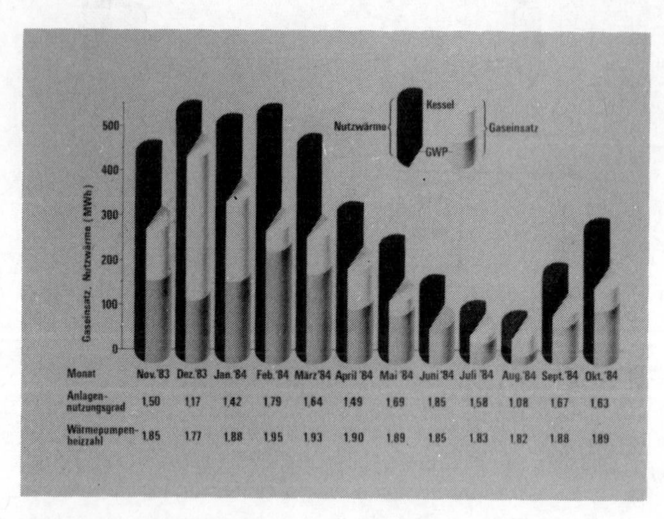

Concluding remarks

The district heating scheme of the city of Göppingen provides an example of a judicious reconciliation between the dictates of ecology and economy, enabling environmental offence to be reduced and rational use made of energy.

This conception also acts as a signpost for other supply areas, offering as it does important ideas for energy experts in industrial undertakings and utilities.

Today more than ever the common interests of everyone involved are obvious: rational utilization of energy and protection of the environment.

These demands are fulfilled excellently by the state-of-the-art district heating scheme at Göppingen.

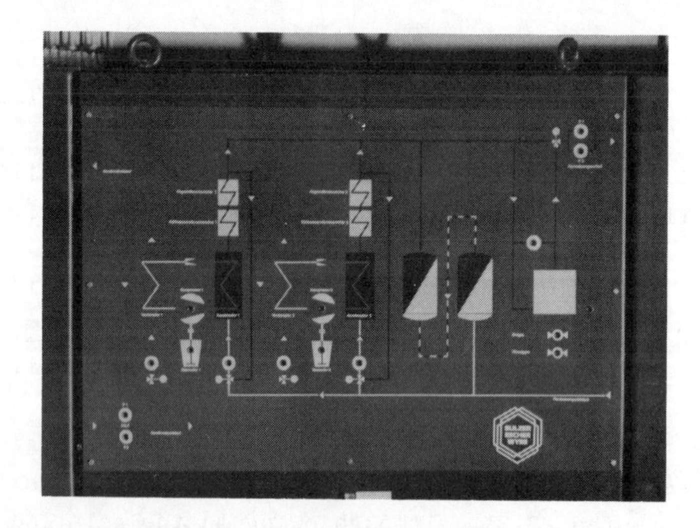

This mimic panel with signal lamps is in the heating station. It allows visual supervision of the individual functions.